This book is devoted to some mathematical methods that arise in two domains of artificial intelligence: neural networks and qualitative physics. The rapid advances in these two areas have left unanswered several mathematical questions that should motivate and challenge mathematicians.

Professor Aubin makes use of control and viability theory in neural networks and cognitive systems, regarded as dynamical systems controlled by synaptic matrices, and set-valued analysis that plays a natural and crucial role in qualitative analysis and simulation. This allows many examples of neural networks to be presented in a unified way. In addition, several results on the control of linear and nonlinear systems are used to obtain a "learning algorithm" of pattern classification problems, such as the back-propagation formula, as well as learning algorithms of feedback regulation laws of solutions to control systems subject to state constraints.

Mathematical models involve many features of a problem that may not be relevant to its solution. Qualitative physics, however, deals with an imperfect knowledge of the problem model. It is therefore more suited to the study of expert systems, which are shallow models and do not require structural knowledge of the problem.

This book should be a valuable introduction to the field for researchers in neural networks and cognitive systems, and should help to expand the range of study for viability theorists.

Neural Networks and Qualitative Physics

Neural Networks and Qualitative Physics

JEAN-PIERRE AUBIN
Ecole Doctorale de Mathématiques de la Décision
Université de Paris - Dauphine

CAMBRIDGE UNIVERSITY PRESS
Cambridge, New York, Melbourne, Madrid, Cape Town,
Singapore, São Paulo, Delhi, Tokyo, Mexico City

Cambridge University Press
The Edinburgh Building, Cambridge CB2 8RU, UK

Published in the United States of America by Cambridge University Press, New York

www.cambridge.org
Information on this title: www.cambridge.org/9781107402843

First published 1996
First paperback edition 2011

A catalogue record for this publication is available from the British Library

ISBN 978-0-521-44532-0 Hardback
ISBN 978-1-107-40284-3 Paperback

Contents

Preface

This book is devoted to some **mathematical methods** that arise in two domains of artificial intelligence: neural networks and qualitative physics (which here we shall call "qualitative analysis"). These two topics are treated independently. Rapid advances in these two areas have left unanswered many mathematical questions that should motivate and challenge a wide range of mathematicians. The mathematical techniques that I choose to present in this book are as follows:

control and viability theory in neural networks and cognitive systems, regarded as dynamical systems controlled by synaptic matrices.

set-valued analysis, which plays a natural and crucial role in qualitative analysis and simulation by emphasizing properties common to a class of problems, data, and solutions. Set-valued analysis also underlies *mathematical morphology*,[1] which provides useful techniques for image recognition.

This allows us to present in a unified way many examples of neural networks and to use several results on the control of linear and nonlinear systems to obtain a *learning algorithm* of pattern-classification problems (including time series in forecasting), such as the *back-propagation formula*, in addition to learning algorithms concerning feedback-regulation laws for solutions to control systems subject to state constraints (inverse dynamics).

These mathematical techniques may also serve to contribute to the various attempts to devise mathematical metaphors for cognitive pro-

[1]See the forthcoming book by Michel Schmitt and Luc Vincent, *Morphological Image Analysis* (Cambridge University Press). The links between mathematical morphology and set-valued analysis and viability theory will be explored in a subsequent book, *Mutational and Morphological Analysis: tools for shape regulation and optimization*.

cesses. I present here a very speculative metaphor under the name of *cognitive systems* based on these mathematical techniques. They go beyond neural networks in the sense that they involve the problem of adaptation to viability constraints. They can *recognize* the state of the environment and *act* on the environment to *adapt* to given viability constraints. Instead of encoding knowledge in synaptic matrices as neural networks do, the knowledge is stored in *conceptual controls*. Given the mechanism of recognition of the state of the environment by conceptual controls, perception and action laws, and viability constraints, the viability theorems allow one to construct *learning rules* that describe how conceptual controls evolve in terms of sensorimotor states to adapt to viability constraints.

There is always a combination of two basic motivations for dealing with formal models of cognition - *neural networks* being content with implementation of "neural-like" systems on computers, and *cognitive systems* attempting to model actual biological nervous systems. Every model lies between these two requirements - the first allowing more freedom in the choice of a particular representation (computing efficiency being the main criterion), and the second constraining the modeling to be closer to biological reality.[2]

The symbolic processing capabilities that neural networks try to achieve are unexpected technological consequences of the use of digital computers, which were not designed for such a purpose at their inception. In the same way, the skills of reasoning logically and solving mathematical problems also represent a kind of unexpected "technological fallout" of the human brain, because they certainly were not among the advantages necessary for the survival of the human species when they appeared.

Expert systems are *shallow models* that do not require any formal and structural knowledge of the problem, whereas a mathematical model might involve too many features that would not be relevant for solving the problem at hand. For many problems, we have only imperfect knowledge of the model, and we may be interested in only a few features (often of a qualitative nature) of the solution, and so we see at once that the concept of *partial knowledge* involves two types of ideas:

[2] Actually, we should say "degree of reality for a social group at a given time," which is understood here in terms of the consensus interpretations of the group members' perceptions of their physical, biological, social, and cultural environments. This concept of reality is thus relative to a social group and is subject to evolution.

1. We require less precision in the results (e.g., signs of the components of vectors instead of their numerical values),
2. We take into account a broader universality or robustness of these results with respect to uncertainty, disturbances, and lack of precision.

As in numerical analysis, which deals both with approximation of problems in infinite-dimensional spaces by problems in finite-dimensional spaces and with the algorithms for solving such approximated problems, the problems of qualitative analysis arise at two levels: the passage from quantitative analysis to qualitative analysis (which deals with the association of discrete problems with continuous problems) and the algorithms to solve discrete problems.[3] In particular, Kuipers's QSIM algorithm for tracking the monotonicity properties of solutions to differential equations is revisited by placing it in a rigorous mathematical framework. This allows us to determine a priori the *landmarks* (i.e., the states at which the monotonicity properties change) instead of discovering them a posteriori by tracking the qualitative evolution of the solutions to the differential equation. These landmarks delineate *qualitative cells*, in which the monotonicity behaviors of the solutions are the same. Once these qualitative cells are computed, the Dordan QSIM algorithm provides the transition laws from one qualitative cell to the others.

This book is divided into 10 chapters. Chapters 1-7 deal with neural networks and some mathematical background needed to treat them (pseudoinverses, tensor products, gradient methods for convex potentials), Chapter 8 deals with cognitive systems, and Chapters 9 and 10 deal with some mathematical questions raised by qualitative physics, in the static and dymanic cases, respectively.

Chapter 1 provides the definitions of neural networks and learning processes (including the perceptron algorithm) and the *heavy learning algorithm*, which allows learning without forgetting.

Chapter 2 deals with some mathematical tools: pseudoinverses of linear operators and tensor products. Indeed, we have to use the specific structure of the space of synaptic matrices as a tensor product to justify mathematically the *connectionist features* of neural networks. Tensor products *explain* the Hebbian nature of many learning algorithms. This is due to the fact that derivatives of a wide class of nonlinear maps de-

[3]The first aspect has been quite neglected, and it is the one we shall emphasize in this book.

fined on spaces of synaptic matrices are tensor products and also to the fact that the pseudoinverse of a tensor product of linear operators is the tensor product of their pseudoinverses.

Chapter 3 is devoted to the case of linear neural networks, also called *associative memories*. We begin by showing that the *heavy learning algorithm* for neural networks that are affine with respect to the synaptic matrices (but nonlinear with respect to the signals) has a Hebbian character. We proceed with purely linear networks with a single layer or with a finite number or a continuum of layers. The chapter ends with an introduction to associative memories with gates, which are well adapted to compute Boolean and fuzzy Boolean functions.

Chapter 4 is devoted to the proof of the convergence of the gradient method for minimization problems involving a convex criterion with or without constraints. We discuss an application to the Minover algorithm of Mézard that replaces the perceptron algorithm. Many more features of convex analysis could be used in the study of a class of neural networks, but such results would go beyond the scope of this book and the common knowledge of its expected audience.

Chapter 5 adapts these results to the case of nonlinear networks and presents two main types of learning rules. The first class consists of algorithms derived from the gradient method and includes in particular the back-propagation rule. The second class is composed of learning rules based on the Newton method.

Chapter 6 is devoted to the use of neural networks for finding viable solutions to control systems, that is, solutions to control systems that will satisfy given viability (or state) constraints. The purpose of this chapter is to derive learning processes for *regulation feedback* for control problems through neural networks. Two classes of learning rules are presented. The first, called the class of *external learning rules*, is based on the gradient method (of optimization problems involving nonsmooth functions). The second deals with *uniform algorithms*.

In Chapter 7, the *internal-learning algorithm* provides learning rules based on viability theory. Two sections are devoted to a short presentation of the main results of viability theory and its application to the regulation of viable solutions to control systems. Applications to the control of cart-pole problems and other benchmark problems have been designed by N. Seube. This algorithm is applied to stabilization problems.

Chapter 8 goes beyond neural networks as they are usually defined. It proposes a very speculative mathematical model of what is called a *cognitive system*. A cognitive system is a dynamical system describing the evolution of sensorimotor states, recognized and controlled by *conceptual controls*, according to perception and action laws, and required to obey some viability constraints. *Adaptive learning processes* associating conceptual controls with sensorimotor states are then obtained by using viability theorems, including the ones that obey the *inertia principle: Change the conceptual controls only when the viability of the cognitive system is at stake.* This chapter is oriented toward mathematical metaphors motivated by cognisciences, of which we present a few relevant facts.

Chapter 9 treats the qualitative resolution of static problems described in the form of both equations and inclusions. It proposes a general framework (*confluence frames*) to link quantitative problems with qualitative ones. In particular, sign confluences are thoroughly investigated.

Chapter 10 is devoted to qualitative simulation of differential equations and to a mathematical treatment of Kuipers's QSIM algorithm to track the monotonicity properties of solutions to differential equations. We also consider Dordan's QSIM algorithm, which provides the qualitative cells delineated by the landmarks, and then the transition map associating with each qualitative cell its successor(s). Dordan's QSIM algorithm was designed to study the qualitative behaviors of a class of differential systems, the *replicator systems*, which play important roles in several domains of biology and biochemistry. We consider several examples obtained by using software designed by O. Dordan.

Two appendixes conclude this book. Appendix A provides a survey of convex optimization and set-valued analysis that goes beyond the minimal survey of Chapter 4. Appendix B describes applications of Nicolas Seube's algorithms, presented in Chapters 6 and 7, to the control of autonomous underwater vehicles (AUVs) tracking the trajectory of an exosystem.

Acknowledgements

I thank Alain Bensoussan for asking me to report to INRIA about neural networks and qualitative physics. That triggered a deeper interest in this topic.

This book owes much to Olivier Dordan and Nicolas Seube, who developed most of the material of this book and the computer applications. I have the pleasure to express here my gratitude. I also thank Luc Doyen and Juliette Mattioli for correcting parts of the text and for developing other applications of set-valued analysis and viability theory to visual control and mathematical morphology.

1
Neural Networks: A Control Approach

Introduction

A *neural network* is a network of subunits, called "formal neurons," processing input signals to output signals, which are coupled through "synapses." The synapses are the nodes of this particular kind of network, the "strength" of which, called the *synaptic weight*, codes the "knowledge" of the network and controls the processing of the signals.

Let us be clear at the outset that the resemblance of a formal neuron to an animal-brain neuron is not well established, but that is not essential at this stage of abstraction. However, this terminology can be justified to some extent, and it is by now widely accepted, as discussed later. Chapter 8 develops this issue.

Also, there is always a combination of two basic motivations for dealing with neural networks - one attempting to model actual biological nervous systems, the other being content with implementation of neural-like systems on computers. Every model lies between these two requirements - the first constraining the modeling, the second allowing more freedom in the choice of a particular representation.

There are so many different versions of neural networks that it is difficult to find a common framework to unify all of them at a rather concrete level. But one can regard neural networks as dynamical systems (discrete or continuous), the states of which are the signals, and the controls of which are the synaptic weights, which regulate the flux of transmitters from one neuron to another. They yield what are also known as *adaptive systems*, controlled by *synaptic matrices*.

We investigate in this chapter the *supervised learning* of a finite set of patterns, called the *training set*, each pattern being a pair of input-output signals. A *learning process* amounts to matching the given input signals of the patterns of this training set with the associated output signals. It thus involves a comparison with desired answers, done by a "supervisor." Hence, the learning problem amounts to finding a synaptic matrix that can "learn" the patterns of a given training set. Any algorithm yielding such a synaptic matrix will then be regarded as a *learning rule*. This seems to exclude neural networks from reasonable models of basic neural functions. However, the problems under investigation here are as follows:

1. Establish the existence of exact (or approximate) synaptic matrices that have learned a given set of patterns,
2. Find algorithms, regarded as learning rules, that will provide a sequence of synaptic matrices converging to a solution of the learning problem.

I propose in this first chapter a short presentation of the learning processes of neural networks in the framework of dynamical systems controlled by synaptic matrices. Section 1.2 explains how neural networks operate to solve pattern-classification problems and, in particular, to extrapolate time series in forecasting problems. These problems are called supervised learning problems, because the patterns to be taught are provided by a supervisor, so to speak.

Biological Comments. Although the results to be derived here will be "formal" and will not necessarily be associated with any biological implementation, it may be useful to provide a crude description of the nature of the processing carried out by some biological neurons (Figure 1.1).

It should be first pointed out that many kinds of neuronal cells evolved during phylogenesis, each of them selected to provide adequate technical solutions to biological or environmental problems. Most neurons in the central nervous system in higher animals can be regarded as *impulse oscillators*. They produce trains or volleys of neural impulses whose average frequency will depend on the input excitation. The processing carried out by neuron ensues from biophysical and biochemical phenomena in the membrane of the neuron, wherein functions the machinery that controls the interaction of the cell and its environment.

Models of neural dynamics that attempt to describe the triggering phenomena, the transmission of the impulses, and their biochemical control have been extensively studied since the Hodgkin-Huxley model was first proposed. The central features are the following: There is a static

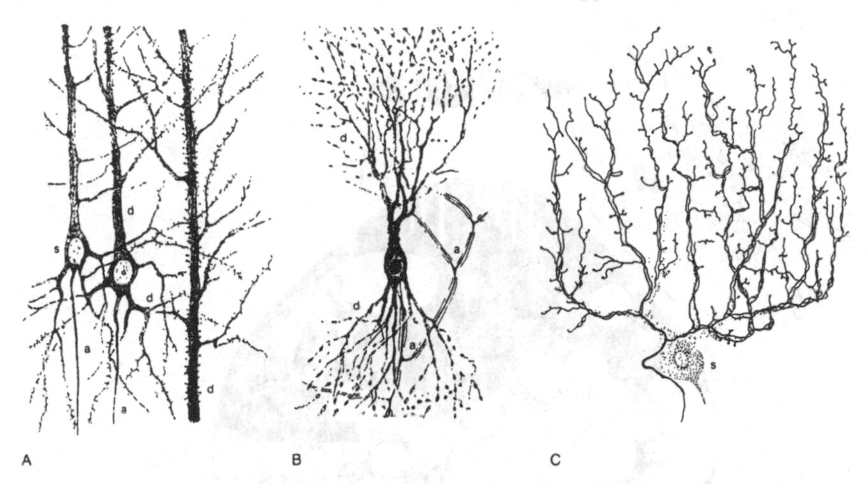

Fig. 1.1. Neurons. Pyramidal and Purkinje neurons

electrical-potential difference of about 70 mV between the inside of the neuron (negatively charged) and its environment; it is maintained by diffusion of ions through the membrane. The synaptic receptors are able to control the ionic conductance of the membrane through a highly sophisticated biochemical and ionical mechanism (Figure 1.2).

When neurotransmitters (excitation signals) arrive at a receptor, the excitatory synapses tend to produce a depolarization of the membrane, and the inhibitory synapses tend to produce a hyperpolarization. When the sum (i.e., algebraic sum) of the depolarizations exceeds a given threshhold, then the membrane's permeability to ions is increased, and the membrane becomes electrically active, sending an output impulse of about 100 mV amplitude during a period of 0.5-2 ms. After each impulse, there is a short refractory period during which the membrane recovers so as to be ready for the next impulse (Figure 1.3).

Because in our framework we are interested in the collective processing by neurons, we shall be content with only a crude analytical description of the processing role of the neuron. We retain only the following features: The impulse frequency oscillates between bounds determined by physical and chemical factors (oscillating between 0 and 500 Hz). Between those bounds, the postsynaptic *average oscillatory frequency* is assumed to be a monotonic function of the net algebraic sum of the presynaptic average frequencies of the inputs afferent to the neurons (inhibitory excitations being regarded as negative excitations). Even though, for simplicity, the integration of the presynaptic inputs is often

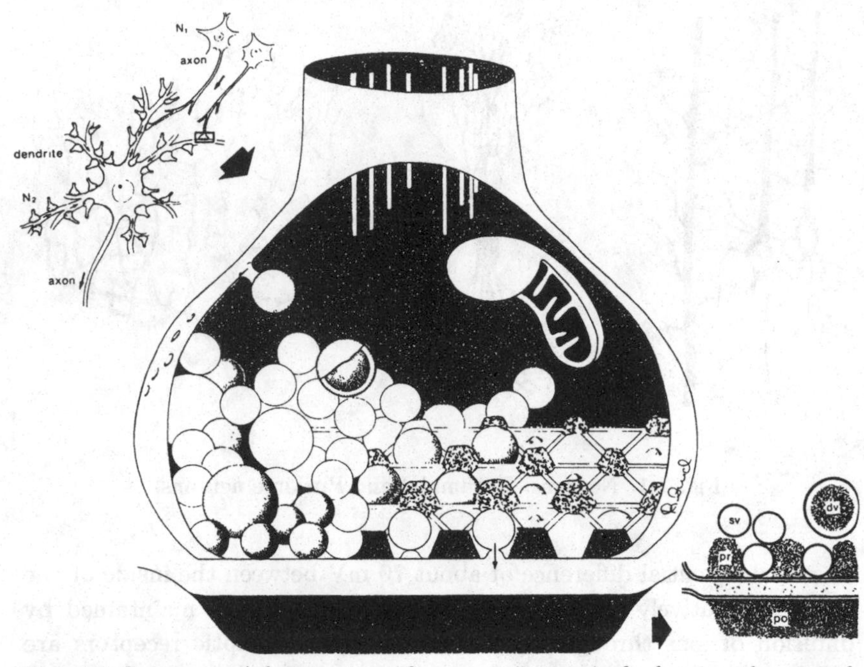

Fig. 1.2 Synapses. A larger view of a synaptic terminal, showing the vesical grid, the vesicles incorporated in the cell membrane, a single mitochondrion and the cleft, but not the postsynaptic detail.

chosen to be linear, biological neurons are somewhat "leaky integrators" of presynaptic inputs, certainly nonlinear.

1.1 Neural Networks: A General Presentation

A way to encompass most of the neural networks - also called *parallel distributed processes* (PDP) - studied in the literature is to regard them as dynamical systems controlled by synaptic matrices.

1.1.1 Formal Neurons

We begin with a set of n "formal neurons" (or abstract neurons, processing units, etc.) labeled by $j = 1, \ldots, n$. The processing of a neural network is carried out by these formal neurons. The formal neuron's job is simply to receive afferent signals from the other formal (presynaptic) neurons (or the input signals) and to provide (process, compute, etc.) an output that is sent to the other (postsynaptic) neurons (or to

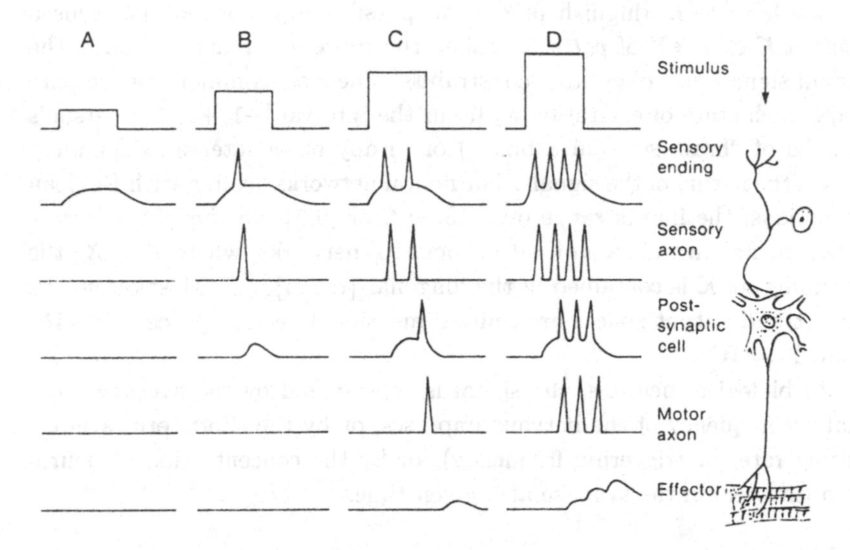

Fig. 1.3 Propagation of the Nervous Influx. A stimulus is applied to a sensory nerve terminal with increasing intensityfrom A to B.

the output of the network). A neural network is *parallel* in that many (if not all) neurons carry out their processing at the same time. This processing is said to be *synchronous* if the duration of the processing of a unit of input is the same for all formal neurons. If not, it is called *asynchronous*. Synchronous models can be regarded as discrete models. Biological neurons are asynchronous and thus require a continuous-time treatment, through differential equations or inclusions.

1.1.2 Signals: The States of the Network

Formal neurons link (directly for one-layer networks, and indirectly for multiple-layer networks) an input space X of signals to an output space Y. Hence, the *state space* of the system is the space $X \times Y$ of input-output pairs (x, y), which are called *patterns* in pattern recognition, data analysis, and classification problems. The set of input signals is often called the *retina* in the framework of image recognition, for obvious reasons. When the inputs are different from the outputs, the system is called *heteroassociative*. When $X = Y$ and when the inputs of the patterns (x, x) coincide with the outputs, we speak of *autoassociative* networks.

We have to distinguish among all possible input-output patterns a subset $\mathcal{K} \subset X \times Y$ of *patterns*, called the *training set*. For instance, the input signals may obey state constraints. The most common ones require that each state of excitation x_i lie in the interval $[-1, +1]$ (Anderson's model of "brain state in a box") or in any other interval, depending upon the nature of the signals. For neural networks dealing with Boolean functions, the inputs range over $\{0,1\}^n$, or $[0,1]^n$ in the case of fuzzy statements. In the case of autoassociative networks, where $Y = X$, the training set $\mathcal{K}$ is contained in the diagonal $\{(x,x)\}_{x \in X}$. Most often, the input and output spaces are finite-dimensional vector spaces $X := \mathbf{R}^n$ and $Y := \mathbf{R}^m$.

In biological neurons, the signal is represented by the average oscillatory frequency of the nervous impulses, or by the short-term average firing rate (or triggering frequency), or by the concentration of neurotransmitters in the synapse at a given time.

1.1.3 Synaptic Matrices: The Controls of the Network

Formal neurons are connected to one another in a neural network. The conventional wisdom among neural-network scientists is that *knowledge* is encoded in the pattern of connectivity of the network. Let N denote the set of neurons, and 2^N or $\mathcal{P}(N)$ the family of subsets of neurons, called *conjuncts* or *coalitions* (of neurons) (Figure 1.4).

The connection links a postsynaptic neuron j to conjuncts or coalitions $S \subset N$ of presynaptic neurons. Each conjunct S of neurons preprocesses (or gates[1]) the afferent signals x_i produced by the presynaptic neurons through a function

$$\varphi_S \;:\; x := (x_j)_{j=1,\ldots,n} \mapsto \varphi_S(x)$$

In most models, the conjuncts $S = \{i\}$ are reduced to individual neurons i. Then the role of the control is played by the synaptic matrix

$$W = \left(w_j^S\right)_{\substack{S \in 2^N \\ j=1,\ldots,n}}$$

the entries w_j^S of which are the synaptic weights. The modulus of the synaptic weight represents the strength, and its sign the direction, of the connection from the conjunct S to the formal neuron j, counted positively if the synapse is excitatory, and counted negatively if it is

[1]This gating process is useful to compute (or extrapolate) Boolean functions.

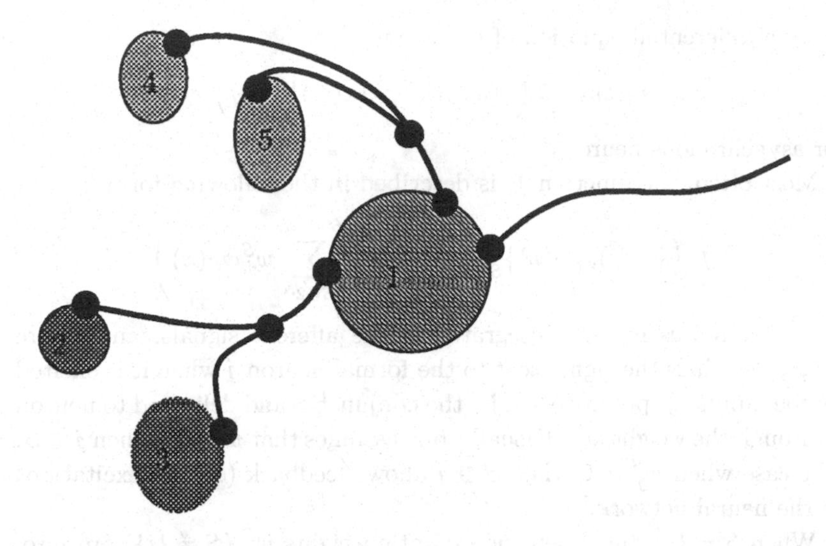

Fig. 1.4 Conjucts of Neurons. Two conjucts of neurons $\{2,3\}$ and $\{4,5\}$ provide inputs to the post-synaptic neuron 1.

inhibitory. Therefore, it is often assumed in this kind of neuromimetic model that the jth neuron is excited by $\sum_{S\in 2^N} w_j^S \varphi_S(x)$.

Some models impose constraints bearing on the structure of the synaptic matrix, which may represent a division of the set of neurons in several layers, with the synaptic matrix as a "product" of elementary synaptic matrices mapping one layer to another (top-down or bottom-up processing systems). There is no need to assume that a synaptic matrix is symmetric. Actually, experimental evidence suggests asymmetric matrices, and perhaps antisymmetric matrices.

1.1.4 Propagation Rules: The Dynamics of the Network

The output of the jth neuron is a function of the neuron potential caused by the preprocessed signals sent by the presynaptic neurons, (possibly) gated by conjuncts of neurons and weighted by the synaptic weights that tune these presynaptic neuron potentials.

The evolution of the average is governed by a discrete dynamical system (for synchronous neurons),

$$y_j := f_j\left(\{(w_j^S), \varphi_S(x)\}_{S\in 2^N}\right)$$

or by a differential equation of the form

$$x'_j(t) = f_j\left(\{(w_j^S)(t), \varphi_S(x(t))\}_{S\in 2^N}\right)$$

for asynchronous neurons.

Most often, the function f_j is described in the following form:

$$f_j\left(\{(w_j^S), \varphi_S(x)\}_{S\in 2^N}\right) := g_j\left(\sum_{S\in 2^N} w_j^S \varphi_S(x)\right)$$

where g_j represents the integrator of the afferent signals, and where $w_j^S\varphi_S$ describes the signal sent to the formal neuron j when it is excited by the inputs x_i, preprocessed by the conjunct S and delivered to neuron j through the weight w_j^S. [Usually, one assumes that $w_j^S = 0$ when $j \in S$. The case when $w_j^S \neq 0$ when $S \ni j$ allows feedback (or autoexcitation) in the neural network.]

When $S = \{j\}$ and when the synaptic weights w_i^S $(S \neq \{i\})$ are zero, we obtain the *loss term* $g_i(0, \ldots, w_j^j\varphi_j(x), \ldots, 0)$, which may represent "forgetting," or at least the decaying of the signal frequency when the neuron is not excited by the other neurons.

To be realistic, delays and, more generally, the history of $x(\cdot)$ should appear in these functions φ_S.

1.2 Examples of Neural Networks

Let us cite several examples of propagation rules:

1. Associative Memories

The richest class of neural networks is obtained when there is no preprocessing by conjuncts of neurons, that is, when $\varphi_S(x) = 0$ if $|S| > 1$ and when $\varphi_{\{i\}}(x) = x_i$ for any afferent signal i (where $|S|$ denotes the number of elements of S) and when the functions g_i are affine: They are governed by the simple law

$$y_j = \sum_{i=1,\ldots,n} w_j^i x_i + c_j$$

Such networks are called *associative memories* by Kohonen. We shall answer most questions in this familiar setting. When the dimension of the output space $Y := \mathbf{R}$ is 1, we find the Widrow *adaline network* (for "adaptive linear element") introduced in 1962.

2. Associative Memories with Gates

When the functions g_i are still affine and there is preprocessing of the inputs of an associative memory through conjuncts of neurons, we obtain the associative memories with gates:

$$y_j := \sum_{S \in 2^N} w_j^S \varphi_S(x) + c_j$$

Example: Boolean Associative Memories

Denoting by $|S|$ the number of elements of a conjunct S (a subset of neurons), an afferent signal $x \in \mathbf{R}^n$ can be gated by the positively homogeneous[2] function

$$\varphi_S(x) := \left(\prod_{i \in S} x_i \right)^{1/|S|}$$

This is quite useful for computing n-variable Boolean functions. In this case, $X := \mathbf{R}^n$, $Y = \mathbf{R}$, and the subset $\mathcal{K}$ of input-output patterns is $\{0,1\}^n \times \{0,1\}$ (and, in the fuzzy case, $[0,1]^n \times [0,1]$). We shall prove that the neural network

$$y = \sum_{S \in 2^N} w^S \left(\prod_{i \in S} x_i \right)^{1/|S|}$$

can compute any Boolean function b defined on $\{0,1\}^n$, in the sense that there exist weights w^S such that

$$\forall\, x \in \{0,1\}^n, \quad \sum_{S \in 2^N} w^S \left(\prod_{i \in S} x_i \right)^{1/|S|} = b(x)$$

This allows us to extrapolate them to *fuzzy statements* $x \in [0,1]^n$. Another (obvious) choice of preprocessing functions having the same property is given by the multilinear function φ_S defined by

$$\varphi_S(x) := \prod_{i \in S} x_i \prod_{j \notin S} (1 - x_j)$$

because we can always write

$$\forall\, x \in \{0,1\}^n, \quad b(x) = \sum_{y \in \{0,1\}^n} b(y) \prod_{\{i | y_i = 1\}} x_i \prod_{\{j | y_j = 0\}} (1 - x_j)$$

This is the analytical version of the standard result from Boolean algebra stating that any Boolean function may be expressed in the *disjunctive*

[2]The choice of positively homogeneous functions allows independence of the normalization rule attributing the value 1 to "true."

normal form. The drawback is that such functions φ_S involve the afferent signals from presynaptic neurons that do not belong to S, contrary to the first example. However, there is nothing to "compute," because we have an explicit formula $w^S = b(y_S)$, where y_S denotes the characteristic function of the subset S. □

3. Nonlinear Automata

The next class of rules of propagation is made of maps f of the form

$$f_j(x, W) \;:=\; g_j\left(\sum_{S\in 2^N} w_j^S \varphi_S(x)\right)$$

where g_j are automata of various forms.

McCulloch-Pitts Neurons These are also called *threshold logic units.* They are associated with functions g_j that are built from the Heaviside function $\mathbf{1}$ defined by $\mathbf{1}(\lambda) = 0$ if $\lambda < 0$, and $\mathbf{1}(\lambda) = 1$ if $\lambda \geq 0$. The rules of propagation require also *thresholds* β_j. Therefore, such a neural network evolves according to the law

$$y_j = \begin{cases} 1 & \text{if } \sum_{S\in 2^N} w_j^S \varphi_S(x) \geq \beta_j \\ 0 & \text{if } \sum_{S\in 2^N} w_j^S \varphi_S(x) < \beta_j \end{cases}$$

Naturally, we can replace the Heaviside function $\mathbf{1}$ by functions g such that $g(x) := A$ when $x \geq 0$, and $g(x) = a$ when $x < 0$.

"Continuous" Automata Because the Heaviside function $\mathbf{1}$ and the forgoing two-valued functions g are not continuous, it may be useful to replace them by continuous and even differentiable approximations in some problems (Figure 1.5). This can be done by using the function $g_{k,\gamma}$:

$$g_{k,\gamma}(\lambda) \;:=\; \frac{A\gamma e^{k\lambda} + a}{\gamma e^{k\lambda} + 1}$$

where $A, a > 0, \gamma$ and $k > 0$ are given parameters representing the automaton. We set $g_k := g_{k,1}$. [the constant $\beta := -\log\gamma/k$ represents a threshold because $g_{k,\gamma}(\lambda) = g_k(\lambda - \beta)$]. The function $g_{k,\gamma}$ maps $\mathbf{R}$ onto $[a, A]$ and has a "sigmoid" shape. The parameter $T := 1/k$ is called the "temperature," by analogy with spin glasses. Its derivative is equal to $k\gamma e^{k\lambda}(A - a)/(\gamma e^{k\lambda} + 1)^2$.

We observe that when k goes to ∞, the function g_k converges to the

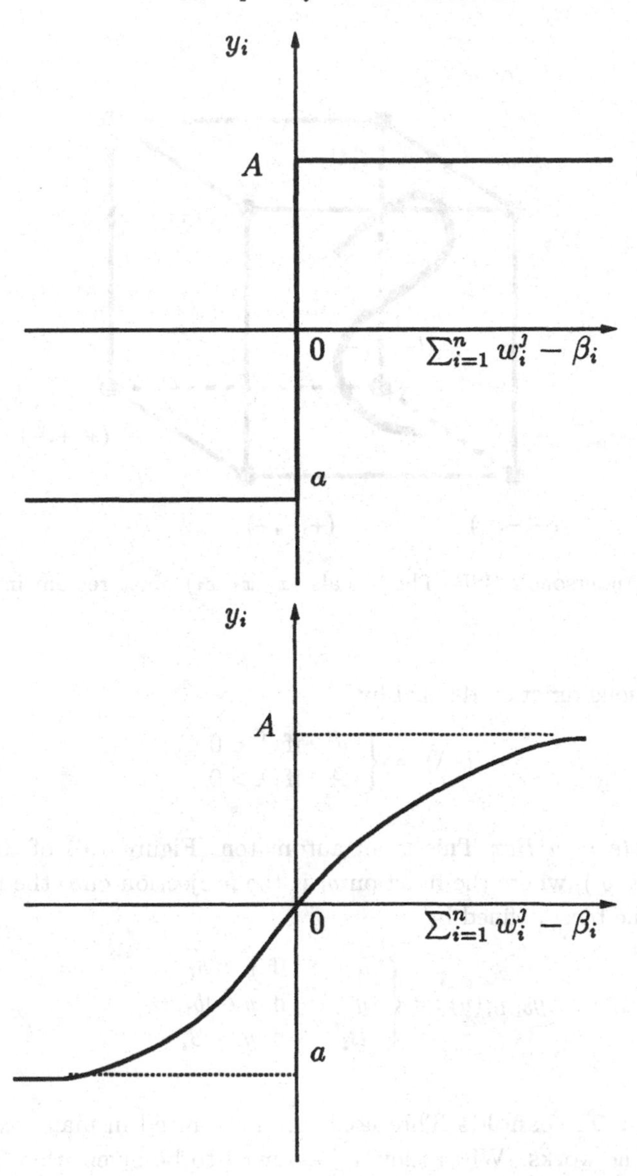

Fig. 1.5 Discrete and Continuous Automata. They are parametrized by the *temperature* $T := 1/k$.

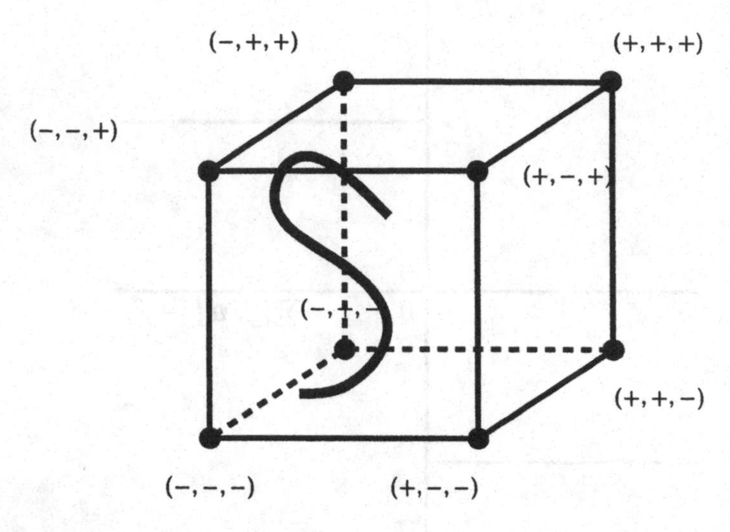

Fig. 1.6 Andersons's BSB. The signals (x_1, x_2, x_3) must remain in a cube (box).

discontinuous function defined by

$$g(\lambda) := \begin{cases} a & \text{if } \lambda < 0 \\ A & \text{if } \lambda > 0 \end{cases}$$

Brain State in a Box This is an automaton (Figure 1.6) of the form $g(y)_i := g_i(y_i)$, where the function g_i is the projection onto the interval $[b_i, B_i]$ (the box) defined by

$$g_{b_i,B_i}(y) := \begin{cases} b_i & \text{if } y < b_i \\ y & \text{if } y \in [b_i, B_i] \\ B_i & \text{if } y > B_i \end{cases}$$

Remark : Thresholds Thresholds are introduced in many examples of neural networks. When they are assumed to be given, they may be integrated in the definition of the processing function g, which then reads $g(\lambda) := h(\lambda - \beta)$, where $\beta \in Y$ is the threshold.

If the level of the threshold is part of the controls to be adjusted in the learning process, then it can be regarded as an entry of an extended synaptic matrix

$$\widehat{W} := (\widehat{w}_{ij})_{i=1,\ldots,m,\ j=0,\ldots,n} \in \mathcal{L}(\mathbf{R} \times X, Y)$$

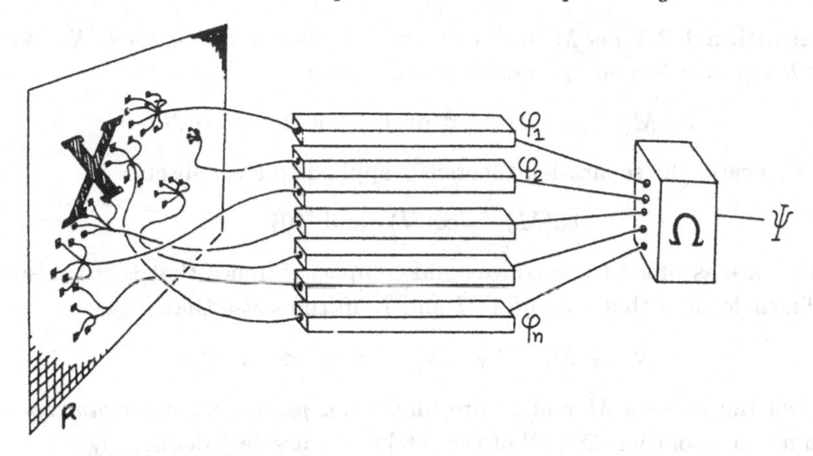

Fig. 1.7 Rosenblatt's Perceptron. The one-layer perceptron analysed by Minsky and Papert (1969, MIT Press)

defined in the following way: $\widehat{w}_{ij} = w_{ij}$ for $j = 1, \ldots, m$, and $\widehat{w}_{i0} = \beta_i$, so that $\widehat{W}(-1, x)_i = \sum_{j=1}^{m} w_{ij} x_j - \beta_i$. □

1.3 Rosenblatt's Perceptron and the Perceptron Algorithm

When the dimension of the output space $Y := \mathbf{R}$ is equal to 1, when the input space is $X := \mathbf{R}^n$, and when there is no preprocessing by conjuncts of neurons [i.e., when $\varphi_S(x) = 0$ if $|S| > 1$, and when $\varphi_{\{i\}}(x) = x_i$ for any afferent signal i], we obtain the celebrated *perceptron* introduced by Rosenblatt in 1961 (Figure 1.7).

The synaptic matrix is then just a linear form $w = (w^i)_{i=1,\ldots,n} \in X^\star$. Hence the perceptron operates according to

$$y = \begin{cases} 1 & \text{if } \sum_{i=1,n} w^i x_i \geq \beta \\ 0 & \text{if } \sum_{i=1,n} w^i x_i < \beta \end{cases}$$

where the threshold β is given.

The ambition of the designers of the perceptron was the computation of Boolean functions, so the set of inputs was the subset $\{0, 1\}^n$. Unfortunately, dimensionality problems prevent the computation of any Boolean function by a perceptron. □

However, the perceptron has been used successfully in classification algorithms.

Definition 1.3.1 *Let M and N be two subsets of a vector space X. We shall say that they are separable if and only if*

$$\overline{\text{co}}(M) \cap \overline{\text{co}}(N) = \emptyset \text{ or } 0 \notin \overline{\text{co}}(M) - \overline{\text{co}}(N)$$

In this case, the separation theorem[3] applied to the subsets

$$\overline{\text{co}}(M) - \overline{\text{co}}(N) \text{ and } \{0\}$$

which are assumed to be convex and compact, implies that there exists a linear form w that *classifies* M and N in the sense that

$$\forall\, x \in M, \quad \forall\, y \in N, \quad \langle w, x \rangle < \langle w, y \rangle$$

When the subsets M and N are finite, the *perceptron algorithm* converges to a solution $\overline{w} \in \mathcal{W}$ of the set $\mathcal{W}$ of classifiers defined by

$$\mathcal{W} := \{ w \in X \mid \forall\, x \in M, \ \forall\, y \in N, \quad \langle w, x \rangle < \langle w, y \rangle \}$$

We assume that

$$M := \{a_1, \ldots, a_P\} \quad \textit{and} \quad N := \{b_1, \ldots, b_Q\}$$

Proposition 1.3.3 *Assume that the finite subsets M and N are separable. We give an infinite sequence x_n of items a_p $(p = 1, \ldots, P)$ and b_q $(q = 1, \ldots, Q)$, each of these occurring infinitely often (this is called the training sequence). We start with $w_1 = 0$ (for instance). If at a given step n, w_n classifies x_n, it remains unchanged. If the classification is not correct, w_n is changed to w_{n+1} according to the following rule:*

$$w_{n+1} := \begin{cases} w_n + x_n & \text{if } x_n \in N \\ w_n - x_n & \text{if } x_n \in M \end{cases}$$

Then this algorithm, called the perceptron algorithm, converges in a finite number of steps to a linear form $\overline{w}$ classifying M and N.

[3]This separation theorem is one cornerstone of functional analysis. It was enunciated by the German mathematician Minkowski at the beginning of this century for finite-dimensional spaces and later extended by Hahn, an Austrian mathematician, and Banach, the Polish founder of linear functional analysis, for Banach spaces. It is then known as the *Hahn-Banach theorem*: **Theorem 1.3.2** *Let K be a nonempty subset of a Banach space X, and $x_0 \notin K$. Then there exist $p \in X^\star$ and $\varepsilon > 0$ such that*

$$\forall\, x \in K, \qquad \langle p, x \rangle \leq \langle p, x_0 \rangle - \varepsilon$$

If X is a finite dimensional vector space and K is only convex (not necessarily closed), we guarantee the existence of $p \in X^\star$ only such that

$$p \neq 0 \quad \textit{and} \quad \forall\, x \in K, \qquad \langle p, x \rangle \leq \langle p, x_0 \rangle$$

Proof We set $K := M - N$, so that we have to separate the finite set $K := \{c_1, \ldots, c_J\}$ from $\{0\}$. We choose any $w^\star \in \mathcal{W}$, which is not empty by assumption. The perceptron algorithm becomes

$$w_{n+1} := \begin{cases} w_n & \text{if } \langle w_n, x_n \rangle < 0 \quad \text{(correct classification)} \\ w_n - x_n & \text{if } \langle w_n, x_n \rangle \geq 0 \quad \text{(incorrect classification)} \end{cases}$$

Let $n_1, \ldots, n_k, \ldots$ be the sequence of trials at which the linear forms w_{n_k} are changed. Hence

$$w_{n_{k+1}} = w_{n_k} - x_{n_k} = -\sum_{l=1}^{k} x_{n_l} \quad \text{and} \quad \langle w_{n_k}, x_{n_k} \rangle \geq 0$$

Since M and N are separable, we can separate K from zero, so that, by the separation theorem, there exists $w^\star$ such that

$$\forall\, x \in K, \quad \langle w^\star, x \rangle < 0$$

Hence $\alpha := -\sup_{x \in K} \langle w^\star, x \rangle > 0$. Let us set $\beta := \sup_{x \in K} \|x\|$. We first infer from the Cauchy-Schwarz inequality that

$$k\alpha \leq \sum_{l=1}^{k} \langle w^\star, -x_{n_l} \rangle = \langle w^\star, w_{n_{k+1}} \rangle \leq \|w^\star\| \left\|w_{n_{k+1}}\right\|$$

and, second, we observe that

$$\left\|w_{n_{k+1}}\right\|^2 = \|w_{n_k}\|^2 + \|x_{n_k}\|^2 - 2\langle w_{n_k}, x_{n_k} \rangle \leq \|w_{n_k}\|^2 + \beta^2 \leq k\beta^2$$

These two inequalities imply that

$$k\alpha \leq \sqrt{k}\, \|w^\star\|\, \beta$$

Since α is positive, we conclude that the procedure must terminate after a finite number of terms. Since every element of the finite set K occurs infinitely often in the training set, the algorithm stops only if a classifier is found, thus proving the theorem. □

1.4 Multilayer Neural Networks

Let us first consider the case of a synchronous (or discrete) network governed by the propagation law $y = g(Wx)$. In the case of synchronous systems, we shall distinguish between one layer networks and multilayer networks:

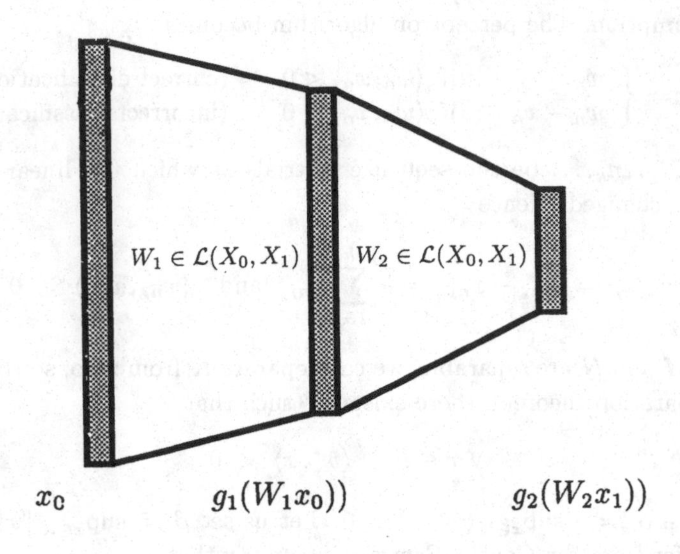

Fig. 1.8 Multi-layer Neural Network. Intermediate layers of hidden neurons process the input signals before sending them to the outputs neurons.

One Layer Networks

The neural network is applied once: It processes an input signal x to the output $y = g(Wx)$.

Multilayer Networks

There are L layers $l = 1, \ldots, L$, and the system is applied L times. Therefore, it is controlled by a sequence $(W_l)_{l=1,\ldots,L}$ of synaptic matrices. This means that we have a number $L + 1$ ($L \geq 2$) of signal spaces X_l, the first one being regarded as the space of inputs, and the last one as the space of outputs. The others are spaces of "hidden units" (Figure 1.8).

For simplicity, we shall neglect the preprocessing by conjuncts of neurons. The signals are first processed by synaptic matrices

$$\forall\, l = 1, \ldots, L, \quad W_l \in \mathcal{L}(X_{l-1}, X_l)$$

They weight the signals coming from one layer to the next before they are processed by the formal neurons of the next layer. We assume that the processing of the lth layer by formal neurons is described by a function $g_l : X_l \mapsto X_l$ and that each formal neuron of the lth layer processes the afferent signals [i.e., that the ith component of $g_l(x_l)$ depends only on the ith component x_{l_i} of the signal x_l]. Starting with an input $x_0 \in X_0$, the neural network weights this input using a synaptic matrix W_1 of the first layer and processes it, mapping it to the signal $x_1 := g_1(W_1 x_0)$. This signal is then weighted by a synaptic matrix W_2 of the second layer and processed, so that it becomes $x_2 := g_2(W_2 x_1)$. Hence, the law of evolution of the signals using a sequence of synaptic matrices W_l $(l = 1, \ldots, L)$ is given by

$$\forall\, l = 1, \ldots, L, \quad x_l = g_l(W_l x_{l-1})$$

We denote by Φ_L the map that associates with a sequence $\vec{W}$ of synaptic matrices W_l and an initial signal x_0 the final state

$$\Phi(x_0, \vec{W}) \;:=\; \Phi_L(x_0, W_1, \ldots, W_l) \;:=\; x_L$$

obtained through this discrete control system.

Naturally, a multilayer neural network can be regarded as a one-layer network controlled by a sequence $\vec{W} = (W_l)_{l=1,\ldots,L}$ of synaptic matrices and mapping input signals $x \in X$ to the output $\Phi_L(x_0, W_1, \ldots, W_L)$.

Infinite-Layer Networks

An asynchronous system governed by the differential equation

$$x'(t) \;=\; f(x(t), W(t))$$

maps a given input signal $x_0 \in X$ to the output signal

$$\Phi_T(x_0, W(\cdot)) \;:=\; x_{W(\cdot)}(T)$$

where $x(\cdot)$ is a solution to the foregoing differential equation controlled by $t \mapsto W(t)$. It can be regarded as an infinite-layer network.

Among the input-output pairs, one can distinguish the *equilibria* $(\bar{x}, \bar{x})$ associated with stationary synaptic matrices $\overline{W}$. They are solutions to the equations

$$f(\bar{x}, \overline{W}) = 0$$

In summary, a neural network maps input signals $x \in X$ to output signals $y = \Phi(x, W)$ after running the network through one-layer or a finite number of layers in the case of discrete networks, or a "continuous" set of layers in the case of asynchronous networks. In this sense, neural networks are instances of what is called an *adaptive system*.

1.5 Adaptive Systems

The general form of an adaptive network is given by a map $\Phi : X \times U \mapsto Y$, where X is the input space, Y the output space, and U a control space (or parameter space).

A *pattern* is an input-output pair $(a^p, b^p) \in X \times Y$. The inputs a^p are often called *keys* (or search argument), and the outputs b^p are the *memorized data*. In the case of neural networks or, more generally, of *connectionnist networks*, the control space is a matrix space (synaptic matrices, fitness matrices, activation matrices, etc., depending on the context and the fields of motivation). A pattern $(x, y) \in \mathcal{K}$ made of an input-output pair is *recognized* (or *discovered, generalized*) by the adaptive system programmed by such a control u if $y = \Phi(x, u)$ is a signal processed by the network excited by the input x. A *teaching set* (or *training set*) $\mathcal{P}$ is a finite set of patterns $(a^p, b^p)_{p\in\mathcal{P}} \subset \mathcal{K}$ of input-output signals.

The choice of such a control is made by *learning* a given number of input-output pairs (a^p, b^p); $(p \in \mathcal{P})$, (the teaching set), that is, by finding a control $u_{\mathcal{P}}$ satisfying

$$\forall\, p \in \mathcal{P}, \quad \Phi(a^p, u_{\mathcal{P}}) \;=\; b^p \tag{1.1}$$

We shall say that such a control $u_{\mathcal{P}}$ has *learned the teaching set*. With such a control u, the system *generalizes* from this teaching set by associating with any input x the output $y = \Phi(x, u_{\mathcal{P}})$. This is called the *generalization phase* of the operation of the adaptive system.

A common feature of adaptive systems is that they are not "programmed" by a sequence of instructions, but operate by making a control u (a synaptic matrix W in the case of neural networks) *learn a set of patterns* that they can reproduce, and thus, it is hoped, discover new patterns associated with such a synaptic matrix. It is nothing else, after all, than an extrapolation procedure set in a new framework.

Remark When $X = Y$ and $a^p = b^p$, the network is said to be *autoassociative*, so that we have to find a control $u_{\mathcal{P}}$ for which the states a^p are equilibria, in the sense that

$$\forall\, p \in \mathcal{P}, \quad \Phi(a^p, u_{\mathcal{P}}) \;=\; a^p \quad \square$$

This includes the *forecasting problem*, when $X = Y$ and when the patterns associated with a time series $a_1, \ldots, a_{T+1}$ are defined by $a^p := a_t$ and $b^p := a_{t+1}$ (the input is the present state, and the output the future state):

$$\forall\, t = 1, \ldots, T, \quad \Phi(a_t, u_{\mathcal{P}}) \;=\; a_{t+1}$$

The record of past states constitutes the teaching set.

Any algorithm converging to a solution $u_{\mathcal{P}}$ to the learning problem (1.1) is regarded as a *learning algorithm*. Since the problem (1.1) is generally a nonlinear problem, the first algorithm that comes to mind is the *Newton algorithm*, which is studied later, both in the case of general adaptive systems and for neural networks.

1.5.1 The Gradient Algorithms

We observe that the learning problem (1.1) is a nonlinear problem, which can be mapped to a minimization problem of the form

$$0 \;=\; \inf_u \left(\sum_{p \in \mathcal{P}} d_p(\Phi(a^p, u), b^p)^\alpha \right)^{1/\alpha}$$

where $\alpha \in [1, \infty]$ and where d_p are various kinds of *distances* on the output space Y. There are two advantages in doing that. The first one is the possibility of defining a solution to such a minimization problem even when there is no solution to the foregoing system of equations: When the infimum is not equal to zero, the minimal solution can be regarded as a quasi solution. The second one is the possibility of using the whole family of minimization algorithms (including variants of the gradient method) converging to a solution to the minimization problem.

When Y is a finite-dimensional vector space, the distances often are associated with *evaluation functions* $E_p : Y \mapsto \mathbf{R}$ vanishing (only) at the origin:

$$\inf_u \left(\sum_{p \in \mathcal{P}} E_p(\Phi(a^p, u) - b^p)^\alpha \right)^{1/\alpha}$$

The simplest example of distance is naturally provided by the Euclidean norm, so that the minimization problem becomes

$$\inf_u \left(\sum_{p \in \mathcal{P}} \|\Phi(a^p, u) - b^p\|^2 \right)^{1/2}$$

We include the important (although nonsmooth) case in which $\alpha = \infty$, where we have to solve the minimization problem

$$\inf_u \left(\sup_{p \in \mathcal{P}} \|\Phi(a^p, u) - b^p\| \right)$$

Definition 1.5.1 (Stationary Learning Problem) *The (stationary) learning problem amounts to finding a control u such that the adaptive system processes the keys a^p to the memorized data b^p for each pattern $b^p = \Phi(a^p, u)$, exactly if possible, or else, in an adequate approximate way, by solving minimization problems of the form*

$$\inf_u \left(\sum_{p \in \mathcal{P}} d_p(\Phi(a^p, u), b^p)^\alpha \right)^{1/\alpha}$$

where $1 \leq \alpha \leq \infty$.

Observe, however, that when the Fermat rule applies, any solution to such a minimization problem is a solution to a system of equations (when the criterion is differentiable) or a system of *inclusions* (when the criterion is nonsmooth) (in this case, the gradient of the criterion is replaced by a *generalized gradient*, which is generally a subset). This happens when $\alpha = \infty$ in the foregoing criteria.

Once we have transformed the learning problem into an optimization problem of the form

$$\inf_{u \in U} H(u)$$

then, as we indicated earlier, the natural algorithms that may lead to the minima are in most cases variants of the *gradient algorithm*. When H is differentiable, then it can be written

$$u^{n+1} - u^n = -\varepsilon_n H'(u^n)$$

The convergence holds true under convexity assumptions; we shall prove this theorem. In this case, we shall see that the differentiability is no longer necessary to implement the Fermat rule of the gradient algorithm, since continuous convex functions are *subdifferentiable*. A brief presentation of convex analysis will be provided to cover this case.

1.5.2 Learning without Forgetting: The Heavy Learning Algorithm

For simplicity, let us set $\mathcal{P} := \{1, \ldots P-1\}$ and $u_{P-1} := u_{\mathcal{P}}$. An algorithm that learns without forgetting is defined as follows. Assume that u_{P-1} has been obtained for learning the teaching set of $(a^p, b^p)_{1 \le p \le P-1}$ made of $P-1$ input-output pairs. At the Pth iteration, we add another pair (a_P, b_P) to the teaching set. We want to find a control u_P that will learn the whole new teaching set $(a^p, b^p)_{1 \le p \le P}$ (and not only the last pattern).

A natural way to design such an algorithm is to use the control u_{P-1} that has learned the teaching set $(a^p, b^p)_{1 \le p \le P-1}$ and to modify it as little as possible for learning the new teaching set $(a^p, b^p)_{1 \le p \le P}$: This is what we call the *heavy algorithm*. Formally, u_P is obtained from u_{P-1} in the following way:

$$\begin{cases} u_P \text{ minimizes } u \mapsto \|u - u_{P-1}\| \\ \text{under the constraints } \quad \forall\, t = 1, \ldots, P, \quad \Phi(a^p, u) \;=\; b^p \end{cases}$$

When the system is affine with respect to the control (but still nonlinear with respect to the state), the heavy algorithm provides very simple formulas. For that purpose, we shall need to recall some basic facts about pseudoinverses.

1.5.3 Hebbian Learning Rules

What, then, is special about neural networks? We shall have to apply these algorithms (Newton's algorithm, gradient algorithms, heavy algorithms) to adaptive systems controlled by synaptic matrices, and thus to use specific properties of the spaces of matrices (regarded as tensor products).

Since Hebb's classic book *The Organization of Behavior* (1949), most studies of neural networks have dealt with learning rules that prescribe a priori the evolution of the synaptic matrix W^n through an algorithm of the form

$$W^{n+1} - W^n = R(t, W^n, (a^p, b^p)_{p \in \mathcal{P}})$$

which converges to a solution to a stationary learning problem. They all are variants and combinations of the same idea: When the pattern to be learned is made of one pair (a, b), the synaptic weight from a neuron i to a neuron j should be strengthened whenever the connection is highly active. The activity of synapse (i, j) is measured in most models as the product of activities $\alpha_i(W, a, b)$ in the presynaptic neuron and activities $\beta_j(W, a, b)$ in the postsynaptic neuron:

$$r_{ij}(W, a, b) := \beta_i(W, a, b)\alpha_j(W, a, b)$$

Such learning rules are often called *Hebbian learning rules* or Δ-rules.

We shall not follow this line of thought here, since our aim is quite opposite: Instead of analyzing the properties of a given learning rule, we shall try to design a posteriori learning rules satisfying given requirements. However, we shall see that the algorithms mentioned earlier yield Hebbian learning rules, because they involve tensor products.

Curiously enough, the history of neural networks is quite similar to that of mathematical economics, in which the role of the synaptic matrix is played by prices, and the role of learning rules is played by price-adjustment rules, describing the law of supply and demand, the most famous one being the Walras tâtonnement. In both cases, price-adjustment rules and learning rules were proposed and studied. Afterward, scientists looked at them as allowing the system to satisfy a given purpose.

We need, for that purpose, some results dealing with the structure of the space of linear operators that we shall present later, in the framework of linear neural networks, known also under the name of "associative memories." We shall also see that in the case of multiple layers ($l = 1, \dots L$), the learning rule derived from the gradient method becomes the *back-propagation formula.*

2

Pseudoinverses and Tensor Products

Introduction

The projection theorem allows construction of the orthogonal right-inverse of a linear surjective operator A, associating with any datum y the solution x to the equation $Ax = y$ with minimal norm. In the same way, it allows construction of the orthogonal left-inverse of a linear injective operator A, associating with any datum y the solution x to the equation $Ax = \bar{y}$, where $\bar{y}$ is the orthogonal projection of y onto the image of A. More generally, when A is any linear operator between finite-dimensional vector spaces, the *pseudoinverse* of A associates with any datum y the solution x (with minimal norm) to the equation $Ax = \bar{y}$, where $\bar{y}$ is the orthogonal projection of y onto the image of A.

These definitions show how useful the concept of the pseudoinverse is in many situations. It is used explicitly or implicitly in many domains of statistics and data analysis. It is then quite natural that the pseudoinverse plays an important role in the use of adaptive systems in learning algorithms of patterns.

This is what we do to construct the heavy algorithm for adaptive systems that are affine with respect to the controls. Because we are looking for synaptic matrices when we deal with neural networks, we have to make a short pause to study the structure of the space of linear operators, of its dual, and of a *tensor product* of linear operators. Actually, these tensor products appear naturally in many formulas (such as the gradients of functionals defined on matrices). The main result, which will be used over and over, is that *the pseudoinverse of a tensor product is the tensor product of the pseudoinverses.*

2.1 Pseudoinverses

2.1.1 Finite-Dimensional Vector Spaces

We denote by $(e^j)_{j=1,\dots,n}$ a basis of X and by $(f^i)_{i=1,\dots,m}$ a basis of Y. Let $X^\star$ and $Y^\star$ denote the dual spaces of X and Y, respectively. The bilinear form $\langle p, x\rangle := p(x)$ on $X^\star \times X$ is called the *duality product.* We associate with the basis $(e^j)_j$ its dual basis $(e_j^\star)_{j=1,\dots,n}$ of $X^\star$ defined by $\langle e_j^\star, e^l\rangle = 0$ when $j \neq l$ and $= \langle e_j^\star, e^j\rangle = 1$. Hence the jth component x_j of x in the basis $(e^j)_j$ is given by $x_j = \langle e_j^\star, x\rangle$, the jth component p^j of $p \in X^\star$ in the dual basis $(e_j^\star)_j$ is given by $p^j = \langle p, e^j\rangle$, and $\langle p, x\rangle := \sum_{j=1}^n p^j x_j$ and $\langle q, y\rangle := \sum_{i=1}^m q^i y_i$.

We recall that the transpose $W^\star \in \mathcal{L}(Y^\star, X^\star)$ is defined by

$$\forall\, q \in Y^\star \quad \text{and} \quad \forall\, x \in X, \quad \langle W^\star q, x\rangle \;=\; \langle q, Wx\rangle$$

Let p belong to $X^\star$. We shall identify the linear form $p \in \mathcal{L}(X, \mathbf{R})$: $x \in X \mapsto < p, x >$ with its transpose $p \in \mathcal{L}(\mathbf{R}, X^\star)$: $\lambda \mapsto \lambda p \in X^\star$. In the same way, we shall identify $x \in X$ with both the maps $p \in X^\star \mapsto < p, x >$ and $\lambda \in \mathbf{R} \mapsto \lambda x \in X$.

Let us introduce the following notations: We associate with any $p \in X^\star$ and $y \in Y$ the linear operator $p \otimes y \in \mathcal{L}(X, Y)$, defined as

$$p \otimes y \;: x \;\mapsto\; (p \otimes y)(x) \;:=\; \langle p, x\rangle y$$

the matrix of which is

$$\left(p^i y_j\right)_{\substack{i=1,\dots,m \\ j=1,\dots,n}}$$

Its transpose $(p \otimes y)^\star \in \mathcal{L}(Y^\star, X^\star)$ maps $q \in Y^\star$ to

$$(p \otimes y)^\star(q) = \langle q, y\rangle p$$

because $\langle q, \langle p, x\rangle y\rangle \;=\; \langle\langle q, y\rangle p, x\rangle$. *Henceforth, we shall identify the transpose* $(p \otimes y)^\star$ *with* $y \otimes p$.

In the same way, if $q \in Y^\star$ and $x \in X$ are given, $q \otimes x$ denotes the map $y \in Y \mapsto < q, y > x \in X$ belonging to $\mathcal{L}(Y, X)$, and its transpose $p \in X^\star \mapsto < p, x > q \in Y^\star$ belonging to $\mathcal{L}(X^\star, Y^\star)$ is identified with $x \otimes q$.

We observe that the map $(p, y) \in X^\star \times Y \mapsto p \otimes y \in \mathcal{L}(X, Y)$ is bilinear. *The operators* $(e_j^\star \otimes f^i)_{\substack{i=1,\dots,m \\ j=1,\dots,n}}$ form a basis of $\mathcal{L}(X, Y)$, because we can write $x = \sum_{j=1}^{n} \langle e_j^\star, x\rangle e^j$ and

$$\begin{aligned} Wx &= \sum_{i=1}^{m} \langle f_i^\star, Wx\rangle f^i = \sum_{i=1}^{m}\sum_{j=1}^{n} \langle f_i^\star, W(e^j)\rangle \langle e_j^\star, x\rangle f^i \\ &= \left(\sum_{i=1}^{m}\sum_{j=1}^{n} \langle f_i^\star, W(e^j)\rangle e_j^\star \otimes f^i\right)(x) \end{aligned}$$

where the entries

$$w_i^j := \langle f_i^\star, W(e^j)\rangle$$

of the matrix W are the components of W in this basis.

2.1.2 Orthogonal Right-Inverses

We shall begin by defining and characterizing the orthogonal right-inverse of a surjective linear operator and then the orthogonal left-inverse of an injective linear operator. These concepts depend, respectively, on the scalar products l and m defined on the finite-dimensional vector spaces X and Y, which are then given once and for all. We shall denote by $L \in \mathcal{L}(X, X^\star)$ and $M \in \mathcal{L}(Y, Y^\star)$ their *duality mappings* defined, respectively, by

$$\forall\, x_1,\ x_2 \in X, \quad \langle Lx_1, x_2\rangle := l(x_1, x_2)$$

$$\forall\, y_1,\ y_2 \in Y, \quad \langle My_1, y_2\rangle := m(y_1, y_2)$$

[The matrix $(l(e^i, e^j)_{i,j=1,\ldots,n})$ of the bilinear form l coincides with the matrix of the linear operator L.]

We shall denote by

$$\lambda(x) := \sqrt{l(x,x)} \quad \text{and} \quad \mu(y) := \sqrt{m(y,y)}$$

the norms associated with these scalar products.

The bilinear form $l_\star(p_1, p_2) := \langle p_1, L^{-1}p_2\rangle = l(L^{-1}p_1, L^{-1}p_2)$ is then a scalar product on the dual of $X^\star$, called the *dual scalar product*.

It is quite important to keep options open for the choice of a scalar product. The choice of the Euclidean scalar product is not always wise, despite its simplicity. We shall see later that we can associate with a scalar product and a linear operator initial and final scalar product.

Another instance of a choice of a scalar product is to make a given basis orthonormal.

Lemma 2.1.1 *Assume that* $\{e^1, \ldots, e^n\}$ *is a basis of a finite-dimensional vector space* X, *and* $B \in \mathcal{L}(\mathbf{R}^n, X)$ *is the associated operator defined by*

$$\forall\, x \in \mathbf{R}^n, \quad Bx \;=\; \sum_{i=1}^{n} x_i e^i$$

Then

$$l(x,y) \;:=\; \langle B^{-1}x, B^{-1}y\rangle$$

is a scalar product for which the basis $\{e^1, \ldots, e^n\}$ *is orthonormal. The associated duality mapping is equal to* $L = (BB^\star)^{-1}$.

Proof Indeed, to say that the sequence $\{e^1, \ldots, e^n\}$ is linearly independent amounts to saying that the associated linear operator B is bijective and maps the canonical basis of $\mathbf{R}^n$ onto the basis $\{e^1, \ldots, e^n\}$. Hence the bilinear form l defined earlier is a scalar product satisfying

$$l(e^i, e^j) \;:=\; \langle B^{-1}e^i, B^{-1}e^j\rangle \;=\; \delta_{i,j}$$

□

Let us consider a surjective linear operator $A \in \mathcal{L}(X, Y)$. Then, for any $y \in Y$, the problem $Ax = y$ has at least a solution. We may select the solution $\bar{x}$ with minimal norm $\lambda(x)$, that is, a solution to the minimization problem with linear equality constraints

$$A\bar{x} = y \quad \text{and} \quad \lambda(\bar{x}) = \min_{Ax=y} \lambda(x)$$

The solution to this problem is given by the formula

$$\bar{x} = L^{-1}A^\star(AL^{-1}A^\star)^{-1}y$$

Indeed, for any $v \in \operatorname{Ker}(A)$ and h, we have $\lambda(\bar{x}) \leq \lambda(\bar{x} + hv)$, so that, taking the limit when $h \longrightarrow 0$ of $(\lambda(\bar{x}+hv)^2 - \lambda(\bar{x})^2)/2h$, we infer that $l(\bar{x}, v) = \langle L\bar{x}, v\rangle = 0$ for any $v \in \operatorname{Ker}(A)$. This means that $L\bar{x}$ belongs to the orthogonal of $\operatorname{Ker}(A)$, which is the image of $A^\star$. Thus there exists some $q \in Y^\star$ such that $L\bar{x} = A^\star q$, so that $A\bar{x} = AL^{-1}A^\star q = y$. But $AL^{-1}A^\star$ is positive-definite, because $A^\star$ is injective (A being surjective): Indeed, for all $q \in Y^\star$,

$$\langle AL^{-1}A^\star q, q\rangle \;=\; \langle L^{-1}A^\star q, A^\star q\rangle \;=\; \lambda_\star(A^\star q)^2 = 0$$

is equivalent to

$$A^\star q \;=\; 0 \iff q \;=\; 0$$

We thus infer that $q = (AL^{-1}A^\star)^{-1}y$.

Definition 2.1.2 *If $A \in \mathcal{L}(X,Y)$ is surjective, we say that the linear operator $A^+ := L^{-1}A^\star(AL^{-1}A^\star)^{-1} \in \mathcal{L}(Y,X)$ is the orthogonal right-inverse of A (associated with the scalar product l on X).*

Indeed, A^+ is obviously a right-inverse of A, because $AA^+y = y$ for any $y \in Y$. We observe that $1 - A^+A$ is the *orthogonal projector onto the kernel of A.*

2.1.3 Quadratic Minimization Problems

Hence, we can write explicitly the solution $\bar{x}$ to the quadratic optimization problem with linear equality constraints

$$A\bar{x} = y \quad \text{and} \quad \frac{1}{2}\lambda(\bar{x} - u)^2 = \min_{Ax=y} \frac{1}{2}\lambda(x - u)^2$$

Proposition 2.1.3 *Let us assume that $A \in \mathcal{L}(X,Y)$ is surjective. Then the unique solution $\bar{x}$ to the foregoing quadratic minimization problem is given by*

$$\begin{cases} \bar{x} = u - L^{-1}A^\star\bar{q} = u - A^+(Au - y) \\ \text{where } \bar{q} := (AL^{-1}A^\star)^{-1}(Au - y) \text{ is the } \textit{Lagrange multiplier} \end{cases}$$

that is, a solution to the dual minimization problem

$$\min_{q \in Y^\star} \left(\frac{1}{2}\lambda_\star(A^\star q - Lu)^2 + \langle q, y \rangle \right)$$

We observe easily that $\xi(q) := u - L^{-1}A^\star(q)$ minimizes over X the function $x \mapsto \frac{1}{2}\lambda(x - u)^2 + \langle q, Ax \rangle$ and that the Lagrange multiplier minimizes $q \mapsto \frac{1}{2}\lambda_\star(\xi(q))^2 + \langle q, y \rangle$.

2.1.4 Projections onto Cones

Let us consider now a closed convex cone $Q \subset Y$, regarded as the cone of nonnegative elements of the partial ordering associated with Q ($x \leq y \Longleftrightarrow y - x \in Q$). We supply the dual $Y^\star$ of Y with the scalar product defined by $l_\star(A^\star q_1, A^\star q_2)$, whose duality mapping is $AL^{-1}A^\star \in \mathcal{L}(Y^\star, Y)$.

Let us consider now the quadratic optimization problem with linear inequality constraints

$$A\overline{x}_{+} \geq y \quad \text{and} \quad \frac{1}{2}\lambda(\overline{x}_{+} - u)^2 = \min_{Ax \geq y} \frac{1}{2}\lambda(x-u)^2$$

We observe that the dual problem can be written

$$\inf_{q \in Q^-} \left(\frac{1}{2}\lambda_\star(A^\star q - Lu)^2 - < q, y > \right)$$

So the solution $\overline{x}_{+}$ of the minimization problem with inequality constraints and the solution $\overline{q}_{-} \in Q^-$ of its dual problem, the Lagrange multiplier, are defined by

$$\begin{cases} \overline{x}_{+} = u - L^{-1}A^\star\overline{q}_{-} \\ \text{where} \quad < \overline{q}_{-}, A\overline{x}_{+} - y >= 0 \end{cases}$$

Since $A\overline{x}_{+} - y \in Q$, then $\langle A\overline{x}_{+} - y, q \rangle \leq 0$ for all $q \in Q^-$, so that we can write

$$\forall\, q \in Q^-, \quad \langle A\overline{x}_{+} - y, q - \overline{q}_{-} \rangle = \langle Au - y - AL^{-1}A^\star\overline{q}_{-}, q - \overline{q}_{-} \rangle$$

$$= \langle AL^{-1}A^\star(\bar{q} - \overline{q}_{-}), q - \overline{q}_{-} \rangle = l_\star(A^\star(\bar{q} - \overline{q}_{-}), A^\star(q - \overline{q}_{-})) \leq 0$$

Therefore, we have proved that the Lagrange multiplier $\overline{q}_{-} \in Q^-$ of the minimization problem with inequality constraints is the orthogonal projection onto Q^- of the Lagrange multiplier $\bar{q} \in Y^\star$ of the minimization problem with equality constraints by the formulas of Proposition 2.1.3. Hence the solution $\bar{x}_{+}$ to the minimization problem with inequality constraints is given by the formula

$$\begin{cases} \overline{x}_{+} \; = \; u - L^{-1}A^\star\overline{q}_{-}, \quad \text{where} \\ \overline{q}_{-} \text{ is the orthogonal projection onto } Q^- \text{ of the Lagrange multiplier} \\ \bar{q} := (AL^{-1}A^\star)^{-1}(Au - y) \text{ of the problem with equality constraints} \\ \text{when } Y^\star \text{ is supplied with the scalar product } \; l_\star(A^\star q_1, A^\star q_2) \end{cases}$$

2.1.5 Projections onto Inverse Images of Convex Sets

Consider now a closed convex subset $M \subset Y$ and the minimization problem

$$\inf_{Ax \in M} \lambda(x - u)$$

When $A \in \mathcal{L}(X,Y)$ is surjective, we can supply the space Y with the *final scalar product*

$$m^A(y_1, y_2) := l(A^+y_1, A^+y_2)$$

its associated *final norm*[1]

$$\mu^A(y) := \lambda(A^+y) = \inf_{Ax=y} \lambda(x)$$

Its duality mapping is equal to $(AL^{-1}A^\star)^{-1} \in \mathcal{L}(Y, Y^\star)$. In this case, we denote by π_M^A the projector of best approximation onto the closed convex subset M when Y is supplied with the norm μ^A.

Proposition 2.1.4 *Assume that $A \in \mathcal{L}(X,Y)$ is surjective and that $M \subset Y$ is closed and convex. Then the unique solution $\bar{x}$ to the minimization problem*

$$\inf_{Ax\in M} \lambda(x-u)$$

is equal to

$$\bar{x} = u - A^+(Au - \pi_M^A(Au))$$

Proof Indeed, we can write

$$\begin{cases} \lambda(\bar{x}-u) = \inf_{Ax\in M} \lambda(x-u) \\ = \inf_{y\in M} \inf_{Ax=y} \lambda(x-u) \\ = \inf_{y\in M} \lambda(A^+(y - Au)) \\ \text{(thanks to Proposition 2.1.3)} \\ = \lambda(A^+(\pi_M^A(Au)) - Au) \end{cases}$$

Hence $\bar{x} = u - A^+(Au - \pi_M^A(Au))$. □

2.1.6 Orthogonal Left-Inverses

Let us consider now an injective linear operator $B \in \mathcal{L}(X,Y)$. Because the problem $Bx = y$ may not have a solution (when y does not belong

[1]This can also be denoted by $\mu^A := A \cdot \mu$, by contrast with the *initial norm* $\lambda \cdot A$. We observe that $A \cdot \mu = \lambda \cdot A^+$.

to the image of B), we shall project y onto this image and take for an approximated solution the inverse of this projection. In other words, the approximated solution to this problem is the element $\bar{x} \in X$ defined by

$$\mu(B\bar{x} - y) = \min_{x \in X} \mu(Bx - y)$$

Since the derivative of the convex function $\frac{1}{2}\mu(\cdot)^2$ at y is equal to $\langle My, \cdot \rangle$, the Fermat rule states that $\bar{x}$ is a solution to the equation

$$\forall\, v \in X, \quad \langle M(B\bar{x} - y), Bv \rangle = \langle B^\star M(B\bar{x} - y), v \rangle = 0$$

The self-transposed operator $B^\star MB$ is positive-definite when B is injective [because $\langle B^\star MBx, x \rangle = \mu(Bx)^2 = 0$ if and only if $Bx = 0$, i.e., if and only if $x = 0$]. Therefore $B^\star MB$ is invertible, and we derive that $\bar{x}$ is equal to $(B^\star MB)^{-1}B^\star My$.

Definition 2.1.5 *If $B \in \mathcal{L}(X, Y)$ is injective, we say that the linear operator $B^- := (B^\star MB)^{-1}B^\star M \in \mathcal{L}(Y, X)$ is the* orthogonal left-inverse *of B.*

Indeed, B^- is obviously a left-inverse of B, because $B^-Bx = x$ for any $x \in X$, and BB^- is the orthogonal projector onto the image of B.

Proposition 2.1.6 *Let us assume that $B \in \mathcal{L}(X, Y)$ is injective. Then*

$$\begin{cases} \text{(i)} & (B^-)^\star = (B^\star)^+ \\ \text{(ii)} & B = (B^-)^+ \\ \text{(iii)} & (B^\star MB)^{-1} = B^-M^{-1}(B^-)^\star \\ \text{(iv)} & \text{If } V \in \mathcal{L}(Z, X) \text{ is invertible, then } (BV)^- = V^{-1}B^- \end{cases}$$

If $A \in \mathcal{L}(X, Y)$ is surjective, then

$$\begin{cases} \text{(i)} & (A^+)^\star = (A^\star)^- \\ \text{(ii)} & A = (A^+)^- \\ \text{(iii)} & (AL^{-1}A^\star)^{-1} = (A^+)^\star LA^+ \\ \text{(iv)} & \text{If } W \in \mathcal{L}(Y, Z) \text{ is invertible, then} (WA)^+ = A^+W^{-1} \end{cases}$$

Example: Orthogonal Projector Let $P := [b^1, \ldots, b^n]$ be the vector space spanned by n linearly independent vectors b^l of a finite-dimensional vector space Y, supplied with the scalar product $m(y_1, y_2)$. If we denote by $B \in \mathcal{L}(\mathbf{R}^n, Y)$ the linear operator defined by $Bx := \sum_{i=1}^n x_i b^i$, which is injective because the vectors b^i are independent, we infer that BB^- is the orthogonal projector on the subspace $P = \mathrm{Im}(B)$.

The entries of the matrix of $B^\star MB$ are equal to $m(b^i, b^j)$ $(i, j =$

$1,\ldots,n$). Let us denote by g_{ij} the entries of its inverse. We infer that $B^- y = (\sum_{j=1}^n g_{ij} m(y, b^j))_{i=1,\ldots,n}$ and therefore that

$$BB^- y = \sum_{i,j=1}^{n} g_{ij} m(y, b^j) b^i$$

When the vectors are m-orthogonal, the formula becomes

$$BB^- y = \sum_{i=1}^{n} \frac{m(b^i, y)}{\mu(b^i)^2} b^i \ \square$$

In particular, we can regard an element $y \in Y$ different from zero as the injective linear operator $y \in \mathcal{L}(\mathbf{R}, Y)$ associating with α the vector αy. It transpose $y^\star \in \mathcal{L}(Y^\star, \mathbf{R}) = Y$ is the map $p \mapsto \langle p, y \rangle$, and its left-inverse $y^- \in \mathcal{L}(Y, \mathbf{R}) = Y^\star$ is equal to

$$y^- = \frac{My}{\lambda(y)^2} \in Y^\star$$

2.1.7 Pseudoinverses

Let us consider now any linear operator $C \in \mathcal{L}(X, Y)$. It can be split as the product $C = BA$ of a surjective linear operator $A \in \mathcal{L}(X, Z)$ and an injective linear operator $B \in \mathcal{L}(Z, Y)$: We can take, for instance, $Z := \mathrm{Im}(C)$, $A := C \in \mathcal{L}(X, Z)$, and $B := I \in \mathcal{L}(Z, Y)$.

Take now two decompositions $C = B_1 A_1 = B_2 A_2$, where $A_i \in \mathcal{L}(X, Z_i)$ is surjective and where $B_i \in \mathcal{L}(Z_i, Y)$ is injective ($i = 1, 2$). Then the images of the B_i's are equal to $\mathrm{Im}(C)$, and the kernels of the A_i's are also equal to the vector subspace $\mathrm{Ker}(C) \subset X$. Let $\varphi \in \mathcal{L}(X, X/\mathrm{Ker}(C))$ denote the canonical surjection from X to the factor space $X/\mathrm{Ker}(C)$. Hence the surjective operators A_i split as $\tilde{A}_i \varphi$, where $\tilde{A}_i \in \mathcal{L}(X/\mathrm{Ker}(C), Z_i)$ are invertible ($i = 1, 2$). Then $V := \tilde{A}_1 \tilde{A}_2^{-1}$ is invertible, and we have the relations

$$A_1 = V A_2 \quad \text{and} \quad B_2 = B_1 V$$

We thus deduce that $A_1^+ B_1^- = A_2^+ B_2^-$ does not depend upon the decomposition of C as a product of an injective operator and a surjective operator, because $A_1^+ B_1^- = A_2^+ V^{-1} V B_2^- = A_2^+ B_2^-$.

Definition 2.1.7 *Let $C = BA \in \mathcal{L}(X, Y)$ be any linear operator split as the product of an injective operator B and a surjective operator A. Then the operator $C^{\ominus 1} := A^+ B^- \in \mathcal{L}(Y, X)$ is called the pseudoinverse of C.*

If $y \in Y$ is given, we can regard $\bar{x} = C^{\ominus 1}y$ as the closest solution with minimal norm. Indeed, by taking $A = C$ and $B = I$, we see that $\bar{x} = A^{+}\bar{y}$, where $\bar{y} := I^{-}(y)$ is the projection of y onto the image of C, and that $\bar{x}$ is the solution with minimal norm to the equation $Cx = \bar{y}$. □

Naturally, if C is surjective, the pseudoinverse $C^{\ominus 1} = C^{+}$ coincides with the orthogonal right-inverse; if C is injective, the pseudoinverse $C^{\ominus 1} = C^{-}$ coincides with the orthogonal left-inverse; and if C is bijective, the pseudoinverse $C^{\ominus 1} = C^{-1}$ coincides with the inverse.

These are the (obvious) properties of the pseudoinverses:

$$\begin{cases} \text{(i)} & C^{\ominus 1}CC^{\ominus 1} = C^{\ominus 1} \\ \text{(ii)} & CC^{\ominus 1}C = C \\ \text{(iii)} & C^{\ominus 1}C \text{ is the orthogonal projector onto } \operatorname{Ker}(C) \\ \text{(iv)} & CC^{\ominus 1} \text{ is the orthogonal projector onto } \operatorname{Im}(C) \end{cases}$$

We also observe that

$$\begin{cases} \text{(i)} & (C^{\ominus 1})^{\ominus 1} = C \\ \text{(ii)} & (C^{\ominus 1})^{\star} = (C^{\star})^{\ominus 1} \\ \text{(iii)} & (C_2C_1)^{\ominus 1} = C_1^{\ominus 1}C_2^{\ominus 1} \end{cases}$$

2.2 The Heavy Algorithm Theorem

We assume that X, Y, and U are finite-dimensional vector spaces, and that the dynamics of the adaptive network Φ are affine with respect to the control

$$\Phi(x,u) := c(x) + G(x)u$$

where

$$\begin{cases} \text{(i)} & c : X \mapsto Y \text{ is continuous} \\ \text{(ii)} & G : X \mapsto \mathcal{L}(U,Y) \text{ is continuous} \\ \text{(iii)} & \forall\, x \in X,\ \ G(x) \in \mathcal{L}(U,Y) \text{ is surjective} \end{cases} \tag{2.1}$$

We supply the control space U with a scalar product $n(u,v) := \langle Nu, v\rangle$, where $N \in \mathcal{L}(U,U^{\star})$ is the duality mapping, which is a symmetric and positive-definite matrix.

Let us consider a finite *teaching set* of patterns $(a^p, b^p)_{p\in\mathcal{P}} \subset X \times Y$ of input-output signals and the *learning problem*: Find a control $u_{\mathcal{P}}$ satisfying

$$\forall\, p \in \mathcal{P},\ \ \Phi(a^p, u_{\mathcal{P}}) = b^p \tag{2.2}$$

Because the continuous linear operators $G(x)$ are surjective, the transposes $G(x)^\star$ are isomorphisms from $Y^\star$ onto their closed images $\mathrm{Im}(G(x)^\star) = (\mathrm{Ker}(G(x)))^\perp$. We shall posit the orthogonality condition

$$\forall\, p,q = 1,\dots,P, \quad p \neq q, \quad G(a^p)N^{-1}G(a^q)^\star = 0 \qquad (2.3)$$

which is equivalent to

$$\forall\, p = 1,\dots,P, \text{ the subspaces } \mathrm{Im}(G(a^p)^\star) \text{ are mutually orthogonal}$$

We recall that the operator

$$G(x)^+ := N^{-1}G(x)^\star \left(G(x)N^{-1}G(x)^\star\right)^{-1}$$

is a right-inverse of $G(x)$, called the *orthogonal right inverse* of $G(x)$, because it associates with any $y \in Y$ the solution $\bar{u}$ with minimal norm to the equation $G(x)u = y$.

Theorem 2.2.1 *We posit the assumptions (2.1) and (2.3). Then the heavy algorithm associates with u_{P-1} the control u_P defined by the formula*

$$u_P = u_{P-1} - G(a^P)^+ \left(c(a^P) + G(a^P)u_{P-1} - b^P\right)$$

Proof Denote by $\mathcal{A}$ the surjective linear operator from U to the space $\mathcal{Y} := Y^P$ defined by

$$\mathcal{A}u := \left(G(a^p)u\right)_{p=1,\dots,P}$$

and set $y := \left(b^p - c(a^p)\right)_{p=1,\dots,P}$. The control u_P we are looking for is given by

$$u_P \text{ minimizes } \|u - u_{P-1}\| \text{ under } \mathcal{A}u = y$$

the solution of which is given by

$$u_P = u_{P-1} - N^{-1}\mathcal{A}^\star \left(\mathcal{A}N^{-1}\mathcal{A}^\star\right)^{-1}\left(\mathcal{A}u_{P-1} - y\right)$$

It remains to observe that $\mathcal{A}^\star \in \mathcal{L}(Y^{\star P}, U^\star)$ is defined by

$$\mathcal{A}^\star \left(\pi_p\right)_{p=1,\dots,P} = \sum_{p=1}^{P} G(a^p)^\star \pi_p$$

so that, by (2.3), $\mathcal{A}N^{-1}\mathcal{A}^\star$ is the diagonal block matrix

$$\mathcal{A}N^{-1}\mathcal{A}^\star = \left(G(a^p)N^{-1}G(a^p)^\star\right)_{p=1,\dots,P}$$

Since

$$\mathcal{A}u_{P-1} - y = (0, \ldots, 0, c(a^P) + G(a^P)u_{P-1} - b_P)$$

because $c(a^p) + G(a^p)u_{P-1} - b^p = 0$ for $p \leq P-1$, we infer that

$$N^{-1}\mathcal{A}^\star \left(\mathcal{A}N^{-1}\mathcal{A}^\star\right)^{-1}(\mathcal{A}u - y) = G(a^P)^+ \left(c(a^P) + G(a^P)u_{P-1} - b^P\right)$$

□

2.3 Duals of Spaces of Linear Operators

Because we deal with synaptic matrices, we devote this section to some results concerning spaces $\mathcal{L}(X, Y)$ of linear operators W from a finite-dimensional vector space X to another one Y and their duals.

2.3.1 The Isomorphism Theorem

We shall deal with linear operators $\mathcal{A}$ mapping a matrix space $\mathcal{L}(X, Y)$ to a finite-dimensional vector space Z. The following isomorphism theorem plays a very important role.

Theorem 2.3.1 *The two spaces*

$$\mathcal{L}(\mathcal{L}(X,Y), Z) \approx \mathcal{L}(X^\star, \mathcal{L}(Y, Z)) \tag{2.4}$$

are isomorphic and will henceforth be identified through the following formula:

$$\forall\, p \in X^\star \quad \text{and} \quad \forall\, y \in Y, \quad \mathcal{A}(p)(y) := \mathcal{A}(p \otimes y)$$

Proof The isomorphism j associates with any linear operator

$$\mathcal{A} \in \mathcal{L}(\mathcal{L}(X,Y), Z)$$

the operator $j(\mathcal{A}) \in \mathcal{L}(X^\star, \mathcal{L}(Y, Z))$ defined in the following way:

$$\forall\, p \in X^\star, \quad j(\mathcal{A})(p) : y \in Y \mapsto j(\mathcal{A})(p)(y) := \mathcal{A}(p \otimes y)$$

This makes sense because $p \otimes y$ is a linear operator from X to Y. The map j is obviously linear. Let k be the operator mapping an operator $\widetilde{\mathcal{A}} \in \mathcal{L}(X^\star, \mathcal{L}(Y, Z))$ to the operator

$$\mathcal{A} = k(\widetilde{\mathcal{A}}) \in \mathcal{L}(\mathcal{L}(X,Y), Z)$$

defined in the following way:

$$\forall\, W := \sum_{i=1}^{m}\sum_{j=1}^{n} w_i^j e_j^\star \otimes f^i, \quad \mathcal{A}(W) := \sum_{i=1}^{m}\sum_{j=1}^{m} w_i^j \widetilde{\mathcal{A}}(e_j^\star) f^i$$

This operator k is clearly the inverse of j. Hence j is an isomorphism. From now on, we shall identify $\mathcal{A}$ with $j(\mathcal{A})$. □

Examples of such maps $\mathcal{A}$ are the maps

$$x \otimes B : W \in \mathcal{L}(X,Y) \mapsto (x \otimes B)(W) := BWx$$

where x is given in X and B is given in $\mathcal{L}(Y,Z)$, which associate with any $W = p \otimes y$ the operator $p \otimes By$.

In particular, when $Z = \mathbf{R}$, we obtain the isomorphism between the dual of $\mathcal{L}(X,Y)$ and $\mathcal{L}(X^\star, Y^\star)$.

Corollary 2.3.2 *We shall identify the dual $(\mathcal{L}(X,Y))^\star$ of $\mathcal{L}(X,Y)$ with the space $\mathcal{L}(X^\star, Y^\star)$. The following formulas hold true:*

$$\begin{cases} \text{(i)} & \langle x \otimes q, W\rangle = \langle q, Wx\rangle \\ \text{(ii)} & \langle x \otimes q, p \otimes y\rangle = \langle p, x\rangle\langle q, y\rangle \end{cases} \tag{2.5}$$

If $(e^j)_{j=1,\ldots,n}$, $(e_j^\star)_{j=1,\ldots,n}$ and $(f^i)_{i=1,\ldots,m}$, $(f_i^\star)_{i=1,\ldots,m}$ are dual bases of X and Y, then the bases $(e_j^\star \otimes f^i)_{i,j}$ and $(e^k \otimes f_l^\star)_{k,l}$ are dual bases of $\mathcal{L}(X,Y)$ and $\mathcal{L}(X^\star, Y^\star)$.

Furthermore, for any pair of dual bases, the duality product can be written in the form

$$\begin{aligned} \langle V^\star, U\rangle &:= \sum_{i=1}^{m}\sum_{j=1}^{n} \langle V^\star e_j^\star, f^i\rangle\langle f_i^\star, Ue^j\rangle \\ &= \sum_{j=1}^{n} \langle V^\star e_j^\star, Ue^j\rangle \\ &= \sum_{i=1}^{m} \langle U^\star f_i^\star, Vf^i\rangle \end{aligned}$$

and does not depend on the choice of the pairs of dual bases.

Proof We have to check the formula for the duality product. Because the bases $(f^i, f_i^\star)_i$ are dual bases, for any $q \in Y^\star$ and $y \in Y$, we have $\langle q, y\rangle = \sum_{i=1}^{m} \langle q, f^i\rangle\langle f_i^\star, y\rangle$. Therefore, by taking $q := V^\star e_j^\star$ and $y :=$

Ue^j, we obtain

$$\sum_{i=1}^{m}\sum_{j=1}^{n} < V^\star e_j^\star, f^i >< f_i^\star, Ue^j > = \sum_{j=1}^{n} < V^\star e_j^\star, Ue^j >$$

The proof of the second formula is analogous. □

2.3.2 Tensor Products of Vector Spaces

It is useful to set

$$X^\star \otimes Y \; := \; \mathcal{L}(X,Y)$$

to recall this structure. We deduce from Theorem 2.3.1 that the tensor product is associative,

$$(X^\star \otimes Y) \otimes Z \; := \; X^\star \otimes (Y \otimes Z)$$

and in particular that

$$(X^\star \otimes Y)^\star = X \otimes Y^\star \;\; \square$$

We also observe that the following spaces are isomorphic:

$$X \otimes \mathbf{R} \; = \; \mathbf{R} \otimes X \; = \; X$$

The operator $W \mapsto W^\star$ is an isometry between $X^\star \otimes Y$ and $Y \otimes X^\star$.

2.4 Tensor Products of Linear Operators

Let us consider two pairs (X, X_1) and (Y, Y_1) of finite-dimensional vector spaces. Let $A \in \mathcal{L}(X_1, X)$ and $B \in \mathcal{L}(Y, Y_1)$ be given. We denote by $A^\star \otimes B$ the linear operator from $\mathcal{L}(X,Y)$ to $\mathcal{L}(X_1, Y_1)$, defined by

$$\forall\, W \in \mathcal{L}(X,Y), \quad (A^\star \otimes B)(W) \; := \; BWA$$

We observe that when $W = p \otimes y$, we have

$$(A^\star \otimes B)(p \otimes y) = A^\star p \otimes By$$

These notations are consistent with the notation

$$x \otimes B \in \mathcal{L}(\mathcal{L}(X,Y), Z)$$

introduced earlier: We set $Y_1 := Z$; we identify $Y_1 = \mathcal{L}(\mathbf{R}, Y_1)$ and take $A := x : \lambda \mapsto \lambda x$. First, we observe that

$$(A^\star \otimes B)^\star \; = \; A \otimes B^\star \tag{2.6}$$

It is enough to check that

$$\langle (A^\star \otimes B)^\star (x_1 \otimes q_1), p \otimes y\rangle \;=\; \langle (x_1 \otimes q_1), (A^\star \otimes B)(p \otimes y)\rangle$$

$$=\; \langle (x_1 \otimes q_1),\, (A^\star p \otimes By)\rangle \;=\; \langle A^\star p, x_1\rangle\langle q_1, By\rangle$$

$$=\; \langle p, Ax_1\rangle\langle B^\star q, x_1\rangle \;=\; \langle (A \otimes B^\star)(x_1 \otimes q_1), p \otimes y\rangle$$

For example, the transpose of $x \otimes B$ is equal to $(x \otimes B)^\star = x \otimes B^\star$. It maps any $q \in Z^\star$ to the linear operator $x \otimes B^\star q \in \mathcal{L}(X^\star, Y^\star)$.

Let $A_1 \in \mathcal{L}(X_2, X_1)$ and $B_1 \in \mathcal{L}(Y_1, Y_2)$. Then it is easy to check that

$$(A_1^\star \otimes B_1)(A^\star \otimes B) \;=\; (A_1^\star A^\star) \otimes (B_1 B)$$

The following formulas are straightforward:

$$\begin{cases} \text{(i)} & \text{If } A \text{ and } B \text{ are invertible, then} \\ & (A^\star \otimes B)^{-1} \;=\; (A^{-1})^\star \otimes B^{-1} \\ \text{(ii)} & \text{If } A \text{ is left-invertible and } B \text{ is right-invertible,} \\ & \text{then } A^\star \otimes B \text{ is right-invertible} \\ \text{iii} & \text{If } A \text{ is right-invertible and } B \text{ is left-invertible,} \\ & \text{then } A^\star \otimes B \text{ is left-invertible} \\ \text{(iv)} & \text{If } A \text{ and } B \text{ are projectors, so is } A^\star \otimes B \end{cases}$$

2.4.1 Gradients of Functionals on Linear Operators

Let us consider three vector spaces X, Y, and Z, an element $x \in X$, and a differential map $g : Y \mapsto Z$, with which we associate the map $H : W \mapsto H(W) := g(Wx)$.

Proposition 2.4.1 *The differential of H at $W \in \mathcal{L}(X,Y)$ is defined by*

$$H'(W) \;=\; x \otimes g'(Wx)$$

Proof Indeed,

$$\lim_{\lambda\to 0} \frac{H(W+\lambda U)-H(W)}{\lambda}$$

$$= \lim_{\lambda\to 0} \frac{g(Wx+\lambda Ux)-g(Wx)}{\lambda}$$

$$= g'(Wx)Ux = (x \otimes g'(Wx))(U)$$

□

More generally, let E be a differentiable functional from Z to $\mathbf{R}$. We set $\Psi(W) := E(g(Wx))$. The chain rule implies the following:

Proposition 2.4.2 *The gradient of* Ψ *at* $W \in \mathcal{L}(X,Y)$ *is defined by*

$$\Psi'(W) = x \otimes g'(Wx)^\star E'(g(Wx)) \in \mathcal{L}(X^\star, Y^\star)$$

Proof Indeed,

$$\Psi'(W)(U) = \langle E'(g(Wx)), g'(Wx)Ux\rangle = \langle g'(Wx)^\star E'(g(Wx)), Ux\rangle$$

□

2.5 Pseudoinverses of Tensor Products

2.5.1 Scalar Products on $\mathcal{L}(X,Y)$

If we supply the spaces X and Y with scalar products $l(\cdot,\cdot)$ and $m(\cdot,\cdot)$, we shall denote by $L \in \mathcal{L}(X,X^\star)$ and $M \in \mathcal{L}(Y,Y^\star)$ their *duality mappings*, defined respectively by

$$\forall\, x_1,\ x_2 \in X, \quad \langle Lx_1, x_2\rangle := l(x_1,x_2)$$

$$\forall\, y_1,\ y_2 \in Y, \quad \langle My_1, y_2\rangle := m(y_1,y_2)$$

[The matrix $(l(e^i,e^j))_{i,j=1,\ldots,n}$ of the bilinear form l coincides with the matrix of the linear operator L.] We shall denote by

$$\lambda(x) := \sqrt{l(x,x)} \quad \text{and} \quad \mu(y) := \sqrt{m(y,y)}$$

the norms associated with these scalar products. The bilinear form $l_\star(p_1,p_2) := \langle L^{-1}p_1, p_2\rangle = l(L^{-1}p_1, L^{-1}p_2)$ is then a scalar product on the dual of $X^\star$, called the *dual scalar product*.

It is clear that if $(e^j)_{j=1,\ldots,n}$ is an l-orthonormal basis of X [i.e., satisfying $l(e^j,e^l)=0$ when $j \neq l$ and $l(e^j,e^j)=1$], then the basis $(e_j^\star := Le^j)_j$

is the dual basis of $(e^j)_j$ because $\langle e_j^\star, e^l\rangle = l(e^j, e^l)$. The matrix of the duality mapping L in these bases is the identity matrix.

Let us consider now an l-orthonormal basis (e^j) of X and an m-orthonormal basis (f^i) of Y. Let U and V be two operators of $\mathcal{L}(X,Y)$. We observe that

$$\sum_{j=1}^{n} m(Ue^j, Ve^j) = \sum_{i=1}^{m} l_\star(U^\star f_i^\star, V^\star f_i^\star) = \sum_{i,j} \langle f_i^\star, Ue^j\rangle\langle f_i^\star, Ve^j\rangle$$

so that this expression is independent of the choice of the l- and m-orthonormal bases of X and Y.

We thus associate with the two scalar products l and m the scalar product $l_\star \otimes m$ on $\mathcal{L}(X,Y)$ defined by

$$\begin{aligned}(l_\star \otimes m)(U,V) &:= \sum_{i,j} \langle f_i^\star, Ue^j\rangle\langle f_i^\star, Ve^j\rangle \\ &= \sum_{j=1}^{n} m(Ue^j, Ve^j) = \sum_{i=1}^{m} l_\star(U^\star f_i^\star, V^\star f_i^\star)\end{aligned}$$

For matrices $p_i \otimes y_i$, this formula becomes

$$(l_\star \otimes m)(p_1 \otimes y_1, p_2 \otimes y_2) = l_\star(p_1, p_2) m(y_1, y_2)$$

We observe that the duality mapping from $\mathcal{L}(X,Y)$ to $\mathcal{L}(X^\star, Y^\star)$ of the scalar product $l_\star \otimes m$ is $L^{-1} \otimes M$, because

$$\begin{aligned}(l_\star \otimes m)(U,V) &= \sum_{i,j} \langle f_i^\star, Ue^j\rangle\langle f_i^\star, Ve^j\rangle \\ &= \sum_{i,j} \langle f^i, MUL^{-1}e_j^\star\rangle\langle f_i^\star, Ve^j\rangle \\ &= \langle MUL^{-1}, V\rangle = \langle (L^{-1} \otimes M)U, V\rangle\end{aligned}$$

If we consider the matrices (u_i^j) and (v_i^j) of the operators U and V for the orthonormal bases $(e^j)_j$ and $(f^i)_i$, then

$$(l_\star \otimes m)(U,V) = \sum_{i=1}^{m}\sum_{j=1}^{n} u_i^j v_i^j$$

We shall denote by

$$(\lambda_\star \otimes \mu)(U) := \sqrt{(l_\star \otimes m)(U,U)}$$

the norm associated with this scalar product.

2.5.2 Pseudoinverse of $A \otimes B$

We begin by characterizing the orthogonal right-inverse of a tensor product:

Proposition 2.5.1 *Let us assume that $A \in \mathcal{L}(X_1, X)$ is injective and that $B \in \mathcal{L}(Y, Y_1)$ is surjective. Then $A^\star \otimes B$ is surjective, and its orthogonal right-inverse is given by the formula*

$$(A^\star \otimes B)^+ = (A^\star)^+ \otimes B^+$$

Then $\overline{W} := B^+ W_1 A^-$ is the solution to the operator equation $BWA = W_1$ with minimal norm [for the norm $\lambda_\star \otimes \mu$ on $\mathcal{L}(X,Y)$].

Proof In order to prove that $(A^\star \otimes B)^+ = (A^\star)^+ \otimes B^+$, we apply the formula for the orthogonal right-inverse: Let us set $\mathcal{A} := A^\star \otimes B$ and recall that the duality map $J \in \mathcal{L}(\mathcal{L}(X,Y), \mathcal{L}(X^\star, Y^\star))$ is equal to $L^{-1} \otimes M$, so that its inverse is equal to $J^{-1} = L \otimes M^{-1}$. Hence,

$$\mathcal{A}^+ = J^{-1}\mathcal{A}^\star \left(\mathcal{A}J^{-1}\mathcal{A}^\star\right)^{-1}$$

But we see that

$$\mathcal{A}J^{-1}\mathcal{A}^\star = A^\star LA \otimes BM^{-1}B^\star$$

so that

$$\mathcal{A}^+ = LA(A^\star LA)^{-1} \otimes M^{-1}B(BM^{-1}B^\star)^{-1}$$

It is enough to observe that $M^{-1}B(BM^{-1}B^\star)^{-1} = B^+$ and that

$$LA(A^\star LA)^{-1} = (A^\star)^+ = \left((A^\star LA)^{-1}A^\star L\right)^\star = (A^-)^\star$$

so that the formula holds true. □

As a corollary, we obtain the following result:

Corollary 2.5.2 *Let us assume that $A \in \mathcal{L}(X_1, X)$ is injective. Then $\overline{W} := W_1 A^-$ is the solution to the operator equation $WA = W_1$ with minimal norm [for the norm $\lambda_\star \otimes \mu$ on $\mathcal{L}(X,Y)$]. This solution does not depend on the choice of the scalar product m on Y.*

In particular, we shall quite often use the following consequence when we take $X_1 := \mathbf{R}$ and $A := x \in \mathcal{L}(\mathbf{R}, X)$:

Corollary 2.5.3 *Let us assume that $B \in \mathcal{L}(Y, Y_1)$ is surjective and take $x \neq 0$. Then $\overline{W} := Lx/\lambda(x)^2 \otimes B^+ y$ is the solution to the operator equation $BWx = y$ with minimal norm [for the norm $\lambda_\star \otimes \mu$ on $\mathcal{L}(X, Y)$].*

By transposition, we obtain a formula for the orthogonal left-inverse of a tensor product of linear operators:

Proposition 2.5.4 *Let us assume that $A \in \mathcal{L}(X_1, X)$ is surjective and that $B \in \mathcal{L}(Y, Y_1)$ is injective . Then $A^\star \otimes B$ is injective, and its orthogonal left-inverse is given by the formula*

$$(A^\star \otimes B)^- = (A^\star)^- \otimes B^-$$

Proof Let us set $\mathcal{A} := A^\star \otimes B$, the transpose of which is

$$\mathcal{A}^\star = A \otimes B^\star$$

Since the transpose is surjective, we deduce from the preceding proposition that

$$(\mathcal{A}^\star)^+ = A^+ \otimes (B^\star)^+ = A^+ \otimes (B^-)^\star$$

Therefore, by taking the transpose and using the fact that the transpose of the orthogonal right-inverse is the orthogonal left-inverse of the transpose, we infer that

$$\mathcal{A}^- = (A^+)^\star \otimes B^- = (A^\star)^- \otimes B^-$$

□

The definition and properties of pseudoinverses imply the following formulas.

Theorem 2.5.5 *Let $C \in \mathcal{L}(X_1, X)$ and $D \in \mathcal{L}(Y, Y_1)$ be given. Then the pseudoinverse of $C^\star \otimes D$ is equal to $(C^{\ominus 1})^\star \otimes D^{\ominus 1}$. In particular, if $W_1 \in \mathcal{L}(X_1, Y_1)$ is given, then the linear operator $\overline{W} := D^{\ominus 1} W_1 C^{\ominus 1} \in \mathcal{L}(X, Y)$ is the closest solution with minimal norm to the operator equation $DWC = W_1$.*

Proof We split the operators $C = B_C A_C$ and $D = B_D A_D$ as products of injective and surjective linear operators, so that $C^\star \otimes D = (A_C^\star B_C^\star) \otimes B_D A_D$ is the product $(A_C^\star \otimes B_D)(B_C^\star \otimes A_D)$ of the injective operator $A_C^\star \otimes B_D$ and the surjective operator $B_C^\star \otimes A_D$. Hence the pseudoinverse of $C^\star \otimes D$ is the product $(B_C^\star \otimes A_D)^+(A_C^\star \otimes B_D)^-$, which is equal to

$$(B_C^\star \otimes A_D)^+(A_C^\star \otimes B_D)^- = ((B_C^-)^\star \otimes A_D^+)((A_C^+)^\star \otimes B_D^-)$$

$$= (B_C^{-\star} A_C^{+\star}) \otimes A_D^+ B_D^- = (A_C^+ B_C^-)^\star \otimes A_D^+ B_D^-$$

$$= (C^{\ominus 1})^\star \otimes (D^{\ominus 1})$$

□

2.5.3 Orthogonal Projection on Subspaces of Matrices

Corollary 2.5.3 and Proposition 2.1.4 imply the following formula for the optimal solution to the quadratic minimization problem under constraints:

$$(\lambda_\star \otimes \mu)(\overline{W} - U) = \inf_{BWx \in K} (\lambda_\star \otimes \mu)(W - U)$$

where $U \in \mathcal{L}(X, Y)$ is given and where $K \subset Z$ is a closed convex subset. Recall that when $B \in \mathcal{L}(Y, Z)$ is surjective, we can supply Z with the final norm μ^B defined by $\mu^B(z) := \mu(B^+ z)$, and we denote by π_K^B the projector of best approximation onto the closed subset K for this final norm.

Theorem 2.5.6 *Let $B \in \mathcal{L}(Y, Z)$ be a surjective operator, $K \subset Z$ be a closed convex subset, and $x \in X$ be different from zero. Then the matrix*

$$\overline{W} := U - \frac{Lx}{\lambda(x)^2} \otimes B^+(BUx - \pi_K^B(BUx))$$

is the best approximation of U in the subset of linear operators W satisfying the constraints

$$BWx \in K$$

Proof By Corollary 2.5.3, we know that the right-inverse of the map $W \mapsto BWx$ is equal to the map $z \mapsto Lx/\lambda(x)^2 \otimes B^+ z$. Therefore, the final norm on the vector space Z associated with this operator is equal to $\mu(B^+ z)\lambda(x)$. Consequently, the orthogonal projector onto K associated with this norm is equal to the projector π_K^B associated with the final norm $\mu(B^+ z)$.

Consequently, by Proposition 2.1.4, we infer that the solution to

$$(\lambda_\star \otimes \mu)(\overline{W} - U) = \inf_{BWx \in K} (\lambda_\star \otimes \mu)(W - U)$$

is equal to $U - (x \otimes B)^+(BUx - \pi_K^B(BUx))$. □

3

Associative Memories

Introduction

We investigate in this chapter the case of linear neural networks, named *associative memories* by T. Kohonen (Figure 3.1). We begin by specializing the heavy algorithm we have studied in the general case of adaptive systems to the case of neural networks, where controls are matrices. It shows how to modify the last synaptic matrix that has learned a set of patterns for learning a new pattern *without forgetting* the previous patterns.

Because right-inverses of tensor products are tensor products of right-inverses, we observe that the heavy algorithm has a Hebbian character: The heavy algorithm states that the correction of a synaptic matrix during learning is the product of activities in both presynaptic and postsynaptic neurons. This added feature that plain vectors do not enjoy justifies the specifics of systems controlled by matrices instead of vectors.

We then proceed with associative memories with postprocessing, with multilayer and continuous-layer associative memories. We conclude this chapter with associative memories with gates, where the synaptic matrices link conjuncts (i.e., subsets) of presynaptic neurons with each postsynaptic neuron. They allow computation of any Boolean function. They require a short presentation of fuzzy sets.

3.1 Optimal Synaptic Matrices of Associative Memories

Let us supply the finite-dimensional vector spaces X and Y with scalar products l and m and the space $\mathcal{L}(X,Y)$ with the associated scalar product $l_\star \otimes m$. Let $\mathcal{P}$ be a finite set of pairs $(a^p, b^p) \in X \times Y$ that defines

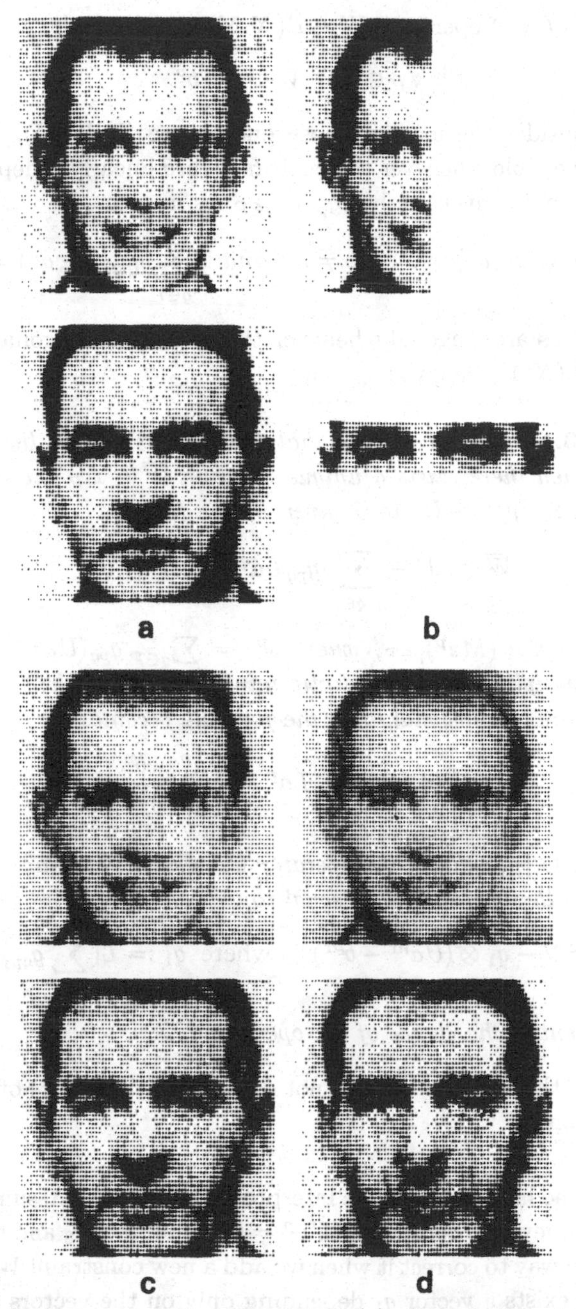

Fig. 3.1 Kohonen's associative memories. (a): Samples of original images. (b): Key patterns. (c): Recollection from a memory with 160 stored images. (d): Recollection from a memory with 500 stored images. From Kohonen's Sel-Organization and Associative Memories.

the subspace $\mathcal{L}_{\mathcal{P}}$ of operators $W \in \mathcal{L}(X, Y)$ satisfying the constraints

$$\forall p \in \mathcal{P}, \quad Wa^p = b^p$$

Let us consider the matrix whose entries are $(l(a^p, a^q))_{p,q \in \mathcal{P}}$. This matrix is invertible whenever the vectors a^p are linearly independent on X. We denote, in this case, by $(g_{pq})_{p,q \in \mathcal{P}}$ its inverse:

$$\sum_{q \in \mathcal{P}} g_{pq} l(a^q, a^r) = 0 \quad \text{if } p \neq r \quad \textit{and} \quad \sum_{q \in \mathcal{P}} g_{pq} l(a^q, a^p) = 1$$

These matrices are diagonal whenever the a^p are l-orthogonal.

Let $U \in \mathcal{L}(X, Y)$ be given.

Theorem 3.1.1 *Let us assume that the vectors a^p are linearly independent. Then there exists a unique solution $\overline{W} \in \mathcal{L}_{\mathcal{P}}$ minimizing the distance $(\lambda_\star \otimes \mu)(W - U)$ to U given by the formula*

$$\overline{W} = U - \sum_{p,q \in \mathcal{P}} g_{pq} La^q \otimes (Ua^p - b^p)$$

and the sequence $(Mz^p)_{p \in \mathcal{P}}$, where $z^p := \sum_{q \in \mathcal{P}} g_{pq}(Ua^q - b^q)$ is the Lagrange multiplier of this quadratic minimization problem.

If the a^p are l-orthogonal, then the formula becomes

$$\overline{W} = U - \sum_{p \in \mathcal{P}} La^p \otimes \frac{Ua^p - b^p}{(\lambda(a^p))^2}$$

If $\mathcal{P} = \mathcal{P}_0 \cup \{p_1\}$ and if the operator U is such that $Ua^p = b^p$ for all $p \in \mathcal{P}_0$, then the operator $\overline{W}$ is equal to

$$\overline{W} = U - q_1 \otimes (Ua^{p_1} - b^{p_1}), \quad \text{where } q_1 := L(\sum_{q \in \mathcal{P}} g_{qp_1} a^q)$$

is independent of the choice of the operator U.

Note that the solution $\overline{W}$ does not depend on the choice of the scalar product m on Y.

Remark The last result can be interpreted as follows: Assume that the operator U obeys the constraints $Ua^p = b^p$ for all $p \in \mathcal{P}_0$ and that we are looking for a way to correct it when we add a new constraint $Wa^{p_1} = b^{p_1}$. Then there exists a vector q_1 depending only on the vectors a^p $(p \in \mathcal{P})$ with which we can correct the former operator U by adding to it the simple operator $q_1 \otimes (b^{p_1} - Ua^{p_1})$. This simple operator is, so to speak, proportional to the error $b^{p_1} - Ua^{p_1}$ and to q_1, which is regarded as a *gain vector*.

It belongs to the class of *corrective methods* of the form

$$W_1 = U + q_1 \otimes (b^{p_1} - Ua^{p_1})$$

where $q_1 \in X^\star$ is any solution to the equations

$$\forall\, p \in \mathcal{P}_0, \quad \langle q_1, a^p\rangle = 0 \quad \text{and} \quad \langle q_1, a^{p_1}\rangle = 1$$

because such an operator W_1 obviously satisfies the constraints $W_1 a^p = b^p$ for all $p \in \mathcal{P}$. □

Proof Let us denote by $\mathcal{A} \in \mathcal{L}(\mathcal{L}(X,Y), Y^{\mathcal{P}})$ the linear operator defined by

$$\mathcal{A}W \;:=\; (Wa^p)_{p\in\mathcal{P}} \;\in Y^{\mathcal{P}}$$

Its transpose $\mathcal{A}^\star \in \mathcal{L}(Y^{\star^{\mathcal{P}}}, \mathcal{L}(X^\star, Y^\star))$ is defined by

$$\mathcal{A}^\star((q_p)_{p\in\mathcal{P}}) \;=\; \sum_{p\in\mathcal{P}} q_p \otimes a^p$$

because

$$\begin{aligned}\langle \mathcal{A}^\star((q_p)_{p\in\mathcal{P}}), W\rangle &= \langle (q_p)_{p\in\mathcal{P}}, \mathcal{A}W\rangle \\ &= \sum_{p\in\mathcal{P}} \langle q_p, Wa^p\rangle = \sum_{p\in\mathcal{P}} \langle q_p \otimes a^p, W\rangle\end{aligned}$$

We recall that the duality mapping of $l_\star \otimes m$ is $L^{-1} \otimes M$.

Therefore, the vectors a^p being linearly independent, the operator $\mathcal{A}$ is surjective. Thus there exists a unique solution $\overline{W}$ to the quadratic minimization problem under equality constraints $\mathcal{A}W = y$:

$$\mathcal{A}\overline{W} = y \quad \text{and} \quad (\lambda_\star \otimes \mu)(\overline{W} - U) = \min_{\mathcal{A}W=y} (\lambda_\star \otimes \mu)(W - U)$$

given by the formula

$$\overline{W} = U - (L^{-1} \otimes M)^{-1}\mathcal{A}^\star \vec{q}$$

where the Lagrange multiplier $\vec{q}$ is a solution to

$$(\mathcal{A}(L^{-1} \otimes M)^{-1}\mathcal{A}^\star)\vec{q} = \mathcal{A}U - y$$

By setting $z^p := M^{-1}q_p$ and observing that $(L^{-1} \otimes M)^{-1} = L \otimes M^{-1}$, the latter equation becomes

$$\forall\, q \in \mathcal{P}, \quad \sum_{p\in\mathcal{P}} \langle La^p, a^q\rangle z^p = Ua^q - b^q$$

which can be written

$$\forall\, q \in \mathcal{P}, \quad \sum_{p\in\mathcal{P}} l(a^q, a^p) z^p = U a^q - b^q$$

and thus

$$\forall\, p \in \mathcal{P}, \quad z^p = \sum_{q\in\mathcal{P}} g_{pq}(U a^q - b^q)$$

The former one becomes

$$\overline{W} = U - \sum_{p\in\mathcal{P}} M^{-1} L a^p \otimes \to q = U - \sum_{p\in\mathcal{P}} L a^p \otimes z^p$$

□

Remark We can deduce Theorem 3.1.1 from Corollary 2.5.2. Let us denote by $A \in \mathcal{L}(\mathbf{R}^{\mathcal{P}}, X)$ the linear operator defined by $Ax := \sum_{p\in\mathcal{P}} x_p a^p$ and by $B \in \mathcal{L}(\mathbf{R}^{\mathcal{P}}, Y)$ the linear operator defined by $By := \sum_{p\in\mathcal{P}} y_p b^p$. Then the constraints

$$\forall\, p \in \mathcal{P}, \quad W a^p \; = \; b^p$$

can be written $WA = B$. Therefore, the solution $\overline{W}$ that minimizes $(\lambda_\star \otimes \mu)(W - U)$ under the constraints $WA = B$ can be written $W = \overline{U} + U$, where $\overline{U}$ minimizes $(\lambda_\star \otimes \mu)(U)$ under the constraints $UA = B - UA$.

By Corollary 2.5.2, the solution $\overline{U}$ is given by the formula $\overline{U} = (B - UA)A^-$, so that $\overline{W} = U + (B - UA)(A^\star LA)^{-1} A^\star L$. The entries of the matrix of $A^\star LA$ (in the canonical basis of $\mathbf{R}^{\mathcal{P}}$) are equal to $l(a^p, a^q)$. Let us denote by $(g_{pq})_{p,q\in\mathcal{P}}$ the entries of the matrix of $(A^\star LA)^{-1}$. Then $(A^\star LA)^{-1} A^\star Lx = \left(\sum_{q\in\mathcal{P}} g_{pq}(Lx, a^q)\right)_{p\in\mathcal{P}}$ and $\overline{W}x = \sum_{p\in\mathcal{P}} \sum_{q\in\mathcal{P}} g_{pq}(La^q, x)(b^p - Ua^p)$. Finally, we have proved that $\overline{W} = U + \sum_{p,q\in\mathcal{P}} g_{pq} La^q \otimes (b^p - Ua^p)$. □

Remark: The Schmidt Algorithm Instead of inverting the matrix of elements $(l(a^p, a^q))_{p,q\in\mathcal{P}}$ for computing the entries g_{pq}, one can first "orthogonalize" the sequence of elements a^p by the Schmidt algorithm and compute the operator $\overline{W}$ using this orthogonal sequence of $\overline{a}^p$ values, because the formula is then given explicitly.

Let us recall that if a sequence of elements a^p is linearly independent, the subspace spanned by the a^p's can also be spanned by the orthogonal sequence $\vec{a}^p$ built recursively through the Schmidt algorithm:

$$\begin{cases} \text{(i)} & \vec{a}^1 = a^1 \\ \text{(ii)} & \vec{a}^p = a^p - \sum_{q=1}^{p-1} \frac{l(\vec{a}^q, a^p)}{\lambda(\vec{a}^q)^2} \vec{a}^q \end{cases}$$

The proof is easy: If this sequence has been built up to $p-1$, the formula giving $\vec{a}^p$ shows that $\vec{a}^p$ belongs to the space spanned by the p first elements of the sequence. Then it is easy to observe that

$$l(\vec{a}^p, \vec{a}^q) = l(a^p, \vec{a}^q) - \sum_{q=1}^{p-1} \frac{l(\vec{a}^q, a^p)}{\lambda(\vec{a}^q)^2} l(\vec{a}^q, \vec{a}^q) = 0$$

for all $q = 1, \ldots, p-1$.

Let us set

$$\vec{b}^p := b^p - \sum_{q=1}^{p-1} \frac{l(\vec{a}^q, a^p)}{\lambda(\vec{a}^q)^2} \vec{b}^q$$

Because the subset $\mathcal{L}_{\mathcal{P}}$ of operators satisfies the constraints

$$\forall\, p \in \mathcal{P}, \quad W\vec{a}^p = \vec{b}^p$$

we can then compute the matrix $\overline{W}$ by the formula

$$\overline{W} = U - \sum_{p \in \mathcal{P}} L\vec{a}^p \otimes \frac{U\vec{a}^p - \vec{b}^p}{(\lambda(\vec{a}^p))^2} \quad \square$$

Remark One can use the algorithms of quadratic programming and translate them in this framework. □

3.2 The Heavy Learning Algorithm for Neural Networks

Let us supply the finite-dimensional vector spaces X and Y with scalar products l and m and the space $\mathcal{L}(X, Y)$ with the associated scalar product $l_\star \otimes m$. Let $\mathcal{P}$ be a finite teaching set of patterns $(a^p, b^p) \in X \times Y$.

Neural networks are adaptive systems when the control space $U := \mathcal{L}(Z_1, Z_2)$ is the space of *synaptic matrices*, where Z_1 and Z_2 are the layers of the network. In general, the output $y = y_1 + y_2$ associated with an input signal $x \in X$ is the sum of two signals:

1. a signal $y_1 = c(x)$ mapped directly to Y through the map $c : X \mapsto Y$
2. a signal $y_2 = \Phi(x) W \psi(x)$, where

$$\begin{cases} \text{(i)} & \psi : X \mapsto Z_1 \text{ maps the input space to the input layer} \\ \text{(ii)} & \text{the synaptic matrix } W \in \mathcal{L}(Z_1, Z_2) \text{ weights the processed signal } \psi(x) \\ \text{(iii)} & \Phi : X \mapsto \mathcal{L}(Z_2, Y) \text{ maps the weighted processed signal to the output space} \end{cases}$$

Hence we have to find a synaptic matrix W_P that can learn the training set in the sense that

$$\forall\, p = 1, \ldots, P, \quad c(a^p) + \Phi(a^p) W_P \psi(a^p) \;=\; b^p$$

We apply Theorem 2.2.1 where $U = \mathcal{L}(Z_1, Z_2)$ and $G(x) := \psi(x) \otimes \Phi(x)$ is the linear operator defined by

$$(\psi(x) \otimes \Phi(x))\,(W) \;:=\; \Phi(x) W \psi(x)$$

The linear operator $G(x)$ is surjective whenever $\Phi(x)$ is surjective and $\psi(x) \neq 0$. Its transpose $(\psi(x) \otimes \Phi(x))^\star$ from $Y^\star$ to

$$(\mathcal{L}(Z_1, Z_2))^\star \;=\; \mathcal{L}(Z_1^\star, Z_2^\star)$$

maps any $p \in Y^\star$ to the operator $\psi(x) \otimes \Phi(x)^\star p$ defined by

$$\forall\, q \in Z_1^\star, \quad (\psi(x) \otimes \Phi(x)^\star p)\,(q) \;=\; \langle q, \psi(x)\rangle \Phi(x)^\star p \;\in\; Z_2^\star$$

Denote by $L_i \in \mathcal{L}(Z_i, Z_i^\star)$ the duality maps of the spaces Z_i. Then the duality map

$$N \in \mathcal{L}\left(\mathcal{L}(Z_1, Z_2), \mathcal{L}(Z_1^\star, Z_2^\star)\right)$$

is equal to $L_1^{-1} \otimes L_2$. We thus observe that

$$G(a^p) N^{-1} G(a^q)^\star \;=\; \langle L_1 \psi(a^q), \psi(a^p)\rangle \Phi(a^p) L_2^{-1} \Phi(a^q)^\star$$

so that

$$\forall\, s \neq t, \quad G(a^p) N^{-1} G(a^q)^\star \;=\; 0$$

whenever the sequence of $\psi(a^p)$ is orthogonal in Z_1 for the scalar product l_1. We have thus proved the following:

Proposition 3.2.1 *We posit the assumptions*

$$\begin{cases} \text{(i)} & c : X \mapsto Y \;\; \textit{and} \;\; \psi : X \mapsto Z_1 \;\; \textit{are continuous} \\ \text{(ii)} & \Phi : X \mapsto \mathcal{L}(Z_2, Y) \;\; \textit{is continuous} \\ \text{(iii)} & \forall\, x \in X, \quad \Phi(x) \in \mathcal{L}(Z_2, Y) \;\; \textit{is surjective} \end{cases}$$

and the elements $\psi(a^p)$ are mutually orthonormal. Then the heavy algorithm associates with the synaptic matrix W_{P-1} the new synaptic matrix W_P defined as follows:

$$W_P = W_{P-1} - \psi(a^P) \otimes \Phi(a^P)^+ \left(c(a^P) + \Phi(a^P) W_{P-1} \psi(a^P) - b^P\right)$$

This a *Hebbian rule*, in the sense that the correction of the synaptic matrix W_{P-1} is obtained by adding to it a matrix $p \otimes u$ whose entries $p_i e^j$ are proportional to the products of activities in the presynaptic and postsynaptic neurons.

3.3 Multilayer Associative Memories

We consider now a linear neural network with L layers. Each layer is described by a vector space X_l, with $X_0 := X$ and $X_L := Y$. The dynamics of the neural network are described by linear operators $A_l \in \mathcal{L}(X_{l-1}, X_l)$ and $B_l \in \mathcal{L}(X_l, X_l)$ and reference signals $y_l \in X_l$. Starting with an input $x_0 = a \in X$, the signal is propagated according to the linear equations

$$\forall\, l = 1, \ldots, L, \quad x_l = A_l x_{l-1} + B_l W_l y_{l-1}$$

An output $b \in X_L$ being given, the problem is to find a sequence $\vec{W} := (W_1, \ldots, W_L)$ of synaptic matrices W_l such that, starting with $x_0 := a$, the neural network arrives at $x_L = b$. Because there may be several solutions to this problem, we shall look for the solution $\vec{W}$ with minimal norm in the space

$$\prod_{l=1}^{L} \mathcal{L}(X_{l-1}, X_l)$$

endowed with the scalar product $\sum_{l=1}^{L} m_{\star l-1} \otimes m_l$ and the duality mapping $J := \left(M_l \otimes M_{l-1}^{-1}\right)_{1 \le l \le L}$. We set

$$\begin{array}{rcll} G(l,k) & = & A_l \cdots A_{k+1} & \text{if } 0 \le k \le l-1 \\ G(l,l) & = & 1 & \text{if } k = l \end{array}$$

This discrete dynamical system can be solved by induction: The formula is given by

$$\begin{aligned} x_l &= G(l,0)x_0 + \sum_{k=1}^{l} G(l,k)B_k W_k y_{k-1} \\ &= \prod_{j=0}^{l-1} A_{l-j}x_0 + B_l W_l y_{l-1} + \sum_{k=1}^{l-1} A_l \cdots A_{k+1} B_k W_k y_{k-1} \end{aligned}$$

Since it is true for $l = 1$, assume that it is true for l, and deduce that it is still true for $l+1$.

Indeed,

$$\begin{aligned} x_{l+1} &= A_{l+1}G(l,0)x_0 + \sum_{k=1}^{l} A_{l+1}G(l,k)B_k W_k y_{k-1} + B_{l+1}W_{l+1}y_l \\ &= G(l+1,0)x_0 + \sum_{k=1}^{l+1} G(l+1,k)B_k W_k y_{k-1} \\ &= A_{l+1}A_l \cdots A_1 x_0 + A_{l+1}B_l W_l y_{l-1} \\ &\quad + \sum_{k=1}^{l-1} A_{l+1}A_l \cdots A_{k+1}B_k W_k y_{k-1} + B_{l+1}W_{l+1}y_l \end{aligned}$$

because $A_{l+1}G(l,k) = G(l+1,k)$ and $A_{l+1}G(l,k) = \mathbf{1}$.

Therefore, the input $a \in X_0$ is mapped to the input x_L:

$$x_L = G(L,0)a + \sum_{k=1}^{L} G(L,k)B_k W_k y_{k-1}$$

Thus, a sequence $\vec{U} := (U_1, \ldots, U_L)$ of synaptic matrices learns the pattern (a,b) if it is a solution to the equation

$$\sum_{k=1}^{L} G(L,k)B_k U_k y_{k-1} = b - G(L,0)a$$

Theorem 3.3.1 *Let us assume that the operator*

$$\sum_{k=1}^{L} \mu_{k-1}(y_{k-1})^2 G(L,k)B_k M_k^{-1} B_k^{\star} G(L,k)^{\star}$$

is invertible. We denote by $\pi_L \in Y^{\star} := X_L^{\star}$ the solution to the equation

$$\left(\sum_{k=1}^{L} G(L,k)B_k M_k^{-1} B_k^{\star} G(L,k)^{\star} \mu_{k-1}(y_{k-1})^2\right) \pi_L = b - G(L,0)a$$

The sequence $\vec{W} := (W_1, \ldots, W_L)$ of synaptic matrices W_l of minimal norm mapping the input a to the output b is given by the formula

$$\forall\, k = 1, \ldots, L, \quad W_k \;=\; M_k^{-1} B_k^\star G(L,k)^\star \pi_L \otimes M_{k-1} y_{k-1}$$

Proof Following introduction of the surjective linear operator $\mathcal{A}$ defined by

$$\mathcal{A}\vec{W} \;:=\; \sum_{k=1}^{L} G(L,k) B_k W_k y_{k-1}$$

a sequence $\vec{U}$ of synaptic matrices learns the pattern (a, b) if it is a solution to the equation

$$\mathcal{A}\vec{U} \;=\; b - G(L,0)a$$

We have to find the sequence $\vec{W}$ of minimal norm satisfying the linear constraints

$$\mathcal{A}\vec{W} \;=\; b - G(L,0)a$$

The transpose of $\mathcal{A}$ is defined by

$$\mathcal{A}^\star \pi \;=\; (B_k^\star G(L,k)^\star \pi \otimes y_{k-1})_{1 \le k \le L}$$

so that the operator $\mathcal{A}J^{-1}\mathcal{A}^\star$ is defined as

$$\mathcal{A}J^{-1}\mathcal{A}^\star \;=\; \left(\sum_{k=1}^{L} G(L,k) B_k M_k^{-1} B_k^\star G(L,k)^\star m_{k-1}(y_{k-1}, y_{k-1}) \right)$$

Hence the Lagrange multiplier of this problem is equal to π_L, so that the optimal solution is equal to

$$\begin{aligned} \vec{W} &= J^{-1}\mathcal{A}^\star \pi_L \\ &= \left(M_k^{-1} B_k^\star G(L,k)^\star \pi_L \otimes M_{k-1} y_{k-1}\right)_{1 \le k \le L} \end{aligned}$$

□

Remark We can regard $b - G(L,0)a$ as the initial discrepancy between the input a and the desired output b. To correct it, we look for π_L and regard the backward adjoint solution $p := (p_k)_{1 \le k \le L} = (G(L,k)^\star \pi_L)_{1 \le k \le L}$. We associate with it the matrices $M_k^{-1} B_k^\star p_k \otimes M_{k-1} y_{k-1}$, and we run forward the neural network. □

3.4 Continuous-Layer Associative Memories

We consider now a linear neural network that maps an input $a \in X$ to an output b through a linear continuous neural network

$$\frac{d}{dt}x(t) = A(t)x(t) + B(t)W(t)y(t)$$

where, for all $t \in [0,T]$, $A(t) \in \mathcal{L}(X,X)$, $B(t) \in \mathcal{L}(Z,X)$, and the synaptic matrices $W(t) \in \mathcal{L}(Y,Z)$. Let us denote by $G(t,s)$ the *fundamental matrix* of $A(t)$. We recall that a solution to the variational equation $w'(t) = A(t)w(t) + f(t)$ starting at time s at the state x_s is given by

$$x(t) = G(t,s)x_s + \int_s^t G(t,r)f(r)dr$$

We also recall that the transpose $G(T,s)^\star$ of the fundamental matrix satisfies the following property:

$$\begin{cases} p(t) := G(T,t)^\star \pi_T \text{ is the solution to the adjoint equation} \\ p'(t) = -A(t)^\star p(t) \quad \text{and} \quad p(T) = \pi_T \end{cases}$$

Indeed, let $x(t) := G(t,s)\xi_s$ be the solution to the differential equation $x'(t) = A(t)x(t)$ starting at time s at ξ_s, and let $p(t)$ be the solution to the differential equation $p'(t) = -A^\star(t)p(t)$ starting at time T at π_T. Then the equation

$$\frac{d}{dt}\langle p(t), x(t)\rangle = \langle p'(t), x(t)\rangle + \langle p(t), x'(t)\rangle = 0$$

implies that

$$\langle \pi_T, G(T,s)\xi_s\rangle = \langle p(T), x(T)\rangle = \langle p(s), x(s)\rangle = \langle p(s), \xi_s\rangle$$

This means that $p(s) = G(T,s)^\star \pi_T$.

Starting with an input $x_0 = a \in X$, the signal is then propagated according to the output

$$x(T) = G(T,0)a + \int_0^T G(T,t)B(t)W(t)y(t)\,dt$$

where $G(t,s)$ denotes the fundamental matrix.

An output $b \in X$ being given, the problem is to find a function $W(\cdot)$ mapping the input $x_0 := a$ to the output $x(T) = b$. Because there may be several solutions to this problem, we shall look for the solution W with minimal norm in the space

$$L^2(0,T;\mathcal{L}(X,Y))$$

endowed with the scalar product $\int_0^T (l_\star \otimes m)(U(t),V(t))\,dt$ and the duality mapping $J := U(\cdot) \mapsto (M \otimes L^{-1})U(\cdot)$. Therefore, a function $U(\cdot)$ learns the pattern (a,b) if it is a solution to the equation

$$\int_0^T G(T,t)B(t)U(t)y(t)\,dt \;=\; b - G(T,0)a$$

Theorem 3.4.1 *Let us assume that the operator*

$$\int_0^T \lambda(y(t))^2 G(T,t)B(t)M^{-1}B(t)^\star G(T,t)^\star\,dt$$

is invertible. We denote by $\pi_T \in Y$ the solution to the equation

$$\left(\int_0^T G(T,t)B(t)M^{-1}B(t)^\star G(T,t)^\star l(y(t),y(t))\,dt\right)\pi_T \;=\; b - G(T,0)a$$

Then the function $W(\cdot)$ of minimal norm mapping the input a to the output b is given by the formula

$$W(t) \;=\; M^{-1}B(t)^\star p(t) \otimes Ly(t)$$

where $p(\cdot)$ is the solution to the adjoint problem

$$-\frac{d}{dt}p(t) \;=\; A(t)^\star p(t)$$

such that

$$p(T) \;=\; \pi_T$$

Proof Following introduction of the surjective linear operator $\mathcal{A}$ defined by

$$\mathcal{A}W(\cdot) \;:=\; \int_0^T G(T,t)B(t)W(t)y(t)\,dt$$

a function $U(\cdot)$ learns the pattern (a,b) if it is a solution to the equation

$$\mathcal{A}U \;=\; b - G(T,0)a$$

We have to find the function W of minimal norm satisfying the linear constraints

$$\mathcal{A}W \;=\; b - G(T,0)a$$

The transpose of $\mathcal{A}$ is defined by

$$\mathcal{A}^\star\pi \;= t \mapsto B(t)^\star G(T,t)^\star \pi \otimes y(t)$$

so that the operator $\mathcal{A}J^{-1}\mathcal{A}^\star$ is defined as

$$\mathcal{A}J^{-1}\mathcal{A}^\star \;=\; \left(\int_0^T G(T,t)B(t)M^{-1}B(t)^\star G(T,t)^\star l(y(t),y(t))\,dt\right)$$

Hence the Lagrange multiplier of this problem is equal to π_T, so that the optimal solution is equal to

$$\begin{aligned} W(\cdot) &= J^{-1}\mathcal{A}^\star\pi_T \\ &= t \mapsto M^{-1}B(t)^\star G(T,t)^\star \pi_T \otimes Ly(t) \end{aligned}$$

□

3.5 Associative Memories with Gates

3.5.1 Fuzzy Sets

Let us consider a set N of neurons and the *power set* 2^N of *conjuncts* or *coalitions* $S \subset N$ of neurons. We shall embed the power set 2^N into the vector space $\mathbf{R}^N$ of real-valued functions defined on N through the familiar concept of the *characteristic function* χ_S of a subset $S \subset N$, defined by

$$\chi_S(i) \;:=\; \begin{cases} 1 & \text{if } i \in S \\ 0 & \text{if } i \notin S \end{cases}$$

We regard such a characteristic function

$$\chi_S \;:\; S \in 2^N \;\mapsto\; \chi_S \in \mathbf{R}^N$$

as a *membership function* of an element i to S, by assigning the value 1 whenever i belongs to S, and 0 whenever i does not belong to S.

We observe that the image of the map $\chi : 2^N \mapsto \mathbf{R}^N$ associating with each subset $S \subset N$ its characteristic function χ_S is equal to

$$\chi\left(2^N\right) \;=\; \{0,1\}^N$$

which is the set of the vertices of the hypercube $[0,1]^N$.

Embedding the power set 2^N into $\mathbf{R}^N$ is quite advantageous, since it allows us to exploit the many properties of the vector space $\mathbf{R}^N$. In this framework, it is quite natural to use the closed convex hull $[0,1]^N$ of the image $\{0,1\}^N$ of the power set 2^N. Because any element $x \in [0,1]^N$ is a function from N into $[0,1]$, we can interpret it as a membership function of a *fuzzy set*[1] (again denoted by x) associating with any element $i \in N$ its membership $x_i \in [0,1]$ in the fuzzy set x_i. Therefore, from now on we shall regard the unit cube $[0,1]^N$ as the set of *fuzzy conjuncts* or *fuzzy coalitions* x of the set N of neurons (Figure 3.2).

Because the set of fuzzy conjuncts is the convex hull of the power set $\{0,1\}^N$, we can write any fuzzy conjunct of neurons in the form

$$x = \sum_{S \in 2^N} m_S \chi_S, \quad \text{where } m_S \geq 0 \quad \text{and} \quad \sum_{S \in 2^N} m_S = 1$$

The memberships are then equal to

$$\forall\, i \in N, \quad x_i = \sum_{S \ni i} m_S$$

Consequently, if m_S is regarded as the probability that the conjunct S will be formed, the membership of the neuron i in the fuzzy conjunct S is the sum of the probabilities for the conjuncts containing the neuron i.

3.5.2 Boolean Functions

We can regard any subset $S \subset N$ as a statement. A Boolean function v is a function from $\{0,1\}^N$ to $\{0,1\}$, associating with any subset $S \subset N$, regarded as a statement, the value 1 (true) or 0 (false). The question arises whether or not a Boolean function v can be *learned* by an associative memory with gates, that is, a neural network of the form

$$\sum_{S \in 2^N} w^S \varphi_S(x)$$

in the sense that

$$\forall\, T \subset N, \quad v(T) = \sum_{S \in 2^N} w^S \varphi_S(\chi_T)$$

A first choice is obtained with multilinear functions γ_S defined by

$$\forall\, S \subset N, \quad \gamma_S(x) := \prod_{i \in S} x_i \prod_{j \notin S} (1 - x_j)$$

because $\gamma_S(T) = 0$ if $S \neq T$ and $\gamma_S(S) = 1$. Therefore,

$$\forall\, T \subset N, \quad v(T) = \sum_{S \subset N} v(S) \gamma_S(T)$$

[1] This concept was introduced by Zadeh in 1968 and has been an area of tremendous development.

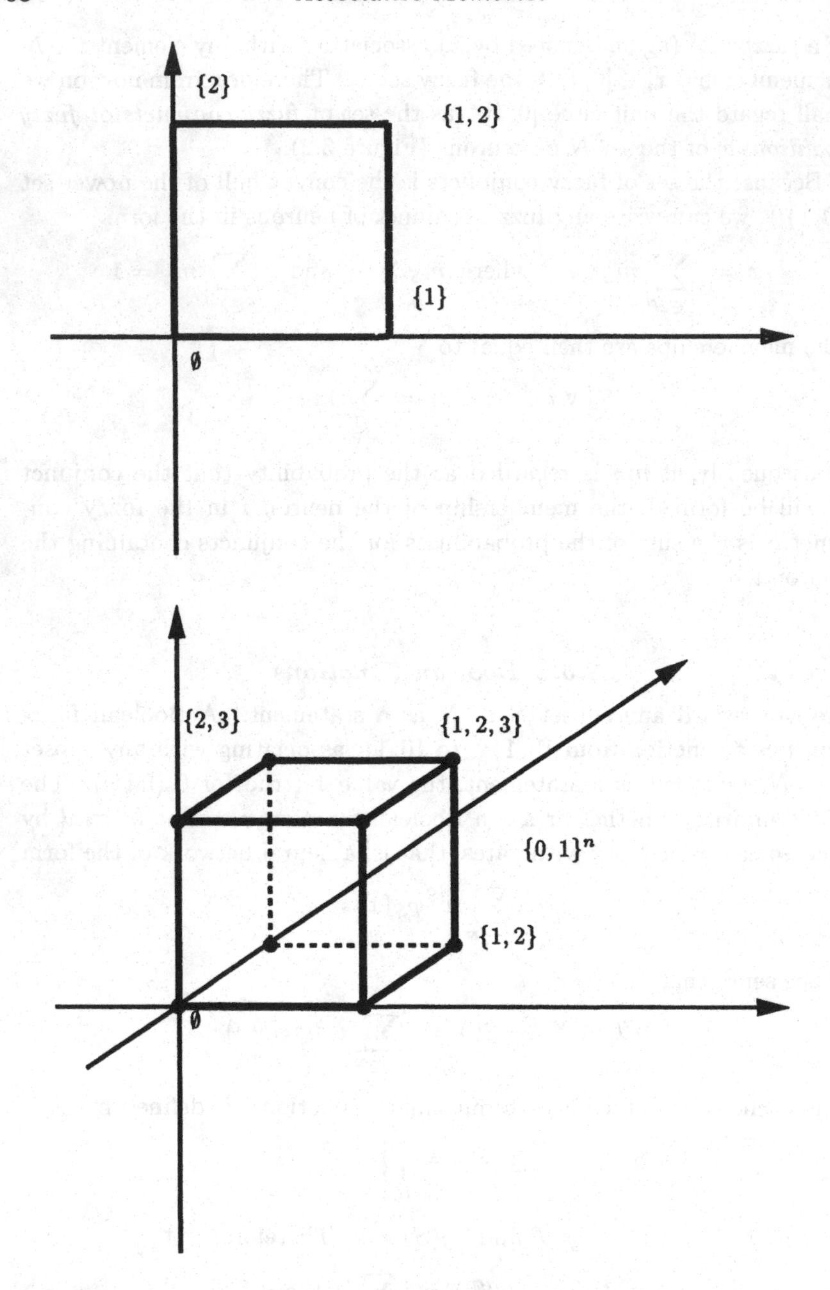

Fig. 3.2 Coalitions and Fuzzy Coalitions. The cube of fuzzy coalitions of a set of two and three neurons; the coalitions are in the vertices of the cube.

One can regard the multilinear function

$$x \in [0,1]^N \mapsto \varpi v(x) := \sum_{S \subset N} v(S)\gamma_S(x)$$

as an extension of the Boolean function v to a fuzzy Boolean function ϖv associating with any fuzzy statement $x \in [0,1]^N$ its value $\varpi v(x)$. It corresponds to the *disjunctive normal form* of a Boolean function.

But the interpretation of this extension procedure in terms of associative memories with gates is not judicious, because it associates with any conjunct S of neurons both signals coming from the neurons of S and signals coming from the neurons of the complement of S. This is the reason we shall rather choose the following functions β_S defined by

$$\forall\, S \subset N, \quad \beta_S(x) := \left(\prod_{i \in S} x_i\right)^{\frac{1}{|S|}}$$

where $|S|$ denotes the cardinal of the subset S. These functions β_S are positively homogeneous and thus are independent of the normalizing choice of 1 for denoting the membership value of an element to a subset. We observe that

$$\beta_S(\chi_T) = \begin{cases} 1 & \text{if } T \supset S \\ 0 & \text{otherwise} \end{cases}$$

Proposition 3.5.1 *The 2^N functionals $v \mapsto \alpha_S(v)$ defined on the space of Boolean functions by*

$$\forall\, S \subset N, \quad \alpha_S(v) := \sum_{T \subset S} (-1)^{|S|-|T|} v(T)$$

satisfy the following interpolation property:

$$\forall\, T \subset N, \quad v(T) = \sum_{S \subset N} \alpha_S(v)\beta_S(\chi_T)$$

Proof The functions

$$\mu_S : T \mapsto \mu_S(T) := \beta_S(\chi_T)$$

are linearly independent on the vector space $\mathbf{R}^{2^N}$. If not, there would exist coefficients $\delta_S \neq 0$ such that $\sum \delta_S \mu_S = 0$. Let S_0 be a subset with minimum cardinality among the subsets S such that $\delta_S \neq 0$. Then we can write

$$\mu_{S_0} = \sum_{S \cap S_0 = \emptyset} \sigma_S \mu_S$$

because, if $S \subset S_0$ and $S \neq S_0$, we should have $|S| < |S_0|$. Since $S \cap S_0 = \emptyset$, we have $\mu_S(S_0) = 0$. On the other hand, $\mu_{S_0}(S_0) = +1$. Therefore, the foregoing equation would imply that $1 = 0$. That is impossible.

The definition of the coefficients μ_S implies that for any $U \in \mathcal{S}$,

$$\begin{aligned} \sum_{S \in \mathcal{S}} \alpha_S(v)\mu_S(U) &= \sum_{S \subset U} \alpha_S(v) = \sum_{S \subset U} \sum_{T \subset S} (-1)^{|S|-|T|} v(T) \\ &= \sum_{T \subset U} \left(\sum_{T \subset S \subset U} (-1)^{|S|-|T|} \right) v(T) \end{aligned}$$

Because there are

$$\binom{|U| - |T|}{|S| - |T|}$$

subsets S between T and U having a given cardinality $|S|$, we deduce that

$$\begin{aligned} \sum_{T \subset S \subset U} (-1)^{|S|-|T|} &= \sum_{|T| \leq |S| \leq |U|} \binom{|U| - |T|}{|S| - |T|} (-1)^{|S|-|T|} \\ &= \sum_{0 \leq i \leq |U|-|T|} \binom{|U| - |T|}{i} (-1)^i = (1-1)^{|U|-|T|} \\ &= \begin{cases} 0 & \text{if } |T| < |U| \\ 1 & \text{if } |T| = |U| \end{cases} \end{aligned}$$

Therefore, the two foregoing equalities imply that

$$\sum_{S \in \mathcal{S}} \alpha_S(v)\mu_S(U) = v(U)$$

□

Remark The foregoing proposition still holds true when the functions β_S are replaced by the functions ψ defined by

$$\forall\, x \in [0,1]^N, \quad \psi_S(x) := \prod_{i \in S} x_i$$

4

The Gradient Method

Introduction

We shall repeatedly use the *gradient algorithm* to devise learning rules. Let $f : X \mapsto \mathbf{R}$ be a numerical function defined on a finite-dimensional vector space that we want to minimize. When f is differentiable, the *Fermat rule* implies that any point $\bar{x}$ achieving a local minimum of f is a solution to the nonlinear problem

$$f'(\bar{x}) = 0$$

Furthermore, the direction $-f'(x)$ points downward, so to speak, whenever we are close to a point $\bar{x}$ that is a local minimum of the function f. Starting from an initial point x_0 close to $\bar{x}$, the gradient algorithm modifies a point x_n by adding to it the direction $-\delta_n f'(x_n)$.

Formally, the gradient algorithm is as follows:

1. (Initialization) Choose $x_0 \in X$, and set $n = 0$.
2. If $f'(x_n) = 0$, then stop. Otherwise, set

$$x_{n+1} := x_n - \delta_n \frac{f'(x_n)}{\|f'(x_n)\|}$$

3. Replace n by $n + 1$, and loop to the previous step.

The convergence of this algorithm is, at best, local. However, it converges when the function f is convex. In this case, we know that the Fermat rule not only provides a necessary condition for a vector $\bar{x}$ to achieve the global minimum of f but also provides a sufficient condition.

Let us consider a sequence of "steps" $\delta_n > 0$ satisfying

$$\lim_{n\to\infty} \delta_n = 0 \quad \text{and} \quad \sum_{n=0}^{\infty} \delta_n = +\infty$$

The following theorem holds true:

Theorem 4.0.2 *Let us assume that the convex function $f : X \mapsto \mathbf{R}$ is differentiable. Assume that the steps of the subgradient algorithm satisfy the assumption*

$$\lim_{n\to\infty} \delta_n = 0 \quad \text{and} \quad \sum_{n=0}^{\infty} \delta_n = +\infty \tag{4.1}$$

and that f is bounded below. Then the decreasing sequence of scalars

$$\theta_k := \min_{n=0,\ldots,k} f(x_n)$$

converges to the infimum $v := \inf f(x)$ of f when $k \to \infty$.

Unfortunately, we shall meet natural optimization problems for which the convex function to minimize is not differentiable, but only continuous. This is the case, for instance, for functions f that are obtained as finite maxima of convex differentiable functions f_i:

$$\forall\, x \in X, \quad f(x) := \max_{i=1,\ldots,n} f_i(x)$$

In the taking of such a maximum, the convexity (and the continuity) is preserved, but the differentiability is destroyed. So we can no longer speak of the Fermat rule and the gradient method for such functions.

Fortunately, in the early 1960s, both Jean-Jacques Moreau and Terry Rockafellar generalized the concept of "gradient" by introducing the concept of the *subdifferential* $\partial f(x)$ of f at x, which is a *subset of subgradients*. If this subset contains only one subgradient, this subgradient coincides with the gradient; it is in this sense that the subdifferential generalizes the gradient.

Going back to our example, one can easily compute the subdifferential of $f := \max_{i=1,\ldots,n} f_i$ in terms of the gradients of the f_i's: Assume that the functions f_i are convex and differentiable. Denote by

$$I(x) := \{i = 1, \ldots, n \,|\, f_i(x) = \max_{j=1,\ldots,n} f_j(x)\,\}$$

the set of "active constraints" at x. Then

$$\partial f(x) = \overline{\text{co}}\,\{f_i'(x)\}_{i \in I(x)}$$

Now, this way of generalizing the concept of differentiability for convex functions allows us to implement the Fermat rule and to adapt the gradient algorithm to the case of continuous convex functions.

The Fermat rule states that $\bar{x}$ achieves the minimum of a convex function f if and only if

$$0 \in \partial f(\bar{x})$$

The subgradient algorithm associates with each x_n an element x_{n+1} defined as

$$x_{n+1} := x_n - \delta_n \frac{p_n}{\|p_n\|}, \quad \text{where } p_n \in \partial f(x_n)$$

We shall prove that it converges whenever the steps δ_n satisfy assumption (4.2).

4.1 Extended Functions and Their Epigraphs

Let us consider an extended function $f : X \mapsto \mathbf{R} \cup \{\pm\infty\}$ whose domain

$$\mathrm{Dom}(f) := \{x \in X \mid f(x) \neq \pm\infty\}$$

is not empty. (Such a function is said to be *proper* in convex and nonsmooth analysis. We shall rather say that it is *nontrivial* in order to avoid confusion with proper maps.)

Any function f defined on a subset $K \subset X$ can be regarded as the extended function f_K, equal to f on K and to $+\infty$ outside of K, whose domain is K. An extended function is characterized by its *epigraph*

$$\mathcal{E}p(f) := \{(x, \lambda) \in X \times \mathbf{R} \mid f(x) \leq \lambda\}$$

An extended function f is convex (respectively positively homogeneous) if and only if its epigraph is convex (resppectively a cone).

The main examples of extended functions are the *indicators ψ_K of subsets K* defined by

$$\psi_K(x) := \begin{cases} 0 & \text{if } x \in K \\ +\infty & \text{if not} \end{cases}$$

The indicator ψ_K is lower-semicontinuous if and only if K is closed, and ψ_K is convex if and only if K is convex. One can regard the sum $f + \psi_K$ as the restriction of f to K.

We recall the convention $\inf(\emptyset) := +\infty$.

Lemma 4.1.1 *Consider a function $f : X \mapsto \mathbf{R} \cup \{\pm\infty\}$. Its epigraph is closed if and only if*

$$\forall\, v \in X, \quad f(v) = \liminf_{v' \to v} f(v')$$

Assume that the epigraph of f is a closed cone. Then the following conditions are equivalent:

$$\begin{cases} \text{(i)} & \forall\, v \in X, \quad f(v) > -\infty \\ \text{(ii)} & f(0) = 0 \\ \text{(iii)} & (0,-1) \notin \mathcal{E}p(f) \end{cases}$$

Proof Assume that the epigraph of f is closed, and pick $v \in X$. There exists a sequence of elements v_n converging to v such that

$$\lim_{n\to\infty} f(v_n) = \liminf_{v'\to v} f(v')$$

Hence, for any $\lambda > \liminf_{v'\to v} f(v')$, there exist N such that, for all $n \geq N, f(v_n) \leq \lambda$, that is, such that $(v_n, \lambda) \in \mathcal{E}p(f)$. By taking the limit, we infer that $f(v) \leq \lambda$ and thus that $f(v) \leq \liminf_{v'\to v} f(v')$. The converse statement is obvious.

Suppose, next, that the epigraph of f is a cone. Then it contains $(0,0)$ and $f(0) \leq 0$. The statements (ii) and (iii) clearly are equivalent. If (i) holds true and $f(0) < 0$, then

$$(0,-1) = \frac{1}{-f(0)}(0, f(0))$$

belongs to the epigraph of f, as do all $(0,-\lambda)$, and (by letting $\lambda \to +\infty$) we deduce that $f(0) = -\infty$, so that (i) implies (ii).

To end the proof, assume that $f(0) = 0$ and that for some v, $f(v) = -\infty$. Then, for any $\varepsilon > 0$, the pair $(v, -1/\varepsilon)$ belongs to the epigraph of f, as does the pair $(\varepsilon v, -1)$. By letting ε converge to zero, we infer that $(0,-1)$ belongs also to the epigraph, since it is closed. Hence $f(0) < 0$, a contradiction. □

4.2 Convex Functions

Convex functions enjoy further properties. We have already seen that an extended function is convex (respectively lower semicontinuous) if and only if its epigraph is convex (respectively closed). We introduce the subdifferential introduced by Moreau and Rockafellar for convex functions:

Definition 4.2.1 *Consider a nontrivial function $f : X \mapsto \mathbf{R} \cup \{+\infty\}$ and $x \in \mathrm{Dom}(f)$. The closed convex subset $\partial f(x)$ defined by*

$$\partial f(x) = \{p \in X^\star \mid \forall\, y \in X, \quad \langle p, y - x\rangle \leq f(y) - f(x)\}$$

(which may be empty) is called the subdifferential of f at x. We say that f is subdifferentiable at x if $\partial f(x) \neq \emptyset$.

From this definition, we see that the Fermat rule follows immediately:

Theorem 4.2.2 *Let $f : X \mapsto \mathbf{R} \cup \{+\infty\}$ be a nontrivial function. Then the following conditions are equivalent:*

$$\begin{cases} \text{(i)} & 0 \in \partial f(x) \quad \text{(the Fermat rule)} \\ \text{(ii)} & x \text{ minimizes } f \end{cases}$$

We also observe that the concept of subdifferential generalizes the concept of gradient in the following sense:

Proposition 4.2.3 *If $f : X \mapsto \mathbf{R} \cup \{+\infty\}$ is convex and differentiable at a point $x \in \text{Int}(\text{Dom}(f))$, then*

$$\partial f(x) = \{f'(x)\}$$

Proof First, the gradient $f'(x)$ belongs to $\partial f(x)$, because, f being convex, the inequalities

$$\frac{f(x + h(y - x))}{h} \leq f(y) - f(x)$$

imply, by letting h converge to zero that

$$\forall y \in X, \quad \langle f'(x), y - x\rangle \leq f(y) - f(x)$$

Conversely, if $p \in \partial f(x)$, we obtain, by taking $y = x + hu$, that

$$\langle p, u\rangle \leq \frac{f(x + hu) - f(x)}{h}$$

By letting h converge to zero, we infer that for every $u \in X, \langle p, u\rangle \leq \langle f'(x), u\rangle$, so that $p = f'(x)$. □

We recall the following important property of convex functions defined on finite-dimensional vector spaces:

Theorem 4.2.4 *A convex function defined on a finite-dimensional vector space is locally Lipschitz and subdifferentiable on the interior of its domain.*

4.3 Subgradient Algorithm

Consider the following *subgradient algorithm*:

1. (Initialization) Choose $x_0 \in \mathrm{Dom}(f)$ and $p_0 \in \partial f(x_0)$. Set $n = 0$.
2. If $p_n = 0$, then stop. Otherwise, set

$$x_{n+1} := x_n - \delta_n \frac{p_n}{\|p_n\|}$$

3. Take $p_{n+1} \in \partial f(x_{n+1})$, replace n by $n+1$, and loop to the previous step.

The following convergence theorem, due to Y. Ermoliev, is taken from the book *Convex Analysis and Minimization Algorithms* by Hiriart-Urruty and Lemaréchal.

Theorem 4.3.1 *Let us assume that a convex function $V : X \mapsto \mathbf{R}$ is bounded below. Assume also that the steps of the subgradient algorithm satisfy the assumption*

$$\lim_{n\to\infty} \delta_n = 0 \quad \text{and} \quad \sum_{n=0}^{\infty} \delta_n = +\infty \tag{4.2}$$

Then the decreasing sequence of scalars

$$\theta_k := \min_{n=0,\dots,k} f(x_n)$$

converges to the infimum $v := \inf_{x\in X} f(x)$ of f when $k \to \infty$.

Proof We prove this theorem by contradiction. If the conclusion is false, then there exists $\eta > 0$ such that $v + 2\eta \leq \theta_k \leq f(x_k)$. Let $\bar{x} \in X$ such that $f(\bar{x}) < v + \eta \leq \theta_k - \eta$. Hence

$$\forall\, k > 0, \quad f(\bar{x}) + \eta < \theta_k \leq f(x_k)$$

We shall contradict this assumption by constructing a subsequence x_{n_k} such that $\lim_{k\to\infty} f(x_{n_k}) \leq f(\bar{x})$. First, we observe that

$$\|x_{n+1} - \bar{x}\|^2 = \|x_n - \bar{x}\|^2 - 2\langle x_n - x_{n+1}, x_n - \bar{x}\rangle + \|x_{n+1} - x_n\|^2$$

so that, by recalling that $\|x_{n+1} - x_n\| = \delta_n$ and that $(x_n - x_{n+1})/\delta_n = p_n/\|p_n\|$, we have

$$\|x_{n+1} - \bar{x}\|^2 = \|x_n - \bar{x}\|^2 - 2\delta_n \left\langle \frac{p_n}{\|p_n\|}, x_n - \bar{x} \right\rangle + \delta_n^2$$

Let us set

$$\alpha_k := \min_{n=0,\ldots,k} \left\langle \frac{p_n}{\|p_n\|}, x_n - \bar{x} \right\rangle$$

By the definition of the subdifferential and the choice of $\bar{x}$, we deduce that

$$\eta < f(x_k) - f(\bar{x}) \leq \langle p_k, x_k - \bar{x} \rangle$$

so that $\alpha_k \|p_k\| > 0$. By summing up the foregoing inequalities from $n = 0$ to k, we obtain

$$\|x_{k+1} - \bar{x}\|^2 \leq \|x_0 - \bar{x}\|^2 - 2\alpha_k \sum_{n=0}^{k} \delta_n + \sum_{n=0}^{k} \delta_n^2 \tag{4.3}$$

On the other hand, we check easily that under assumption (4.2),

$$\frac{\sum_{n=0}^{k} \delta_n^2}{\sum_{n=0}^{k} \delta_n} \text{ converges to } 0 \tag{4.4}$$

Indeed, set $\gamma_k := \sum_{n=0}^{k} \delta_n^2$, $\tau_k := \sum_{n=0}^{k} \delta_n$, and $K(\varepsilon)$ the integer such that $\delta_k \leq \varepsilon$ whenever $k \geq K(\varepsilon)$. Then

$$\gamma_k = \gamma_{K(\varepsilon)-1} + \sum_{k=K(\varepsilon)}^{k} \delta_n^2 \leq \gamma_{K(\varepsilon)-1} + \varepsilon \sum_{k=K(\varepsilon)}^{k} \delta_n = \gamma_{K(\varepsilon)-1} + \varepsilon \tau_k$$

so that

$$\forall\, k \geq K(\varepsilon), \quad \frac{\gamma_k}{\tau_k} \leq \frac{\gamma_{K(\varepsilon)-1}}{\tau_k} + \varepsilon$$

Since $\tau_k \to \infty$, we infer that

$$\limsup_{k \to \infty} \frac{\gamma_k}{\tau_k} \leq \varepsilon$$

By letting ε converge to zero, we have checked (4.4).

Properties (4.3) and (4.4) imply

$$\lim_{k \to \infty} \alpha_k = 0 \tag{4.5}$$

Let n_k be the index such that

$$\left\langle \frac{p_{n_k}}{\|p_{n_k}\|}, x_{n_k} - \bar{x} \right\rangle := \alpha_k := \min_{n=0,\ldots,k} \left\langle \frac{p_n}{\|p_n\|}, x_n - \bar{x} \right\rangle$$

Let us set

$$\bar{x}_{n_k} := \bar{x} + \frac{\langle p_{n_k}, x_{n_k} - \bar{x} \rangle}{\|p_{n_k}\|^2} p_{n_k}$$

We see at once that

$$\langle p_{n_k}, \bar{x}_{n_k} \rangle = \langle p_{n_k}, x_{n_k} \rangle$$

$$\|\bar{x}_{n_k} - \bar{x}\| = \left\langle \frac{p_{n_k}}{\|p_{n_k}\|}, x_{n_k} - \bar{x} \right\rangle = \alpha_k$$

The first inequality implies that

$$f(x_{n_k}) - f(\bar{x}_{n_k}) \le \langle p_{n_k}, x_{n_k} - \bar{x}_{n_k} \rangle = 0$$

by the definition of the subdifferential, and the second implies that there exists $l > 0$ such that, for k large enough,

$$f(\bar{x}_{n_k}) - f(\bar{x}) \le l \|\bar{x}_{n_k} - \bar{x}\| \le l\alpha_k$$

since a convex function defined on a finite-dimensional vector space is locally Lipschitz on the interior of its domain.

Therefore, $f(x_{n_k}) \le f(\bar{x}) + l\alpha_k$, so that, passing to the limit, we obtain the contradiction $\lim_{k\to\infty} f(x_{n_k}) \le f(\bar{x})$ for which we were looking. □

Remark The continuous version of the gradient method is the *differential inclusion*:

$$\text{for almost all } t \ge 0, \quad x'(t) \in -\partial f(x(t))$$

When f is convex and lower-semicontinuous and is bounded below, one can prove that for any initial state $x_0 \in \mathrm{Dom}(f)$, there exists a unique solution to this differential inclusion starting at x_0 such that

$$\lim_{t\to\infty} f(x(t)) = \inf_{x\in X} f(x)$$

The solution $x(\cdot)$ is *slow* in the sense that for almost any t, the norm of the velocity $x'(t)$ is the smallest one. (See, for instance, *Differential Inclusions* by Aubin and Cellina.) □

4.4 Support Functions and Conjugate Functions

There is more to that: Lower-semicontinuous convex functions enjoy duality properties. In the same way that we associate with subspaces their orthogonal spaces, with linear operators their transposes, we can, following Fenchel, associate with lower semicontinuous convex functions *conjugate functions*, for the same reasons, and with the same success.

In order to devise other learning rules for the classification problem, it is useful to recall the following:

Definition 4.4.1 *Let K be a nonempty subset of a Banach space X. We associate with any continuous linear form $p \in X^\star$*

$$\sigma_K(p) := \sigma(K,p) := \sup_{x \in K} \langle p, x \rangle \in \mathbf{R} \cup \{+\infty\}$$

The function $\sigma_K : X^\star \mapsto \mathbf{R} \cup \{+\infty\}$ is called the support function of K. We say that the subsets of $X^\star$ defined by

$$\begin{cases} \text{(i)} & K^- := \{p \in X^\star \mid \sigma_K(p) \le 0\} \\ \text{(ii)} & K^\perp := \{p \in X^\star \mid \forall\, x \in K, \quad \langle p, x \rangle = 0\} \end{cases}$$

are the (negative) polar cone and the orthogonal of K, respectively.

Examples

- When $K = \{x\}$, then $\sigma_K(p) = < p, x >$
- When $K = B_X$, then $\sigma_{B_X}(p) = \|p\|_\star$
- If K is a cone, then

$$\sigma_K(p) = \begin{cases} 0 & \text{if} \quad p \in K^- \\ +\infty & \text{if} \quad p \notin K^- \; \square \end{cases}$$

When $K = \emptyset$, we set $\sigma_\emptyset(p) = -\infty$ for every $p \in X^\star$. We observe that

$$\forall\, \lambda,\, v > 0, \quad \sigma_{\lambda L + \mu M}(p) = \lambda \sigma_L(p) + \mu \sigma_M(p)$$

and, in particular, that if P is a cone, then

$$\sigma_{M+P}(p) = \begin{cases} \sigma_M(p) & \text{if} \quad p \in P^- \\ +\infty & \text{if} \quad p \notin P^- \; \square \end{cases}$$

The separation theorem can be stated in the following way:

Theorem 4.4.2 (Separation Theorem) *Let K be a nonempty subset of a Banach space X. Its closed convex hull is characterized by linear-constraint inequalities in the following way:*

$$\overline{\text{co}}(K) = \{x \in X \mid \forall\, p \in X^\star, \quad \langle p, x \rangle \le \sigma_K(p)\}$$

Furthermore, there is a bijective correspondence between nonempty closed convex subsets of X and nontrivial lower-semicontinuous positively homogeneous convex functions on $X^\star$.

Because the epigraph of a lower-semicontinuous convex function is a closed convex subset, it is tempting to compute its support function and, in particular, to observe that

$$\sigma_{\mathcal{E}p(f)}(p,-1) = \sup_{x\in X,\ \lambda\geq f(x)} (\langle p,x\rangle - \lambda) = \sup_{x\in X}(\langle p,x\rangle - f(x))$$

Definition 4.4.3 *Let $f : X \mapsto \mathbf{R}\cup\{+\infty\}$ be any nontrivial extended function defined on a finite-dimensional vector space X. We associate with it its conjugate function $f^\star : X^\star \mapsto \mathbf{R}\cup\{+\infty\}$ defined on the dual of X by*

$$\forall\, p \in X^\star, \quad f^\star(p) := \sup_{x\in X}(< p,x > - f(x))$$

Its biconjugate $f^{\star\star} : X \mapsto \mathbf{R}\cup\{\pm\infty\}$ is defined by

$$f^{\star\star}(x) := \sup_{p\,\in\, X^\star}(< p,x > - f^\star(p))$$

We see at once that *the conjugate function of the indicator ψ_K of a subset K is the support function σ_K.*

We deduce from the definition the convenient inequality

$$\forall\, x \in X, \forall p \in X^\star, \quad \langle p,x\rangle \leq f(x) + f^\star(p)$$

known as *Fenchel's inequality.* The epigraphs of the conjugate and biconjugate functions being closed convex subsets, the conjugate function is lower-semicontinuous and convex, and so is its biconjugate when it never takes the value $-\infty$. We observe that

$$\forall\, x \in X, \quad f^{\star\star}(x) \leq f(x)$$

If equality holds, then f is convex and lower-semicontinuous. The converse statement, a consequence of the Hahn-Banach separation theorem, is the first basic theorem of convex analysis:

Theorem 4.4.4 *A nontrivial extended function $f : X \mapsto \mathbf{R}\cup\{+\infty\}$ is convex and lower-semicontinuous if and only if it coincides with its biconjugate. In this case, the conjugate function $f^\star$ is nontrivial.*

So the Fenchel correspondence associating with any function f its conjugate $f^\star$ is a one-to-one correspondence between the sets of nontrivial lower-semicontinuous convex functions defined on X and its dual $X^\star$. This fact is at the root of duality theory in convex optimization.

Proof (a) Suppose that $a < f(x)$. Since the pair (x, a) does not belong to $\mathcal{E}p(f)$, which is convex and closed, there exist a continuous, linear form $(p, b) \in X^\star \times \mathbf{R}$ and $\varepsilon > 0$ such that

$$\forall y \in \operatorname{Dom} f, \quad \forall \lambda \geq f(y), \quad \langle p, y\rangle - b\lambda \leq \langle p, x\rangle - ba - \varepsilon \tag{4.6}$$

by virtue of the *separation theorem* (Theorem 2.4).

(b) We note that $b \geq 0$. If not, we take y in the domain of f, and $\lambda = f(y) + \mu$. We then have

$$-b\mu \leq \langle p, x - y\rangle + b(f(y) - a) - \varepsilon < +\infty$$

Then we obtain a contradiction if we let μ tend to $+\infty$.

(c) We show that if $b > 0$, then $a < f^{\star\star}(x)$. In fact, we can divide the inequality (4.6) by b, whence, setting $\bar{p} = p/b$ and taking $\lambda = f(y)$, we obtain

$$\forall y \in \operatorname{Dom} f, \quad \langle \bar{p}, y\rangle - f(y) \leq \langle \bar{p}, x\rangle - a - \varepsilon/b$$

Then, taking the supremum with respect to y, we have

$$f^\star(\bar{p}) < \langle \bar{p}, x\rangle - a$$

This implies that

$$\begin{cases} \text{(i)} & \bar{p} \text{ belongs to the domain } f^\star \\ \text{(ii)} & a < \langle \bar{p}, x\rangle - f^\star(\bar{p}) \leq f^{\star\star}(x) \end{cases} \tag{4.7}$$

(d) We consider the case in which x belongs to the domain of f. In this case, b is always strictly positive. To see this, it is sufficient to take $y = x$ and $\lambda = f(x)$ in formula (4.6) to show that

$$b \geq \varepsilon/(f(x) - a)$$

since $f(x) - a$ is a strictly positive real number. Then, from part (b), we deduce the existence of $\bar{p} \in \operatorname{Dom} f^\star$ and that $a \leq f^{\star\star}(x) \leq f(x)$ for all $a < f(x)$. Thus, $f^{\star\star}(x)$ is equal to $f(x)$.

(e) We consider the case in which $f(x) = +\infty$ and a is an arbitrarily large number. Either b is strictly positive, in which case part (b) implies that
$a < f^{\star\star}(x)$, or $b = 0$. In the latter case, (4.6) implies that

$$\forall y \in \operatorname{Dom} f, \quad \langle p, y - x\rangle + \varepsilon \leq 0 \tag{4.8}$$

Let us take $\bar{p}$ in the domain of $f^\star$ (we have shown that such an element exists, since $\operatorname{Dom} f$ is non empty). Fenchel's inequality implies that

$$\langle \bar{p}, y\rangle - f^\star(\bar{p}) - f(y) \leq 0 \tag{4.9}$$

We take $\mu > 0$, multiply the inequality (4.8) by μ, and add the product to the inequality (4.9) to obtain

$$\langle \bar{p} + \mu p, y\rangle - f(y) \leq f^\star(\bar{p}) + \mu\langle p, x\rangle - \mu\varepsilon$$

Taking the supremum with respect to y, we obtain

$$f^\star(\bar{p} + \mu p) \leq f^\star(\bar{p}) + \mu\langle p, x\rangle - \mu\varepsilon$$

which can be written in the form

$$\langle p, x\rangle + \mu\varepsilon - f^\star(\bar{p}) \leq \langle \bar{p} + \mu p, x\rangle - f^\star(\bar{p} + \mu p) \leq f^{\star\star}(x)$$

Taking $\mu = (a + f^\star(\bar{p}) - \langle \bar{p}, x\rangle)/\varepsilon$, which is strictly positive, we have again proved that $a \leq f^{\star\star}(x)$. Thus, since $f^{\star\star}(x)$ is greater than an arbitrary finite number, we deduce that $f^{\star\star}(x) = +\infty$. □

We deduce at once the following characterization of the subdifferential:

Proposition 4.4.5 *Let $f : X \mapsto \mathbf{R} \cup \{+\infty\}$ be a nontrivial extended convex function defined on a finite-dimensional vector space X. Then*

$$p \in \partial f(x) \iff \; < p, x > = f(x) + f^\star(p)$$

If, moreover, the function f is lower-semicontinuous, then the inverse of the subdifferential $\partial f(\cdot)$ is the subdifferential $\partial f^\star(\cdot)$ of the conjugate function*:*

$$p \in \partial f(x) \iff x \in \partial f^\star(p)$$

Because $-f^\star(0) = \inf_{x \in X} f(x)$, the Fermat rule becomes the following:

Theorem 4.4.6 *Let $f : X \mapsto \mathbf{R} \cup \{+\infty\}$ be a nontrivial lower-semicontinuous convex extended function defined on a finite-dimensional vector space X. Then $\partial f^\star(0)$ is the set of minimizers of f.*

As an example, we obtain the following:

Corollary 4.4.7 *Let $K \subset X$ be a closed convex subset. Then*

$$\begin{cases} \text{(i)} & \partial\psi_K(x) = \{p \in X^\star \;\; \textit{such that } \langle p, x\rangle = \sup_{y \in K}\langle p, y\rangle\} \\ \text{(ii)} & \partial\sigma_K(p) = \{x \in K \;\; \textit{such that } \langle p, x\rangle = \sup_{y \in K}\langle p, y\rangle\} \end{cases}$$

The first subset is called the *normal cone to* K *at* x, and the second is the *support zone of* K *at* p.

Definition 4.4.8 *The negative polar cone of the normal cone* $N_K(x)$ *to a convex subset is called the tangent cone to* K *at* x *and is denoted by*

$$T_K(x) := N_K(x)^-$$

It can be easily characterized by

$$T_K(x) = \overline{S_K(x)}$$

where

$$S_K(x) := \bigcup_{h>0} \frac{K-x}{h} \qquad \Box$$

4.5 The Minover Algorithm

We return to the classification problem that was solved by the perceptron algorithm. When K is a closed, bounded convex subset disjoint from zero, the classifiers $\overline{w}$ are the solutions to the inequalities

$$\sigma_K(\overline{w}) < 0$$

where σ_K is the *support function* of K. Because K is compact, this support function achieves its minimum at some $w^\star$, which is then a classifier. This classifier can be approximated by the *gradient method*:

$$w_{n+1} - w_n \in -\varepsilon_n \partial\sigma_K(w_n)$$

where $\partial\sigma_K(w)$ denotes the *subdifferential* of the support function σ_K at w. We recall that

$$\partial\sigma_K(w) := \{y \in K \mid \langle w, y\rangle = \sigma_K(w)\}$$

When $K := \mathrm{co}(\{a_1, \dots, a_J\})$ is the convex hull of a finite subset, let us set

$$J_n := \{j = 1, \dots, J \mid \langle w_n, a_j\rangle = \max_{k=1,\dots,J} \langle w_n, a_k\rangle\}$$

and denote by $S^J \subset \mathbf{R}^J$ the probability simplex. Then the gradient method can be written in the form

$$w_{n+1} - w_n = -\varepsilon_n \sum_{j\in J_n} \lambda_n^j a_j, \quad \text{where } \lambda_n \in S^{J_n}$$

Indeed, the support function can be written

$$\sigma_K(w) = \sup_{\lambda \in S^J} \sum_{j=1}^{J} \lambda^j \langle w, a_j \rangle$$

In this case,

$$\partial \sigma_K(w_n) := \left\{ \lambda_n \in S^J \mid \sum_{j=1}^{J} \lambda_n^j \langle w_n, a_j \rangle = \sup_{\lambda \in S^J} \sum_{j=1}^{J} \lambda^j \langle w_n, a_j \rangle \right\}$$

Then any $\lambda_n \in S^{J_n}$ achieves the maximum of the linear form

$$\lambda \mapsto \sum_{j=1}^{J} \lambda^j \langle w_n, a_j \rangle \quad \text{on } S^J$$

Hence

$$\partial \sigma_K(w_n) := \left\{ \sum_{j \in J_n} \lambda_n^j a_j \right\}_{\lambda_n \in S^{J_n}}$$

so that the gradient method amounts to

$$w_{n+1} - w_n = -\varepsilon_n \sum_{j \in J_n} \lambda_n^j a_j, \quad \text{where } \lambda_n \in S^{J_n} \; \Box$$

This algorithm was introduced directly by Marc Mézard under the name of the *Minover algorithm*. □

5
Nonlinear Neural Networks

Introduction

We devote this chapter to supervised learning processes for nonlinear neural networks, discrete as well as continuous (since biological neurons are asynchronous). We study learning rules based on gradient methods in the first section, and Newton's methods in the second section. In both cases, we consider successively the cases of one-layer, multilayer, and continuous-layer neural networks.

Because neural networks are dynamical systems controlled by synaptic matrices that match a given set of patterns, we recall the basic facts on discrete and continuous nonlinear control systems that we need. Applied to neural networks, they yield the so-called back-propagation algorithm for multilayer networks, as well as for their continuous versions (continuous-layer networks).

Indeed, in the last analysis, the problem of finding a synaptic matrix W amounts to solving the system of nonlinear equations

$$\Phi(x, W) = y$$

for different kinds of maps Φ depending on whether the neural network has one, two, or an infinite number of layers. We shall adapt to each of these maps Φ

- the gradient method for the derived optimization problem

$$\min_W E\left(\Phi(x, W) - y\right)$$

- Newton-type algorithms

Hence, the task at hand is to compute the derivatives of these maps Φ

and their transposes, in order to write down the algorithms. The nonlinearity of these maps excludes, for the time, the proofs of the convergence of these algorithms (except maybe in their continuous versions). The validity of these algorithms rests on simulation and experimentation. We now summarize the results of this chapter in the simple case in which the network has to learn one pattern.

In the case of a one-layer neural network described by a propagation function $g : Y \mapsto Y$, the map Φ is defined by

$$\Phi(a, W) \ := \ g(Wa) - b$$

Given an evaluation function E on the output space Y and a pattern (a, b), we look for a synaptic matrix that will minimize the function

$$W \ \mapsto \ H(W) \ := \ E(g(Wa) - b)$$

Then the gradient of H is given by

$$H'(W) \ = \ a \otimes g'(Wa)^\star E'(g(Wa) - b)$$

the entries of which are equal to

$$\frac{\partial H}{\partial w_{ij}} = a_j \left(\sum_{k=1}^{m} \frac{\partial g_k}{\partial y_i}(Wa) \frac{\partial E}{\partial y_k}(g(Wa) - b) \right)$$

The gradient method can then be written

$$W^{n+1} - W^n \ = \ -\varepsilon_n a \otimes g'(W^n a)^\star E'(g(W^n a) - b)$$

It yields, for each entry,

$$w_{ij}^{n+1} = w_{ij}^n - \varepsilon_n a_j \left(\frac{\partial g_i}{\partial y_i}(W^n a) \frac{\partial E}{\partial y_i}(g(W^n a) - b) \right)$$

It belongs to the class of *Hebbian learning rules*: The synaptic weight from a neuron j to a neuron i should be *strengthened* whenever the connection is highly active, in proportion of the activities of the presynaptic and postsynaptic neurons of the synapse.

In the case of the multilayer network described by the vector spaces $X_0 := X$, X_l $(l = 1, \ldots, L-1)$, and $X_L := Y$ and the differentiable propagation rules $g_l : X_l \mapsto X_l$, the map Φ_L associates with the sequence $\vec{W}$ of synaptic matrices

$$\vec{W} \ := \ (W_1, \ldots, W_L) \in \prod_{l=1}^{L} \mathcal{L}(X_{l-1}, X_l)$$

the final state x_L the sequence of states starting from $x_0 := a$ according to

$$\forall\, l = 1, \ldots, L, \quad x_l \;=\; g_l(W_l x_{l-1})$$

Given an evaluation function E defined on the output space X_L, we are looking for a sequence $\vec{W} := (W_l)_{l=1,\ldots,L}$ of synaptic matrices minimizing the function

$$\vec{W} \mapsto H(\vec{W}) \;:=\; E(\Phi_L(a, W_1, \ldots, W_L))$$

The back-propagation learning rule is nothing else than the gradient method applied to this function H. When the maps g_l and the evaluation function E are differentiable, the gradient of H is given by the formula

$$H'(W) \;=$$

$$\left(x_{k-1} \otimes g_k'(x_k)^\star W_{k+1}^\star g_{k+1}'(x_{k+1})^\star \cdots W_L^\star g_L'(x_L)^\star E'(x_L - b)\right)_{1 \leq k \leq L}$$

In other words,

$$\frac{\partial H(W)}{\partial W_k} = x_{k-1} \otimes g_k'(x_k)^\star W_{k+1}^\star g_{k+1}'(x_{k+1})^\star \cdots W_L^\star g_L'(x_L)^\star E'(x_L - b)$$

The gradient method provides the celebrated *back-propagation algorithm*: Starting with a synaptic matrix $\vec{W}^0$, we define $\vec{W}^{n+1}$ from the synaptic matrix $\vec{W}^n$ according to the rule

$$W_l^{n+1} - W_l^n \;=\; -\varepsilon_n x_{l-1}^n \otimes g_l'(W_l^n x_l^n)^\star p_l^n$$

where $x_l^n = g_l(W_l^n x_{l-1}^n)$ (starting at $x_0^n = a$) and where p_l^n is given by the formula

$$p_l^n \;=\; W_{l+1}^{n^\star} g_{l+1}'(x_{l+1}^n)^\star \cdots W_L^{n^\star} g_L'(x_L^n)^\star E'(x_L^n - b) \;\square$$

We begin by modifying the synaptic weights of the synaptic matrix of the last layer L by computing $p^{L-1} := W_L^{n^\star} g_L'(x_L^n)^\star E'(x_L^n - b)$: The new matrix is obtained through the Hebbian rule

$$W_L^{n+1} - W_L^n \;=\; -\varepsilon_n x_{L-1}^n \otimes g_L'(W_L^n x_L^n)^\star p_L^n$$

Then the gradient is back-propagated to modify successively the synaptic matrices W_l^n of each layer l from the last one to the first one.

Continuous-Layer Network The evolution of the signals is governed by

$$x'(t) = g(W(t)x(t))$$

We denote by $\Phi_T(a, W(\cdot))$ the map that associates with a time-dependent synaptic matrix $W(t)$ and an initial signal a the final state $x(T)$ obtained through this control system.

Find a time-dependent synaptic matrix $W(\cdot)$ minimizing the function

$$W(\cdot) \mapsto H(W(\cdot)) := E(\Phi_T(a, W(\cdot)) - b)$$

The *continuous back-propagation learning rule* is nothing else than the gradient method applied to this function H.

We associate with a time-dependent synaptic matrix $W(\cdot)$ the solution $x(\cdot)$ to the continuous neural network starting at $x(0) = a$ and the solution $p_W(\cdot)$ to the *adjoint differential equation*

$$p'_W(t) = -W(t)^\star g'(W(t)x(t))^\star p_W(t)$$

starting at $p_W(T) = E'(x(T) - b)$. Then the gradient of H is

$$H'(W(\cdot))(t) = x(t) \otimes g'(W(t)x(t))^\star p_W(t)$$

The gradient method provides the *continuous back-propagation algorithm*. Starting with a synaptic matrix $W^0(\cdot)$, we define $W^{n+1}(\cdot)$ from the synaptic matrix $W^n(\cdot)$ according to this rule: For all $t \in [0, T]$,

$$W^{n+1}(t) - W^n(t) = -\varepsilon_n x^n(t) \otimes g'(W^n(t)x^n(t))^\star p^n(t)$$

where $x^n(\cdot)$ is the solution to the differential equation

$$\frac{d}{dt}x^n(t) = g(W^n(t)x^n(t))$$

starting at $x^n(0) = a$ and the solution $p^n(\cdot)$ to the adjoint differential equation

$$\frac{d}{dt}p^n(t) = -W^n(t)^\star g'(W^n(t)x^n(t))^\star p^n(t)$$

starting at $p^n(T) = E'(x^n(T) - b)$.

The Newton Algorithm The Newton algorithm is

$$W^{n+1} = W^n + \delta a \otimes g'(W^n a)^+ \left(\frac{b - g(W^n a)}{l(a, a)}\right)$$

5.1 Gradient Learning Rules

The simplest nonlinear neural networks, including the celebrated "perceptron," can be regarded as discrete control systems that map an input space $X := \mathbf{R}^n$ to an output space $Y := \mathbf{R}^m$ and that are controlled by synaptic matrices: The network first "weights" an input signal $x \in X := \mathbf{R}^m$ through a synaptic matrix $W \in \mathcal{L}(X, Y)$ and thus processes this signal by mapping it to the output signal $f(x, W) := g(Wx)$, where $g : Y \mapsto Y$. It is also reasonable to assume that for neural networks, the kth component g_k of the map g depends only on the kth component y_k of the incoming signal y. We supply the spaces X and Y with the usual Euclidean scalar product, so that the duality mappings L and M are the identity maps.

Given an evaluation function E on the output space Y and a pattern (a, b), we look for a synaptic matrix that will minimize the function

$$W \mapsto H(W) := E(g(Wa) - b)$$

A *learning rule* associated with this evaluation function is just the gradient method

$$W^{n+1} = W^n - \varepsilon_n H'(W^n)$$

Since the gradient of the function $\Psi : y \mapsto \Psi(y) := E(g(y))$ is equal to $g'(y)^\star E'(g(y))$, we deduce from Proposition 2.4.2 that

$$H'(W) = a \otimes g'(Wa)^\star E'(g(Wa) - b)$$

the entries of which are equal to

$$\frac{\partial H}{\partial w_{ij}} = a_j \left(\sum_{k=1}^{m} \frac{\partial g_k}{\partial y_i}(Wa) \frac{\partial E}{\partial y_k}(g(Wa) - b) \right)$$

Remark: The Need for Nonsmooth Optimization We may not always have the possibility of choosing the evaluation function E nor, for instance, the opportunity to choose the simplest ones, as in the case of *least-squares methods*. In the next chapter, where we deal with viable solutions to control systems, the evaluation function is the function measuring the distance from a point to a set, which is not differentiable. However, nonsmooth analysis and, in particular, convex analysis can provide ways to define *generalized gradients* of any function. Generalized gradients (which boil down to *subdifferentials* in the case of convex functions) are no longer elements, but subsets. They allow us to extend the Fermat rule (stating that zero belongs to the generalized gradient of a function at a point where it reaches its minimum) and the gradient methods. We refer to Chapter 6 of *Set-Valued Analysis* (Aubin and Frankowska, 1990) for an introduction to nonsmooth analysis. For instance, if E is locally Lipschitz and g is differentiable, we deduce that the generalized gradient of the function H is given by

$$\partial H(W) \;=\; a \otimes g'(Wa)^{\star}\partial E(g(Wa)-b)$$

since the generalized gradient of the function $\Psi : y \mapsto \Psi(y) := E(g(y))$ is equal to $g'(y)^{\star}\partial E(g(y))$.

For the sake of simplicity, we shall not elaborate on nonsmooth analysis at this time, but leave to the reader who knows either convex analysis or nonsmooth analysis the task of adapting the following results to the nonsmooth case. □

In the case of propagation rules for neural networks, it is assumed that the component g_i of g depends only on the ith component y_i of y. Then the preceding formula becomes

$$\frac{\partial H}{\partial w_{ij}} = a_j\left(\frac{\partial g_i}{\partial y_i}(Wa)\frac{\partial E}{\partial y_i}(g(Wa)-b)\right)$$

This formula states that the marginal increase in the synaptic weight is proportional to the product of the strength of the input signal a_j sent by the "jth line," the marginal contribution

$$\frac{\partial g_i}{\partial y_i}(Wa)$$

of the ith neuron, and the marginal increase in the evaluation function.

Therefore, we deduce that if a synaptic matrix $\overline{W}$ achieves the minimum of H, then it is a solution to the system of nm equations

$$\forall\, i,j,\quad a_j\left(\frac{\partial g_i}{\partial y_i}(\overline{W}a)\frac{\partial E}{\partial y_i}(g(\overline{W}a)-b)\right)=0$$

and the associated learning rule of the network is given by

$$w_{ij}^{n+1} = w_{ij}^{n} - \varepsilon_n a_j\left(\frac{\partial g_i}{\partial y_i}(W^n a)\frac{\partial E}{\partial y_i}(g(W^n a - b)\right)$$

It belongs to the class of *Hebbian learning rules*: The weight of the synapse from a neuron j to a neuron i should be *strengthened* whenever the connection is highly active.

Example Let us consider the case in which

$$g_i(y_i) := \frac{A_i \gamma_i e^{k_i y_i} + a_i}{\gamma_i e^{k_i y_i} + 1}$$

(where $\gamma_i = -\log \beta_i / k_i$) and $E(y) := \|y - b\|^2/2$. Then

$$H(W) = \frac{1}{2} \sum_{i=1}^{m} \left| \frac{A_i \gamma_i \exp k_i \sum_{j=1}^{n} w_i^j a_j + a_i}{\gamma_i \exp k_i \sum_{j=1}^{n} w_i^j a_j + 1} - b_i \right|^2$$

and

$$\frac{\partial H}{\partial w_{ij}} = \frac{k_i \gamma_i (A_i - a_i) \exp k_i \sum_{j=1}^{n} w_j^i a_j}{(\gamma_i \exp k_i \sum_{j=1}^{n} w_j^i a_j + 1)^2} \left(\frac{A_i \gamma_i \exp k_i \sum_{j=1}^{n} w_i^j a_j + a_i}{\gamma_i \exp k_i \sum_{j=1}^{n} w_i^j a_j + 1} - b_i \right)$$

□

In the case of multiple patterns $(a^j, b^j)_{1 \le j \le J}$ and differentiable evaluation functions E^j that vanish at zero, we minimize the function $H(W)$ defined by

$$H(W) := \sum_{j=1}^{J} E^j(g(Wa^j) - b^j)$$

We infer that

$$H'(W) = \sum_{j=1}^{J} a^j \otimes g'(Wa^j)^\star E'^j(g(Wa^j) - b^j)$$

so that the Hebbian learning rule derived from the gradient method reads

$$W^{n+1} = W^n - \varepsilon_n \sum_{j=1}^{J} a^j \otimes g'(W^n a^j)^\star E'^j(g(W^n a^j) - b^j)$$

When Y is supplied with scalar products m^j whose matrices are denoted by M^j, the preceding formula becomes

$$W^{n+1} = W^n - \varepsilon_n \sum_{j=1}^{J} a^j \otimes g'(W^n a^j)^\star M^j(g(W^n a^j) - b^j)$$

5.2 Back-Propagation Algorithm

We begin by explaining this celebrated algorithm in the general case of adaptive systems and then apply it to the specific case of neural networks. Multilayer adaptive systems are nothing other than discrete control systems.

5.2.1 Discrete Control Systems

Consider the discrete control problem

$$x_l = f_l(x_{l-1}, u_l), \qquad l = 1, \ldots, L \text{ and } x_0 \text{ given}$$

where f_l maps $X_{l-1} \times U_l$ to X_l. We denote by $\vec{u}$ the sequence of controls $(u_1, \ldots, u_L) \in \prod_{l=1}^{L} U_l$, by Φ the map associating with the control sequence $\vec{u}$ and the initial state x_0 the sequence $\Phi(x_0, \vec{u}) := (x_1, \ldots, x_L)$ of states starting from x_0, and by Φ_L the map associating with $(x_0, \vec{u})$ the last component x_L of $\Phi(x_0, \vec{u}) := (x_1, \ldots, x_L)$. Let us denote by

$$A_l := f'_{x_{l-1}}(x_{l-1}, u_l) \quad \text{and } B_l := f'_{u_l}(x_{l-1}, u_l)$$

the derivatives of f with respect to the state and the control, respectively.

Proposition 5.2.1 *Let us assume that the functions f_l are differentiable. Consider a control sequence $\vec{u}$ and the associated sequence $\Phi(x_0, \vec{u})$ of states starting from x_0. Let us set*

$$\begin{array}{rcll} G(l,k) & = & A_l \cdots A_{k+1} & \text{if } 0 \leq k \leq l-1 \\ G(l,l) & = & 1 & \text{if } k = l \end{array}$$

Then the transpose of the derivative of Φ, which maps $\prod_{l=1}^{L} X_l^\star$ to $X_0^\star \times \prod_{k=1}^{L} U_k^\star$, is given by the formula

$$\Phi'(x_0, \vec{u})^\star(p_1, \ldots, p_L)$$

$$= \left(\sum_{l=1}^{L} G(l,0)^\star p_l, \left(\sum_{l=k}^{L} B_k^\star G(l,k)^\star p_l \right)_{1 \leq k \leq L} \right)$$

$$= \left(\sum_{l=1}^{L} A_1^\star \cdots A_l^\star p_l, \left(B_k^\star p_k + \sum_{l=k+1}^{L} B_k^\star A_{k+1}^\star \cdots A_l^\star p_l \right)_{1 \leq k \leq L-1}, B_L^\star p_L \right)$$

In particular, the transpose of the derivative of the map $u \mapsto \Phi_L(x_0, \vec{u})$ is given by the formula

$$\begin{aligned} &\Phi'_L(x_0, \vec{u})^\star(p_L) \\ &= (B_k^\star G(L,k)^\star p_L)_{1\le k\le L} \\ &= \left(B_k^\star A_{k+1}^\star \cdots A_L^\star p_L\right)_{1\le k\le L} \end{aligned}$$

Remark We did not mention explicitly that these matrices G depend on the pair $(x_0, \vec{u})$ to keep the notation simple, but one should not forget this. □

Proof First, it is quite easy to compute the derivative of $\Phi(x_0, \vec{u})$ in the direction $y_0, v_1, \ldots, v_L$. Indeed,

$$(y_1, \ldots, y_L) = \Phi'(y_0, v_1, \ldots, v_L)$$

if and only if for any l,

$$y_l = \lim_{h\to 0} \frac{f(x_{l-1} + hy_{l-1}, u_l + hv_l) - f(x_{l-1}, u_l)}{h}$$

Therefore, $(y_1, \ldots, y_L)$ is the solution of the linearized discrete control system

$$\forall\, l = 1, \ldots, L, \quad y_l := A_l y_{l-1} + B_l v_l$$

This discrete dynamical system can be solved by induction. The formula is given by

$$\begin{aligned} y_l &= G(l,0)y_0 + \sum_{k=1}^{l} G(l,k)B_k v_k \\ &= \prod_{j=0}^{l-1} A_{l-j}y_0 + B_l u_l + \sum_{k=1}^{l-1} A_l \cdots A_{k+1}B_k v_k \end{aligned}$$

Since it is true for $l = 1$, assume that it is true for l, and deduce that it is still true for $l + 1$. Indeed,

$$\begin{aligned} y_{l+1} &= A_{l+1}G(l,0)y_0 + \sum_{k=1}^{l} A_{l+1}G(l,k)B_k v_k + B_{l+1}v_{l+1} \\ &= G(l+1,0)y_0 + \sum_{k=1}^{l+1} G(l+1,k)B_k v_k \\ &= A_{l+1}A_l \cdots A_1 y_0 + A_{l+1}B_l v_l \\ &+ \sum_{k=1}^{l-1} A_{l+1}A_l \cdots A_{k+1}B_k v_k + B_{l+1}v_{l+1} \end{aligned}$$

since $A_{l+1}G(l,k) = G(l+1,k)$ and since $G(l+1,l+1) = \mathbf{1}$. Then

$$\Phi'(\vec{u}^L, x_0)(y_0, v_1, \ldots, v_L) = \left(G(l,0)y_0 + \sum_{k=1}^{l} G(l,k)B_k v_k \right)_{l=1,\ldots,L}$$

$$= \left(A_l \cdots A_1 y_0 + B_l v_l + \sum_{k=1}^{l-1} A_l \cdots A_{k+1} B_k v_k \right)_{l=1,\ldots,L}$$

Second, we transpose this formula, by applying $\Phi'(x_0, \vec{u})(y_0, v_1, \ldots, v_L)$ to $(p_1, \ldots, p_L)$:

$$\begin{aligned} &\sum_{l=1}^{L} \langle p_l, y_l \rangle \\ &= \sum_{l=1}^{L} \langle p_l, G(l,0)y_0 + \sum_{k=1}^{l} G(l,k)B_k v_k \rangle \\ &= \sum_{l=1}^{L} \langle G(l,0)^\star p_l, y_0 \rangle + \sum_{l=1}^{L} \sum_{k=1}^{l} \langle B_k^\star G(l,k)^\star p_l, v_k \rangle \\ &= \sum_{l=1}^{L} \langle G(l,0)^\star p_l, y_0 \rangle + \sum_{k=1}^{L} \sum_{l=k}^{L} \langle B_k^\star G(l,k)^\star p_l, v_k \rangle \\ &= \sum_{l=1}^{L} \langle A_1^\star \cdots A_l^\star p_l, y_0 \rangle \\ &+ \sum_{k=1}^{L-1} (\langle B_k^\star p_k + \sum_{l=k+1}^{L} B_k^\star A_{k+1}^\star \cdots A_l^\star p_l, v_k \rangle) + \langle B_L^\star p_L, v_L \rangle \end{aligned}$$

from which we deduce the proposition. □

Let us consider now an evaluation function $\vec{E}$ defined on the state space $\vec{X} := \prod_{l=1}^{L} X_l$ and the function $H := \vec{E} \circ \Phi$ defined on $X_0 \times \prod_{l=1}^{L} U_l$. We deduce that if $\vec{E}$ is differentiable, the gradient of H at $(x_0, u_1, \ldots, u_L)$ is given by

$$H'(x_0, \vec{u}) = \left(\sum_{l=1}^{L} G(l,0)^\star E'_{x_l}(x_1, \ldots, x_L), \left(\sum_{l=k}^{L} B_k^\star G(l,k)^\star E'_{x_l}(x_1, \ldots, x_L) \right)_{1 \le k \le L} \right)$$

$$= \left(\sum_{l=1}^{L} A_1^\star \cdots A_l^\star E'_{x_l}(x_1, \ldots, x_L), \ldots, B_k^\star E'_{x_k}(x_1, \ldots, x_L) + \sum_{l=k+1}^{L} B_k^\star A_{k+1}^\star \cdots A_l^\star E'_{x_l}(x_1, \ldots, x_L), \ldots, B_L^\star E'_{x_L}(x_1, \ldots, x_L) \right)$$

In particular, if the evaluation function depends only on the final state, we infer that

$$H'(x_0, \vec{u}) = \left(G(L,0)^\star E'(x_L), (B_k^\star G(L,k)^\star E'(x_L))_{1 \le k \le L} \right)$$

$$= (A_1^\star \cdots A_L^\star E'(x_L), \ldots, B_k^\star A_{k+1}^\star \cdots A_L^\star E'(x_L), \ldots, B_L^\star E'(x_L))$$

Finally, we single out the following consequence:

Corollary 5.2.2 *Assume that the maps $f_l : X_{l-1} \times U_l \mapsto X_l$ and the evaluation function $E : X_L \mapsto \mathbf{R}$ are differentiable. Then the gradient of the function $\vec{u} \mapsto H(x_0, \vec{u})$ is given by the formula*

$$\frac{\partial}{\partial \vec{u}} H(x_0, \vec{u}) = (B_k^\star G(L,k)^\star E'(x_L))_{1 \le k \le L}$$

$$= \left((B_k^\star A_{k+1}^\star \cdots A_L^\star E'(x_L))_{1 \le k \le L-1}, B_L^\star E'(x_L) \right)$$

5.2.2 Multilayer Discrete Networks: Back-Propagation Formula

Let us consider a multilayer discrete network described by the vector spaces $X_0 := X$, X_l $(l = 1, \ldots, L-1)$, and $X_L := Y$ and differentiable

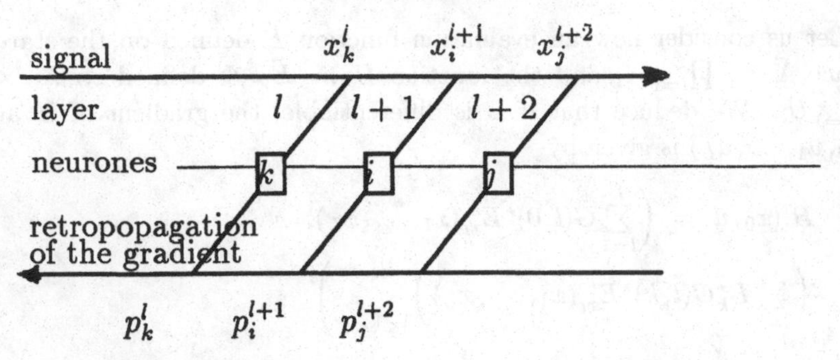

Fig. 5.1 Back-propagation of the gradient. The correction of the synaptic matrix W_l^n involves the signals x_k^n propagated forward and the signals p_k^n propagated backward.

propagation rules[1] $g_l : X_l \mapsto X_l$. We denote by $\vec{W}$ the sequence of synaptic matrices

$$\vec{W} := (W_1, \ldots, W_L) \in \prod_{l=1}^{L} \mathcal{L}(X_{l-1}, X_l)$$

and by Φ the map associating with the sequence $\vec{W}$ and the initial state x_0 the sequence $\Phi(x_0, \vec{W}) := (x_1, \ldots, x_L)$ of states starting from x_0 according to

$$\forall\, l = 1, \ldots L, \quad x_l = g_l(W_l x_{l-1})$$

An initial signal a being given, we shall denote by Φ_L the map associating with the sequence of synaptic matrices W_l the last component x_L of $\Phi(a, \vec{W}) := (x_1, \ldots, x_L)$.

Given an evaluation function E defined on the output space X_L, we are looking for a sequence $\vec{W} := (W_l)_{l=1,\ldots,L}$ of synaptic matrices minimizing the function

$$\vec{W} \mapsto H(\vec{W}) := E(\Phi_L(a, W_1, \ldots, W_L) - b)$$

The back-propagation learning rule is nothing else than the gradient method applied to this function H. Hence, we have to compute the gradient of such a function (Figure 5.1).

[1]Propagation rules such as those whose components are functions defined by

$$g_j(\lambda) := \frac{A_j \gamma_j e^{k_j \lambda} + a_j}{\gamma_j e^{k_j \lambda} + 1}$$

Proposition 5.2.3 *Assume that the maps g_l and the evaluation function E are differentiable. Then the gradient of H is given by the back-propagation formula*

$$H'(W) =$$

$$(x_{k-1} \otimes g_k'(x_k)^\star G(L,k)^\star E'(x_L - b))_{1 \leq k \leq L} =$$

$$\left(x_{k-1} \otimes g_k'(x_k)^\star W_{k+1}^\star g_{k+1}'(x_{k+1})^\star \cdots W_L^\star g_L'(x_L)^\star E'(x_L - b)\right)_{1 \leq k \leq L}$$

In other words,

$$\frac{\partial H(W)}{\partial W_k} = x_{k-1} \otimes g_k'(x_k)^\star W_{k+1}^\star g_{k+1}'(x_{k+1})^\star \cdots W_L^\star g_L'(x_L)^\star E'(x_L - b)$$

Observe that the matrices $g_l'(x_l)$ are diagonal whenever we assume that $g_{l_i}(x_l) = g_{l_i}(x_{l_i})$, so that the entries of the matrices $A_l^\star = W_l^\star g_l'(x_l)^\star$ are given by the formulas $w_{l_{ij}} \partial g_{l_j}(x_{l_j})/\partial x_{l_j}$.

Proof We apply Corollary 5.2.2 to the case of the discrete control system where $f_l(x_{l-1}, W_l) := g_l(W_l x_{l-1})$. Therefore, we know that $A_l := f_{x_{l-1}}'(x_{l-1}, W_l) = g_l'(x_l) W_l$ and that $B_l := f_w'(x_{l-1}, W_l) = (x_{l-1} \otimes g_l'(x_l))$, so that $A_l^\star = W_l^\star g_l'(x_l)^\star$ and $B_l^\star = x_{l-1} \otimes g_l'(x_l)^\star$, which earlier we computed explicitly. We thus obtain

$$\begin{aligned} G(l,k) &= g_l'(x_l) W_l \cdots g_{k+1}'(x_{k+1}) W_{k+1} \\ G(l,k)^\star &= W_{k+1}^\star g_{k+1}'(x_{k+1})^\star \cdots W_l^\star g_l'(x_l)^\star \end{aligned}$$

Replacing A_l, B_l, and $G(l,k)^\star$ by their values in the formula stated in Corollary 5.2.2, we obtain the *back-propagation formula.* □

Consequently, we obtain the back-propagation formula for learning the input-output pair $(a, b) \in X_0 \times X_L$. If a solution $\vec{W}$ to the problem

$$\Phi_L(a, \vec{W}) = b$$

exists, and if the output space X_L is supplied with an evaluation function E, the gradient method provides the celebrated *back-propagation algorithm*: Starting with a synaptic matrix $\vec{W}^0$, we define $\vec{W}^{n+1}$ from the synaptic matrix $\vec{W}^n$ according to the rule

$$W_l^{n+1} - W_l^n = -\varepsilon_n x_{l-1}^n \otimes g_l'(W_l^n x_l^n)^\star p_l^n$$

where $x_l^n = g_l(W_l^n x_{l-1}^n)$ (starting at $x_0^n = a$) and where p_l^n is given by the formula

$$p_l^n = W_{l+1}^{n^\star} g'_{l+1}(x_{l+1}^n)^\star \cdots W_L^{n^\star} g'_L(x_L^n)^\star E'(x_L^n - b) \quad \Box$$

We begin by modifying the synaptic weights of the synaptic matrix of the last layer L by computing $p^{L-1} := W_L^{n^\star} g'_L(x_L^n)^\star E'(x_L^n - b)$: The new matrix is obtained through the Hebbian rule

$$W_L^{n+1} - W_L^n = -\varepsilon_n x_{L-1}^n \otimes g'_L(W_L^n x_L^n)^\star p_L^n$$

Then the gradient is back-propagated to modify successively the synaptic matrices W_l^n of each layer l from the last one to the first one.

Consider now the case in which we have to learn Q patterns $(a^q, b^q)_{1 \le q \le Q}$. If a solution $\vec{W}$ to the learning problem

$$\forall\, q = 1, \ldots, Q, \quad \Phi_L(a^q, \vec{W}) = b^q$$

exists, and if the output space X_L is supplied with Q evaluation functions E_q, then the gradient method provides the following algorithm. Starting with a synaptic matrix $\vec{W}^0$, we define $\vec{W}^{n+1}$ from the synaptic matrix $\vec{W}^n$ according to the rule

$$W_l^{n+1} - W_l^n = h \sum_{q=1}^{Q} x_{l-1}^{n,q} \otimes g'_l(W_l^n x_l^{n,q})^\star p_l^{n,q}$$

where $x_l^{n,q} = g_l(W_l^n x_{l-1}^{n,q})$ (starting at $x_0^{n,q} = a^q$) and where $p_l^{n,q}$ is given by the formula

$$p_l^{n,q} = W_{l+1}^{n^\star} g'_{l+1}(x_{l+1}^{n,q})^\star \cdots W_L^{n^\star} g'_L(x_L^{n,q})^\star E'^q(x_L^{n,q} - b^q) \quad \Box$$

5.3 Continuous Back-Propagation Algorithm

We begin by explaining the continuous version of the back-propagation algorithm in the general case of adaptive systems and, as for the discrete case, then apply it to the specific case of neural networks. Continuous-layer adaptive systems are nothing other than control systems.

5.3.1 Continuous Control Systems

Consider the control problem

$$x'(t) = f(x(t), u(t)) \tag{5.1}$$

where f maps $X \times Z$ to X describes the dynamics of the system on the state space X controlled by controls $u \in Z$ of the control space Z. We denote by $u(\cdot)$ the time-dependent control function $t \in [0,T] \mapsto u(t) \in Z$ and by Φ the map associating with the control function $u(\cdot)$ and the initial state x_0 the solution $\Phi(x_0, u(\cdot))(\cdot) := x(t)$ to the differential equation

$$x'(t) = f(x(t), u(t))$$

starting from x_0.

We shall denote by Φ_T the map associating with $(x_0, u(\cdot))$ the final value $x(T)$ of $\Phi(x_0, u(\cdot))(T) := x(T)$. Our purpose is to compute the derivative of Φ_T and its transpose.

Let us assume that the function f is differentiable. We consider the solution $x(\cdot)$ to the differential equation (5.1), to which we associate the (linear) *variational equation*

$$w'(t) = f'_x(x(t), u(t))w(t)$$

We set

$$A(t) := f'_x(x(t), u(t)) \quad \text{and} \quad B(t) := f'_u(x(t), u(t))$$

and denote by $G(t,s)$ the *fundamental matrix* of $A(t)$. We recall that a solution to the variational equation $w'(t) = A(t)w(t) + f(t)$ starting at time s at the state x_s is equal to

$$x(t) = G(t,s)x_s + \int_s^t G(t,\tau)f(\tau)d\tau$$

We did not mention explicitly that these matrices G depend on the pair $(x_0, u(\cdot))$ to keep the notation simple, but one should not forget this.

We recall that the transpose $G(T,s)^\star$ of the fundamental matrix satisfies the following property:

$$\begin{cases} p(t) := G(T,t)^\star \pi_T \text{ is the solution to the adjoint equation} \\ \\ p'(t) = -A(t)^\star p(t) \quad \text{and} \quad p(T) = \pi_T \end{cases}$$

Indeed, let $x(t) := G(t,s)\xi_s$ be the solution to the differential equation $x'(t) = A(t)x(t)$ starting at time s at ξ_s, and let $p(t)$ be the solution to the differential equation $p'(t) = -A^\star(t)p(t)$ starting at time T at π_T. Then the equation

$$\frac{d}{dt}\langle p(t), x(t)\rangle = \langle p'(t), x(t)\rangle + \langle p(t), x'(t)\rangle = 0$$

implies that

$$\langle \pi_T, G(T,s)\xi_s\rangle = \langle p(T), x(T)\rangle = \langle p(s), x(s)\rangle = \langle p(s), \xi_s\rangle$$

This means that $p(s) = G(T,s)^\star \pi_T$.

Proposition 5.3.1 *Let us assume that the function f is continuously differentiable. Consider a control function $u(\cdot)$ and the associated solution $\Phi(x_0, u(\cdot))$ starting from x_0. Then the transpose of the derivative of Φ_T, which maps $X^\star$ to $X^\star \times \mathcal{C}(0,T;Z)^\star$, is given by the formula*

$$\Phi'_T(x_0, u(\cdot))^\star(p) = (G(T,0)^\star p, t \mapsto B(t)^\star G(T,t)^\star p)$$

Proof First, it is quite classical to derive the variational equation. We begin by associating with any initial state ξ_0 and any continuous control $v(\cdot)$ the solution $y_h(\cdot)$ to the differential equation

$$y'_h(t) = f(y_h(t), u(t) + hv(t))$$

starting at $x_0 + h\xi_0$, and we set

$$w_h(t) = \frac{y_h(t) - x(t)}{h}$$

Hence we can write

$$\begin{aligned} w'_h(t) &= \tfrac{1}{h}\left(f(x(t) + hw_h(t), u(t) + hv(t)) - f(x(t), u(t))\right) \\ &= f'_x(x(t), u(t))w_h(t) + f'_u(x(t), u(t))v(t) + \varepsilon(t,h) \end{aligned}$$

Therefore, w_h can be written

$$w_h(t) = G(t,0)\xi_0 + \int_0^t G(t,s) f'_u(x(s), u(s))v(s)ds + \int_0^t G(t,s)\varepsilon(s,h)ds$$

Since the derivative f' is uniformly continuous, one can prove that

$$\forall\, \alpha > 0, \quad \exists\, \eta > 0 \text{ such that } \|\varepsilon(t,h)\| \leq \alpha \text{ when } h \leq \eta$$

Consequently, by letting h converge to zero, we infer that

$$\Phi'(x_0, u(\cdot))(\xi_0, v(\cdot)) = G(t,0)\xi_0 + \int_0^t G(t,s) f'_u(x(s), u(s))v(s)ds$$

We conclude by observing that for any $\xi_0 \in X$ and $v(\cdot)$,

$$\begin{aligned} &\langle \Phi'_T(x_0, u(\cdot))^\star p, (\xi_0, v(\cdot))\rangle = \langle p, \left(G(T,0)\xi_0 + \textstyle\int_0^T G(T,s)B(s)v(s)ds\right)\rangle \\ &= \langle G(T,0)^\star p, \xi_0\rangle + \left\langle \textstyle\int_0^T B(t)^\star G(T,t)^\star p, v(t)\right\rangle dt \end{aligned}$$

□

Let us consider an evaluation functional $x(T) \mapsto E(x(T)-b)$ depending only on the final state, and let us set

$$H(x_0, u(\cdot)) \;:=\; E(\Phi_T(x_0, u(\cdot)) - b)$$

We infer that

$$H'(x_0, u(\cdot)) = \;(G(T,0)^\star E'(x(T)-b), t \mapsto B(t)^\star G(T,t)^\star E'(x(T)-b))$$

We then single out the following consequence:

Corollary 5.3.2 *Assume that the map f and the evaluation function $E : X \mapsto \mathbf{R}$ are continuously differentiable. Then the gradient of the function $u(\cdot) \mapsto H(x_0, u(\cdot))$ is given by the formula*

$$\frac{\partial}{\partial u(\cdot)} H(x_0, u(\cdot)) = \; t \mapsto B(t)^\star G(T,t)^\star E'(x(T)-b)$$

We can write also that

$$\frac{\partial}{\partial u(\cdot)} H(x_0, u(\cdot))(t) \;=\; B(t)^\star p(t)$$

where the function $p(\cdot)$ is the solution to the adjoint differential equation

$$p'(t) \;=\; -A(t)^\star p(t)$$

starting at

$$p(T) \;=\; E'(x(T)-b)$$

5.3.2 Continuous Neural Networks: Back-Propagation Formula

Let us consider a continuous neural network associated with differentiable propagation rules $g : Y \mapsto X$ and controlled by synaptic matrices $W(t) \in \mathcal{L}(X,Y)$, where X and Y are finite-dimensional vector spaces. The evolution of the signals is thus governed by the differential equation

$$x'(t) \;=\; g(W(t)x(t))$$

We denote by $\Phi_T(a, W(\cdot))$ the map that associates with a time-dependent synaptic matrix $W(t)$ and an initial signal a the final state $x(T)$ obtained through this control system. The same question as for the multilayer neural networks is asked: Given an evaluation function E defined on the state space X, find a time-dependent synaptic matrix $W(\cdot)$ that will minimize the function

$$W(\cdot) \mapsto H(W(\cdot)) \;:=\; E(\Phi_T(a, W(\cdot)) - b)$$

The *continuous back-propagation learning rule* is nothing else than the gradient method applied to this function H. Hence, we have to compute the gradient of such a function.

Proposition 5.3.3 *Assume that the map g and the evaluation function E are continuously differentiable. We associate with a time-dependent synaptic matrix $W(\cdot)$ the solution $x(\cdot)$ to the continuous neural network*

$$x'(t) = g(W(t)x(t))$$

starting at $x(0) = a$ and the solution $p_W(\cdot)$ to the adjoint differential equation

$$p_W'(t) = -W(t)^\star g'(W(t)x(t))^\star p_W(t)$$

starting at

$$p_W(T) = E'(x(T) - b)$$

Then the gradient of H is given by the continuous back-propagation formula

$$H'(W(\cdot))(t) = x(t) \otimes g'(W(t)x(t))^\star p_W(t)$$

Proof We deduce this result from Corollary 5.3.2, where the controls $u(\cdot)$ are the synaptic matrices $W(\cdot)$. We easily check that

$$\begin{array}{rcl} A_W(t) & := & g'(W(t)x(t))W(t) \in \mathcal{L}(X, X) \quad \text{and} \\ A_W^\star(t) & = & W(t)^\star g'(W(t)x(t))^\star \end{array}$$

and that

$$B_W(t) = x(t) \otimes g'(W(t)x(t)) \quad \text{and} \quad B_W^\star(t) = x(t) \otimes g'(W(t)x(t))^\star$$

□

Consequently, the continuous back-propagation formula for learning the input-output pair (a, b) can be made explicit. If a solution $\overline{W}$ to the problem

$$\Phi_T(a, \overline{W}(\cdot)) = b$$

exists, and if the state space X is supplied with an evaluation function E, then the gradient method provides the following algorithm. Starting with a synaptic matrix $W^0(\cdot)$, we define $W^{n+1}(\cdot)$ from the synaptic matrix $W^n(\cdot)$ according to this rule: For all $t \in [0, T]$,

$$W^{n+1}(t) - W^n(t) = -\varepsilon_n x^n(t) \otimes g'(W^n(t)x^n(t))^\star p^n(t)$$

where $x^n(\cdot)$ is the solution to the differential equation

$$\frac{d}{dt}x^n(t) = g(W^n(t)x^n(t))$$

starting at $x^n(0) = a$ and $p^n(\cdot)$ is the solution to the adjoint differential equation

$$\frac{d}{dt}p^n(t) = -W^n(t)^\star g'(W^n(t)x^n(t))^\star p^n(t)$$

starting at

$$p^n(T) = E'(x^n(T) - b)$$

Consider now the case in which we have to learn Q patterns $(a^q, b^q)_{1\leq q\leq Q}$. If a solution $\overline{W}$ to the learning problem

$$\forall\, q \in \{1, \ldots, Q\}, \quad \Phi_T(a^q, \overline{W}(\cdot)) = b^q$$

exists, and if the state space X is supplied with Q evaluation functions E_q, then the gradient method provides the following algorithm. Starting with a synaptic matrix $W^0(\cdot)$, we define $W^{n+1}(\cdot)$ from the synaptic matrix $W^n(\cdot)$ according to this rule: For all $t \in [0, T]$,

$$W^{n+1}(t) - W^n(t) = -\varepsilon_n \sum_{q=1}^{Q} x^{n,q}(t) \otimes g'(W^n(t)x^{n,q}(t))^\star p^{n,q}(t)$$

where $x^{n,q}(\cdot)$ is the solution to the differential equation

$$\frac{d}{dt}x^{n,q}(t) = g(W^n(t)x^{n,q}(t))$$

starting at $x^{n,q}(0) = a^q$ and $p^{n,q}(\cdot)$ is the solution to the adjoint differential equation

$$\frac{d}{dt}p^{n,q}(t) = -W^n(t)^\star g'(W^n(t)x^{n,q}(t))^\star p^{n,q}(t)$$

starting at

$$p^{n,q}(T) = E'_q(x^{n,q}(T) - b^q)$$

5.4 Newton-type Methods

We recall that in the preceding analysis, a neural network with one layer, several layers of a continuous set of layers map inputs x to outputs $y = \Phi(x, W)$. Let us consider a family of patterns $(a^j, b^j)_{1\leq j\leq J}$. We have to find synaptic matrices $\overline{W}$ learning these patterns, that is, satisfying the equation

$$\forall\, j = 1, \ldots, J, \quad \Phi(a^j, W) \;=\; b^j$$

Instead of replacing these nonlinear problems by optimization problems to which we apply gradient methods, we investigate in this section the use of Newton algorithms for finding synaptic matrices learning such patterns. We shall study successively the cases of perceptron-type networks with one layer, with several layers, and with a continuous set of layers.

5.4.1 Newton Learning Rules

Consider a nonlinear neural network that maps an input space $X := \mathbf{R}^n$ to an output space $Z := \mathbf{R}^m$ through an intermediate layer Y. The network first "weights" an input signal $x \in X := \mathbf{R}^m$ through a synaptic matrix $W \in \mathcal{L}(X, Y)$ and thus processes this signal by mapping it to the output signal $f(x, W) := g(Wx)$, where $g : Y \mapsto Z$.

Let us consider a family of patterns $(a^j, b^j)_{1 \le j \le J}$. We have to find synaptic matrices $\overline{W}$ learning these patterns, that is, satisfying the equation

$$\forall\, j = 1, \ldots, J, \quad g(Wa^j) \;=\; b^j$$

which can be written in the more compact form

$$\Phi(\overline{W}) \;:=\; \left(g(\overline{W}a^j) - b^j\right)_{1 \le j \le J} \;=\; 0$$

Instead of using the gradient method, we can think of using the Newton algorithm

$$\forall\, n \ge 0, \quad \Phi'(W^n)\left(W^{n+1} - W^n\right) \;=\; -\delta\Phi(W^n)$$

starting at some initial matrix W^0, where $\delta > 0$ is given. Because the linear operators $\Phi'(W^n)$ are not necessarily injective, we shall use the *slow* Newton algorithm (i.e., choose W^{n+1} such that the norm of the *discrete velocity* $W^{n+1} - W^n$ is minimal among the solutions of the foregoing equation).

Theorem 5.4.1 *Assume that the transfer function $g : Y \mapsto Z$ is differentiable and that it satisfies the hypothesis*

$$\forall\, j = 1, \ldots, J, \quad \forall\, W \in \mathcal{L}(X, Y), \quad g'(Wa^j) \text{ is surjective} \tag{5.2}$$

Assume also that the inputs a_j are orthogonal for the scalar product l of X.

Then the slow solutions of the Newton algorithm are given by the following Hebbian learning rules:

$$W^{n+1} = W^n + \delta \sum_{j=1}^{J} La^j \otimes (g'(W^n a^j))^+ \left(\frac{b^j - g(W^n a^j)}{l(a^j, a^j)} \right) \quad (5.3)$$

Proof Because the directional derivative of the map Φ at W^n in the direction U is equal to

$$\Phi'(W^n)U = g'(W^n a^j) U a^j$$

the Newton algorithm can be written in the form

$$\forall j = 1, \ldots, J, \quad g'(W^n a^j)(W^{n+1} - W^n) a^j = -\delta (g(W^n a^j) - b^j)$$

By Theorem 3.1.1, with $U = 0$, $B^j = g'(W^n a^j)$, and $y^j = \delta (b^j - g(W^n a^j))$ the minimal discrete velocity is given by

$$W^{n+1} - W^n = \delta \sum_{j=1}^{J} La^j \otimes g'(W^n a^j)^+ \left(\frac{b^j - g(W^n a^j)}{l(a^j, a^j)} \right)$$

□

5.4.2 Newton Algorithm for Multilayer Neural Networks

Let us consider a multilayer discrete network described by the vector spaces $X_0 := X$, X_l $(l = 1, \ldots, L-1)$, and $X_L := Y$ and differentiable propagation rules $g_l : X_l \mapsto X_l$ and controlled by sequences of synaptic matrices $W_l \in \mathcal{L}(X_{l-1}, X_l)$ by

$$\forall l = 1, \ldots L, \quad x_l = g_l(W_l x_{l-1})$$

We denote by $\vec{W}$ the sequence of synaptic matrices

$$\vec{W} := (W_1, \ldots, W_L) \in \prod_{l=1}^{L} \mathcal{L}(X_{l-1}, X_l)$$

and by Φ the map associating with the sequence $\vec{W}$ and the initial state x_0 the sequence $\Phi(x_0, \vec{W}) := (x_1, \ldots, x_L)$ of states starting from x_0 according to the foregoing law.

An initial signal a being given, we shall denote by Φ_L the map associating with the sequence of synaptic matrices W_l the last component x_L of $\Phi(a, \vec{W}) := (x_1, \ldots, x_L)$.

We are looking for a sequence of synaptic matrices learning a pattern $(a,b) \in X_0 \times X_L$, that is, a solution $\vec{W} := (W_1, \ldots, W_L)$ to the nonlinear equation

$$\Phi_L(a, W_1, \ldots, W_L) = b$$

We shall approximate a solution to this problem by the slow Newton algorithm

$$\forall\, n \geq 0, \quad \vec{W}^{n+1} - \vec{W}^n = \delta \left(\Phi'_L(a, \vec{W}^n)\right)^+ \left(b - \Phi_L(a, \vec{W}^n)\right)$$

starting at some initial matrix $\vec{W}^0$, where $\delta > 0$ is given. In other words, the *slow Newton algorithm* chooses a solution $\vec{W}^{n+1}$ such that the norm of the *discrete velocity* $\vec{W}^{n+1} - \vec{W}^n$ is minimal among the solutions of the Newton algorithm

$$\forall\, n \geq 0, \quad \Phi'_L(a, \vec{W}^n)\,(W^{n+1} - W^n) = \delta \left(b - \Phi_L(a, \vec{W}^n)\right)$$

We shall make explicit the slow Newton algorithm associated with this problem.

If $\vec{W}^n$ is a given sequence of synaptic matrices, we denote by x_l^n the solutions to the problem

$$\forall\, l = 1, \ldots L, \quad x_l^n = g_l(W_l^n x_{l-1}^n)$$

starting from $x_0 := a$. We set $A_l^n := g'_l(x_l^n) W_l^n$ and

$$\begin{aligned} G^n(l,k) &= g'_l(x_l^n) W_l^n \cdots g'_l(x_{k+1}^n) W_{k+1}^n \\ G^n(l,k)^\star &= W_{k+1}^{n\star} g'_{k+1}(x_{k+1}^n)^\star \cdots W_l^{n\star} g'_l(x_l^n)^\star \end{aligned}$$

Theorem 5.4.2 *Let us assume that the propagation maps g_l are differentiable and that the derivatives g'_l that satisfy*

$$\forall\, n \geq 0, \quad \sum_{k=1}^{L} \mu_{k-1}(x_{k-1}^n)^2 G^n(L,k) g'_k(x_k^n) M_k^{-1} g'_k(x_k^n)^\star G^n(L,k)^\star$$

are invertible. We denote by $\pi_L^n \in Y := X_L$ the solution to the equation

$$\left(\sum_{k=1}^{L} G^n(L,k) g'_k(x_k^n) M_k^{-1} g'_k(x_k^n)^\star G^n(L,k)^\star \mu_{k-1}(x_{k-1}^n)^2\right) \pi_L^n = \delta\,(b - x_L^n)$$

The sequence $\vec{W}^n := (W_1^n, \ldots, W_L^n)$ of synaptic matrices W_l^n determined *by the slow Newton algorithm for mapping the input a to the output b is given by the formula*

$$\forall\, k = 1, \ldots, L, \quad W_k^{n+1} - W_k^n \;=\; M_{k-1} x_{k-1}^n \otimes M_k^{-1} g_k'(x_k^n)^\star G^n(L,k)^\star \pi_L^n$$

Proof First, we check that a sequence

$$(y_0, V_1, \ldots, V_L, y_1, \ldots, y_L)$$

belongs to the tangent space to the graph of Φ at

$$(x_0^n, W_1^n, \ldots, W_L^n, x_1^n, \ldots, x_L^n)$$

if and only if

$$\forall\, l = 1, \ldots, L, \quad y_l := A_l^n y_{l-1} + g_l'(x_l^n) V_l x_{l-1}^{n-1}$$

This discrete dynamical system can be solved by induction: The formula is given by

$$y_l \;=\; G^n(l,0) y_0 + \sum_{k=1}^{l} G^n(l,k) g_k'(x_k^n) V_k x_{k-1}^n$$

Therefore,

$$\Phi_L'(a, \vec{W}^n)\vec{V} \;=\; \sum_{k=1}^{L} G^n(L,k) g_k'(x_k^n) V_k x_{k-1}^n$$

The slow Newton algorithm amounts to finding a sequence of matrices $\vec{V}$ with minimal norm among the solutions to the linear system

$$\sum_{k=1}^{L} G^n(L,k) g_k'(x_k^n) V_k x_{k-1}^n \;=\; \delta\,(b - x_L^n)$$

We then apply Theorem 3.3.1 to conclude the proof. □

5.4.3 Newton Algorithm for Continuous-Layer Neural Networks

Let us consider a continuous neural network associated with differentiable propagation rules $g : Y \mapsto X$ and controlled by synaptic matrices $W(t) \in \mathcal{L}(X,Y)$, where X and Y are finite-dimensional vector spaces. The evolution of the signals is thus governed by the differential equation

$$x'(t) \;=\; g(W(t)x(t))$$

We denote by $\Phi_T(a, W(\cdot))$ the map that associates with a time-dependent synaptic matrix $W(t)$ and an initial signal a the final state $x(T)$ obtained through this discrete control system.

An output $b \in X$ being given, we are looking for a function

$$W(\cdot) \in L^2(0,T;\mathcal{L}(X,Y))$$

[endowed with the scalar product $\int_0^T (l_\star \otimes m)(U(t), V(t))dt$ and the duality mapping $J := U(\cdot) \mapsto (M \otimes L^{-1})U(\cdot)$] that can learn the pattern (a, b), that is, a solution to the problem

$$\Phi_T(a, W(\cdot)) = b$$

We shall approximate a solution to this problem by the slow Newton algorithm

$$\forall\, n \geq 0, \quad W^{n+1}(\cdot) - W^n(\cdot) = \delta\,(\Phi'_T(a, W^n(\cdot)))^+\,(b - \Phi_T(a, W^n(\cdot)))$$

starting at some initial matrix $W^0(\cdot)$, where $\delta > 0$ is given. We shall make explicit the slow Newton algorithm associated with this problem.

If $W^n(\cdot)$ is a given function, we denote by $x^n(\cdot)$ the solutions to the differential equation

$$\frac{d}{dt}x_n(t) = g(W^n(t)x_n(t))$$

starting from $x(0) := a$. We set

$$\begin{array}{rcl} A^n(t) & = & g'(W^n(t)x^n(t))W^n(t) \in \mathcal{L}(X,X) \\ A^{n\star}(t) & = & W(t)^\star g'(W^n(t)x^n(t))^\star \end{array}$$

and denote by $G^n(t,s)$ the fundamental matrix associated with $A^n(t)$.

Theorem 5.4.3 *Let us assume that the propagation map g is differentiable and that the operators*

$$\int_0^T \lambda(x^n(t))^2 G^n(T,t) g'(W^n(t)x^n(t)) M^{-1} g'(W^n(t)x^n(t))^\star G^n(T,t)^\star dt$$

are invertible. We denote by $\pi_T^n \in X$ the solution to the equation

$$\left(\int_0^T G^n(T,t)g'(W^n(t)x^n(t))M^{-1}g'(W^n(t)x^n(t))^\star G^n(T,t)^\star \lambda(x^n(t))^2 dt\right)\pi_T^n$$
$$= \delta\,(b - x_T^n)$$

The function $W^{n+1}(\cdot)$ given by the slow Newton algorithm for mapping the input a to the output b satisfies

$$W^{n+1}(t) - W^n(t) = Lx^n(t) \otimes M^{-1} g'(W^n(t)x^n(t))^\star G^n(T,t)^\star \pi_T^n$$

Proof First, we check, through the variational equation, that

$$\Phi'_T(a, W^n(\cdot))V(\cdot) = \int_0^T G^n(T,t)g'(W^n(t)x^n(t))V(t)x^n(t)dt$$

The slow Newton algorithm amounts to finding a function $V(\cdot)$ with minimal norm among the solutions to the linear system

$$\int_0^T G^n(T,t)g'(W^n(t)x^n(t))V(t)x^n(t)dt = \delta\,(b - x_T^n)$$

We then apply Theorem 3.4.1 to conclude the proof. □

6

External Learning Algorithm for Feedback Controls

Introduction

We devote this chapter to the learning processes[1] of neural networks that allow them to find feedback maps for control problems with state constraints. The results presented in this chapter are due to Nicolas Seube.

We shall study *external learning algorithms* for discrete (or discretized) control systems. Let us consider a discrete control problem defined as follows:

$$x_{n+1} = f(x_n, u_n), \qquad u_n \in U$$

where f is a $\mathcal{C}^1$ mapping from $X \times U$ to X, x_n is the state of the system, u_n is the control of the system, and U is the space of feasible controls. We also assume that $X = \mathbf{R}^n$, and $U \subset \mathbf{R}^m$.

Viability conditions (or state constraints) are described by a subset K of the state space: They are written in the form

$$\forall\, n \geq 0, \quad x_n \in K$$

The viability condition amounts to saying that $\forall\, n \geq 0$, $d_K(x_{n+1}) = 0$, so that viable solutions to the discrete control problem are also solutions to the optimization problem

$$\forall\, n \geq 0, \quad \inf_{u_n \in U} d_K(f(x_n, u_n)) \;=\; 0$$

[1] Standard learning rules (back-propagation and Widrow-Hoff learning rules, etc.) have been tested elsewhere for solving control problems. In the following sections we shall compare such methods with learning rules that involve knowledge of the dynamics of the system.

The strategy underlying the external algorithm is to apply the gradient method for minimization of $d_K(f(x,\cdot))$, where the control is provided by a given neural network (acting as a new controller). Because the function d_K is equal to zero inside K, the gradient methods lead to an "external minimization algorithm": The network will learn only when the system lies outside of the domain K.

For other problems, such as a target problem (we shall give an example of application of our algorithm to such a problem), we do not minimize a distance to a set, but simply a criterion defined on the state of the system (e.g., the distance to a point).

We shall look for a feedback law in the class of L-layer neural networks defined by layer spaces $X_0 := X, \ldots, X_L := X_L = U$, by the synaptic matrices $W_k \in \mathcal{L}(X_{k-1}, X_k)$, and by the propagation rules $g_k : X_{k+1} \to X_{k+1}$.

$$\forall n \geq 0, \quad u(x_n) = \Phi_L(x_n, W_1, W_2, \ldots, W_L) \tag{6.1}$$

where $\Phi_L(x, \vec{W})$ denotes the propagation of a signal x in a neural network:

$$\begin{aligned} x^0 &= x \\ x^{k+1} &= g_k(W_{k+1}x^k), \qquad k = 0, \ldots, L-1 \\ \Phi_L(x, W_1, W_2, \ldots, W_L) &= x^L \end{aligned}$$

Our aim is then to find a sequence of synaptic matrices minimizing a criterion defined on the state of the system. This criterion describes implicitly a viability condition:

$$\min_{(W_1,\ldots,W_L)\in\prod_{i=1}^{L-1}\mathcal{L}(X_i,X_{i+1})} d_K(x_{n+1})^2$$

Conclusion Instead of the coding of patterns for use of the classical back-propagation algorithm, the external learning algorithm integrates knowledge of the dynamics of the control system:

- The learning of the feedback is far more efficient than for other learning algorithms for neural methods of the same kind (e.g., Barto and Sutton and Widrow and Smith for the cart-pole system).
- The feedback control being discovered is stable.
- The time efficiency of a learning step enables the system to discover a control law in real time.

6.1 Construction of the External Learning Algorithm

Let us consider, for simplicity, a closed convex subset K of the state space. The viability condition

$$\forall\, t \geq 0, \quad x_{t+1} := f(x_t, u_t) \in K$$

is satisfied if and only if

$$\forall\, n \geq 0, \quad d_K(x_t) = 0$$

If a viable regulation law exists (i.e., the foregoing condition is true), the gradient method, whenever it converges, can be used to compute that law as a solution of the following minimization problem:

$$\min_{(W_1,\ldots,W_L)\in\prod_{i=0}^{L-1}\mathcal{L}(X_i,X_{i+1})} d_K\left(f\left(x_t, \Phi_L(x_t, W_1,\ldots,W_L)\right)\right)$$

where $d_K(x) = \inf_{y\in K} \| x - y \|$ denotes the distance from x to the set K.

Let us set

$$\begin{aligned} \vec{W} &= (W_1, W_2, \ldots, W_L) \\ E(\vec{W}) &= d_K\left(x, f\left(x, \Phi_L(x, \vec{W})\right)\right)^2 \\ G(\vec{W}) &= f(x, \Phi_L(x, \vec{W})) \end{aligned}$$

The gradient of E at $\vec{W}$ is defined by

$$E'(\vec{W}) = G'(\vec{W})^\star \left(d_K\left(x, \Phi_L(x, \vec{W})\right)^2\right)'$$

with

$$G'(\vec{W})^\star = \Phi_L'(\vec{W}, x)^\star \frac{\partial f}{\partial u}\left(x, \Phi_L(x, \vec{W})\right)^\star$$

In order to compute $(\Phi_L'(\vec{W}, x))^\star$, we can apply Proposition 5.2.3. We obtain

$$G'(\vec{W})^\star = \begin{pmatrix} x^0 \otimes g_1'(x^1)^\star R_1 \\ \vdots \\ x^{L-2} \otimes g_{L-1}'(x^{L-1})^\star R_{L-1} \\ x^{L-1} \otimes g_L'(x^L)^\star R_L \end{pmatrix}$$

where

$$R_k = W_{k+1}^\star g_{k+1}'(x^{k+1})^\star \ldots W_L^\star g_L'(x^L)^\star \frac{\partial f}{\partial u}(x, \Phi_L(x, \vec{W}))^\star$$

is the back-propagation expression. We recall that the back-propagation algorithm is based on the foregoing formula.

Because we have to compute the derivative of d_K^2, we must make assumptions on the domain K. A complete proof of the following results can be found in *Set-Valued Analysis* (Aubin and Frankowska 1990).

Proposition 6.1.1 *Let us assume that K is closed and convex. Then $d_K^2 \in \mathcal{C}^1$, and $d_K^{2'}(x) = 2(x - \Pi_K(x))$, where Π_K is the best approximation projector onto the convex set K.*

We observe that $(x - \Pi_K(x))$ is always contained in the normal cone $N_K(x)$ when $x \notin K$. The foregoing proposition allows us to define a gradient learning rule that can find viable control laws for problem $\mathcal{P}$:

Proposition 6.1.2 *Assume that K is closed and convex. Consider a L-layer neural network described as before. Then the following learning rule allows us to find, whenever they exist, viable regulation laws for the control problem*

$$\vec{W}^{n+1} - \vec{W}^n \in -\varepsilon \begin{pmatrix} x^0 \otimes g_1'(x^1)^\star R_1 \\ \vdots \\ x^{L-2} \otimes g_{L-1}'(x^{L-1})^\star R_{L-1} \\ x^{L-1} \otimes g_L'(x^L)^\star R_L \end{pmatrix} F_K\left(\vec{W}^n, x^L\right)$$

where

$$F_K(\vec{W}, x) = \left\{(I - \Pi_K)\left(f\left(x, \Phi_L\left(\vec{W}, x\right)\right)\right)\right\}$$

and the back-propagation expression

$$R_k = W_{k+1}^\star g_{k+1}'(x^{k+1})^\star \ldots W_L^\star g_L'(x^L)^\star \frac{\partial f}{\partial u}(x, \Phi_L(x, \vec{W}))^\star$$

and where ε is a step vector.

This learning rule is a *generalized back-propagation learning rule.* As a matter of fact, we can recognize in Proposition 6.1.2 the back-propagation algorithm, consisting in adjusting weights from the last layer L to the first one, in order to minimize a cost function defined on the last layer. In the case of our learning rule, the cost function has been chosen such that the viability condition for the control problem is satisfied.

We can deal with other problems by replacing the distance function

d_K by any criterion function J on the state space and considering the minimization problem

$$\min_{(W_1,\dots,W_L)\in\prod_{i=0}^{L-1}\mathcal{L}(X_i,X_{i+1})} J\left(f\left(x_t,\Phi_L(W_1,\dots,W_L,x_t)\right)\right)$$

Let us set $\vec{W}=(W_1,W_2,\dots,W_L)$, $E(\vec{W})=J\left(G(\vec{W})\right)$, and $G(\vec{W})=f(x,\Phi_L(x,\vec{W}))$. The gradient of E is

$$E'(\vec{W})=(G'(\vec{W}))^\star J'\left(f\left(x,\Phi_L\left(x,\vec{W}\right)\right)\right)$$

with

$$(G'(\vec{W}))^\star=\Phi'_L(\vec{W},x)^\star\frac{\partial f}{\partial u}\left(x,\Phi_L(x,\vec{W})\right)^\star.$$

We deduce from Proposition 5.2.3 the following result:

Proposition 6.1.3 *Let us consider an L-layer neural network defined by L synaptic matrices* $(W_i)_{i=1,\dots,L}$ *and L differentiable threshold functions* $(g_i)_{i=0,\dots,L-1}$. *Let us set* $\vec{W}=(W_1,W_2,\dots,W_L)$. *Assume that the function J is differentiable. Then the following learning rule allows us to find, whenever they exist, viable regulation laws for the problem* $\mathcal{P}$:

$$\vec{W}^{n+1}-\vec{W}^n\in -\varepsilon\begin{pmatrix} x^0\otimes g_1'(x^1)^\star R_1\\ \vdots\\ x^{L-2}\otimes g_{L-1}'(x^{L-1})^\star R_{L-1}\\ x^{L-1}\otimes g_L'(x^L)^\star R_L\end{pmatrix} J'\left(f\left(x_t,\Phi_L\left(x_t,\vec{W}\right)\right)\right)$$

where the back-propagation expression R_k *is*

$$R_k=W_{k+1}^\star g_{k+1}'(x^{k+1})^\star\dots W_L^\star g_L'(x^L)^\star\frac{\partial f}{\partial u}(x_t,\Phi_L(x_t,\vec{W}))^\star,$$

$k=1,\dots,L-1$ *and*

$$R_L=\frac{\partial f}{\partial u}(x_t,\Phi_L(x_t,\vec{W}))^\star$$

We thus propose a self-organization algorithm based on the back-propagation formula given by Proposition 6.1.2.

In all simulations we have chosen to set the threshold functions to the sigmoid function defined by:

$$g(x)=A\frac{e^{kx}-1}{e^{kx}+1}.$$

Let K_m, be a domain where the state of the system can live: If the state lies outside of this domain, we consider that the system is no longer controllable. In this case, our trial-and-error algorithm is the following:

1. Initialize the synaptic matrix (e.g., $W^0 = 0$).
2. Set an initial condition for the control problem: $x(0) = x_0 \in K_m$.
3. The system is controlled by the current synaptic matrix W^n:

$$x_{t+1} = f(x_t, u_t), \qquad u_t = \Phi_L(\vec{W^n}, x_t)$$

   At each step of the process, the learning rule defined in Proposition 6.1.3 is activated.
4. $v_t = \Phi_L(\vec{W}^{n+1}, x_t), \quad x_{t+1} = f(x_t, v_t)$
5. If x_t lies outside of K_m, store the previous modifications of $\vec{W^n}$, and go to step 2; otherwise, go to step 3.

6.2 Simulation Results: The Cart Pole

Nicolas Seube has tested the foregoing algorithm numerically for benchmark problems that have been studied with different kinds of neural methods. For each of these problems, his results are compared with those from other methods. A generator of Fortran source code for the external learning rule, based on the symbolic computation software MAPLE, has been developed by N. Seube. This generator needs only a formal description of the network and its dynamics and the criterion J to produce automatically a set of Fortran subroutines that can simulate the algorithm corresponding to one step of our learning rule. It is therefore very easy to simulate different learning algorithms based on the external learning rule. This algorithm has been tested for many nonlinear control problems, neural-network structures, and viability constraints (or criterion functions J).

6.2.1 Viable control of a cart-pole system

This problem (which is strongly nonlinear and unstable) provides a good example for testing the robustness of our method. Standard learning algorithms have been used by Barto and Sutton and Widrow and Smith (their results do not seem to have been published). We give only the main results concerning application of the external learning rule to this problem.

The cart-pole dynamics are given by

$$\frac{dX}{dt} = (\omega, v, \alpha \sin(\theta) + \beta \cos(\theta)u, u), \qquad X = (\theta, x, \omega, v)$$

where θ and ω are the angle and angular velocity with respect to a vertical axis, x and v are the position of the cart and its velocity, u is the control (acceleration given to the cart), and α and β are parameters that depend on the features of the cart pole. We consider the discrete control problem provided by the Euler approximation of the continuous problem. In the case of a viable control problem, the criterion function J is

$$J(x) = d_K(x)^2$$

where $d_K(x) := \min_{y \in K}(\| x - y \|)$ denotes the distance from x to K. When K is closed and convex, its gradient can be easily computed:

$$\frac{1}{2} d_K^{2'}(x) = x - \Pi_K(x)$$

where Π_K denotes the best approximation projection onto the set K.

N. Seube has tested the external learning algorithm with a very simple network structure: One layer of four cells is connected to the state X of the system used as an input.[2] The output of the network is defined by[3]

$$u_n := g(w_1\theta + w_2 x + w_3 \omega + w_4 v)$$

For the cart-pole problem, we choose the following domain $K_m : \{X \mid \ \mid \theta \mid < \Pi/2\}$. In a first learning phase, we set constraints only for angular variables: $K = K_{\theta,\omega} = \{X \mid \theta^2 + \omega^2 < \frac{1}{2}\}$. The number of trials needed to compute a viable control law (in the sense of the domain $K_{\theta,\omega}$) will depend on the temperature of the threshold function (i.e., the parameter k): For $k = 1$, approximately 100 trials are necessary, but for k=4, 23 trials are sufficient (for $k \geq 4$, the learning time grows). After this learning phase, we set position constraints too. Only one trial[4] is necessary to compute a viable control law with respect to the set $K_{\theta,\omega,x,v} = \{X \mid \theta^2 + \omega^2 + x^2 + v^2 \leq \frac{1}{2}\}$.

The graphs show two-dimensional projected trajectories for the cart-pole system after 10 trials (Figure 6.1), and 24 trials (Figure 6.2). In all figures, a two-dimensional projection of the ball of radius $\frac{1}{2}$ is plotted.

[2] Barto and Sutton used two networks and a Hebbian learning rule with memory. They used 300 weights to compute their law.

[3] This means that the control is bounded by the sigmoid parameter $A : \ U = [-A, A]$.

[4] This learning time is much shorter than the Barto-Sutton learning time (300 trials) or the Widrow learning time (thousands of trials).

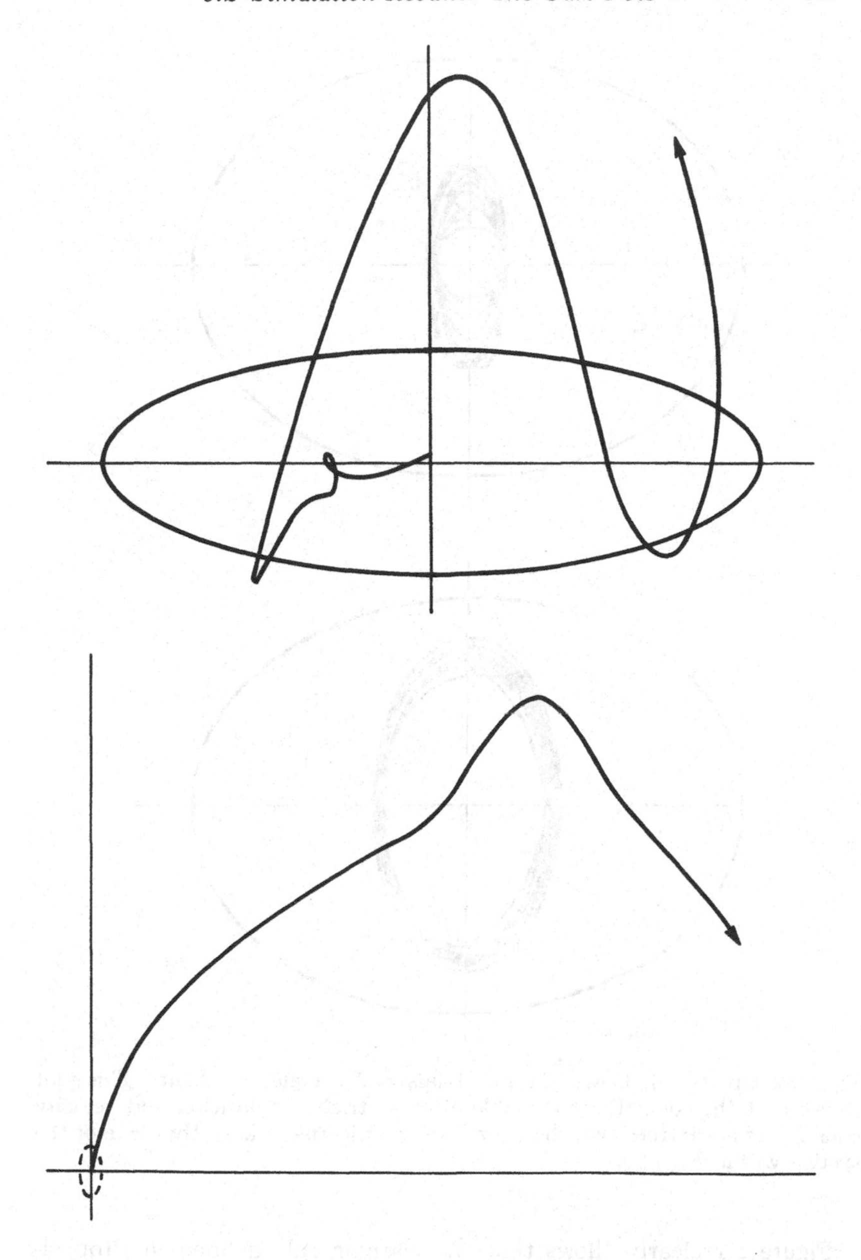

Fig. 6.1 Up: (θ, ω), Down: (x, v). Trials = 10; Time = 20 sec. The plot shows the viability set (centered ball of radius 1/2), and the trajectory of the system. After 10 trials, the control is not stable. The plot shows the ball of radius 1/2, but during learning, we did not set constraints on the variables x and v. Compare this trajectory with plot of Figure 6.2.

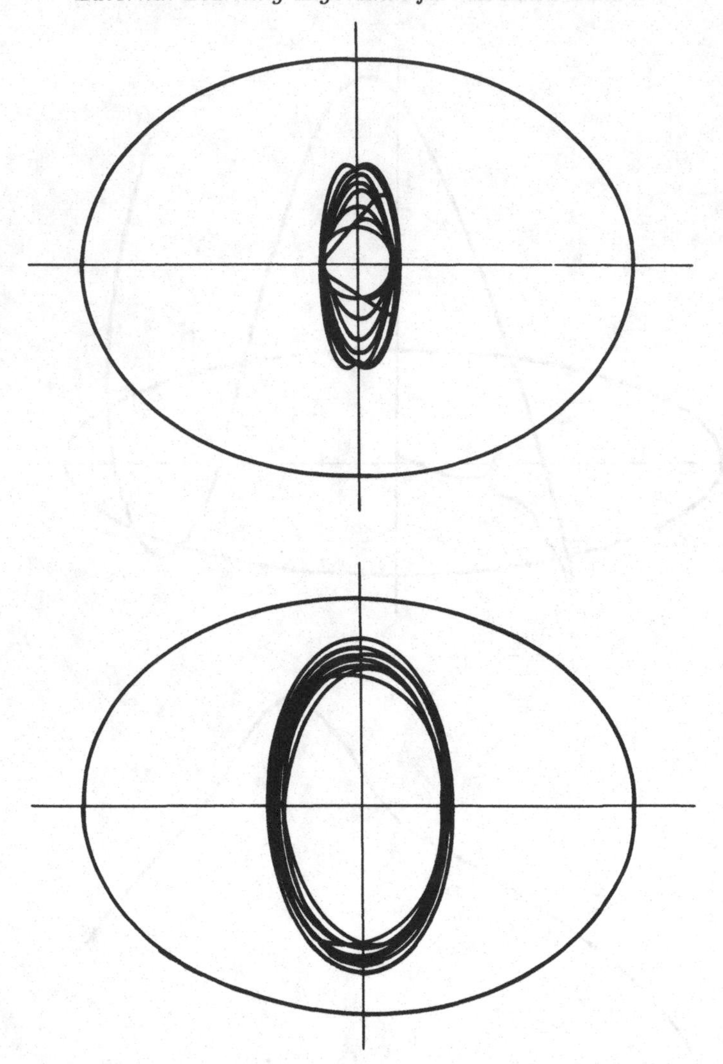

Fig. 6.2 Up: (θ, ω), Down: (x, v). Trials = 23; Time $\geq$ 1 hour. This plot shows that the control law is viable after 24 trials. If position and velocity constraints are introduced, the neural network learns to keep the state of the system within $K_{\theta,\omega,x,v}$.

Figure 6.1 clearly shows that the learning rule defined in Proposition 6.1.3 computes a control law that bends the trajectory toward $K := K_{\theta,\omega}$. This is not surprising, because the learning rule forces the control u to set the velocity of the system in the direction of $(x - \Pi_K(x))$.

Figure 6.2 shows that it is possible to introduce position and velocity

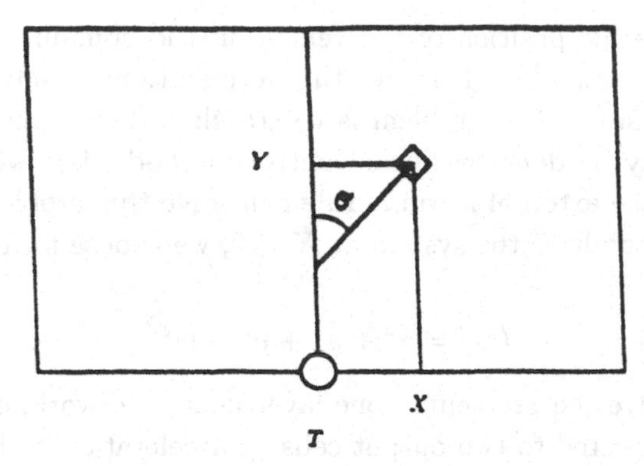

Fig. 6.3 Target problem. The car is placed in the domain K, and has to reach the target T. At each time, sensors give the position (x, y) of the car, and the wheel angle with respect to a vertical axis.

constraints, by setting $K_{\theta,x,\omega,v} = B(0, \frac{1}{2})$ (the plot represents more than an hour of simulated real time).

Conclusion We have shown that a one-layer neural network can solve the difficult problem of finding a viable feedback law for the cart-pole system. Note that the nonlinearity improves the performance of our system.[5] Our control law is discovered in real time.

The external learning algorithm is self organizing. [*We have no problems concerning the coding of patterns (choice of pieces of trajectories with appropriate control) as in supervised learning control systems.*]

6.3 Simulation Results: A Target Problem

We consider now the following target problem: A vehicle is placed in a closed domain and must reach a given point in the domain (Figure: 6.3).

6.3.1 Linearized Model of the Problem

The discrete dynamics of the vehicle are:

$$X = (x, y, v, w), \qquad X_{n+1} = X_n + h(v_n, w_n, u_n^1, u_n^2)$$

[5] As a matter of fact, the learning time will depend on the sigmoid temperature. Note that the control is not a pseudolinear feedback; Figure 6.2 clearly shows that there is no stabilization.

where (x, y) is the position of the vehicle in the domain, v and w are the velocities, and u^1 and u^2 are the accelerations (controls); h is a discretization step. The problem is to stabilize the system at $X = 0$. This can easily be done by linear control methods, but we want only to show how the external learning rule can solve this problem. Because we want to "stabilize" the system at $X = 0$, we choose for criterion the function J:

$$J(x) = x^2 + y^2 + v^2 + w^2$$

In order to solve the problem, a one-layer neural network, composed of four cells, connected to two output cells (x acceleration and y acceleration), was used. A sigmoid function was used as propagation rule. The domain K_m was $[-10, 10] \times [0, 10]$. During the first trial, the learning rule was able to compute a feedback law that allowed the system to reach the domain $K = B(0, 10^{-2})$; at the time it was reached, the velocity was zero (Figure 6.4). Figure 6.4 shows that the trajectory of the system follows the direction of the gradient of J (straight line from the initial-condition position to the center of the target).

6.3.2 Nonlinear Model

When the velocity of the vehicle is constant, the dynamics of the previous system can be approximated (at order 2) by

$$X_{n+1} = X_n + h\left(\sin\theta_n + \frac{u_n h \cos\theta_n}{2}, -\cos\theta_n - \frac{u_n h \sin\theta_n}{2}, u_n\right)$$

where (x, y) is the position of the vehicle, θ is the angle formed by the axis of the vehicle and the vertical Y axis, and $X = (x, y, \theta)$ (Figure 6.3). Now the control is the steering-angle variation.[6] We want to find a control law such that the vehicle will reach the point $X = 0$. A criterion J corresponding to this problem is, for instance,

$$J(x) = \lambda x^2 + y^2 + \theta^2$$

A large value for parameter λ has been chosen because the criterion is stiff with respect to the variable y (caused by the term θ^2).

Figure 6.5 shows trajectories corresponding to the first six trials of the learning algorithm. One can see that the learning rule forces the control to curve the trajectories in the direction of the target $B(0, 10^2)$. During

[6] Note that the control is bounded by the amplitude of the sigmoid function, as in the case of the cart-pole system.

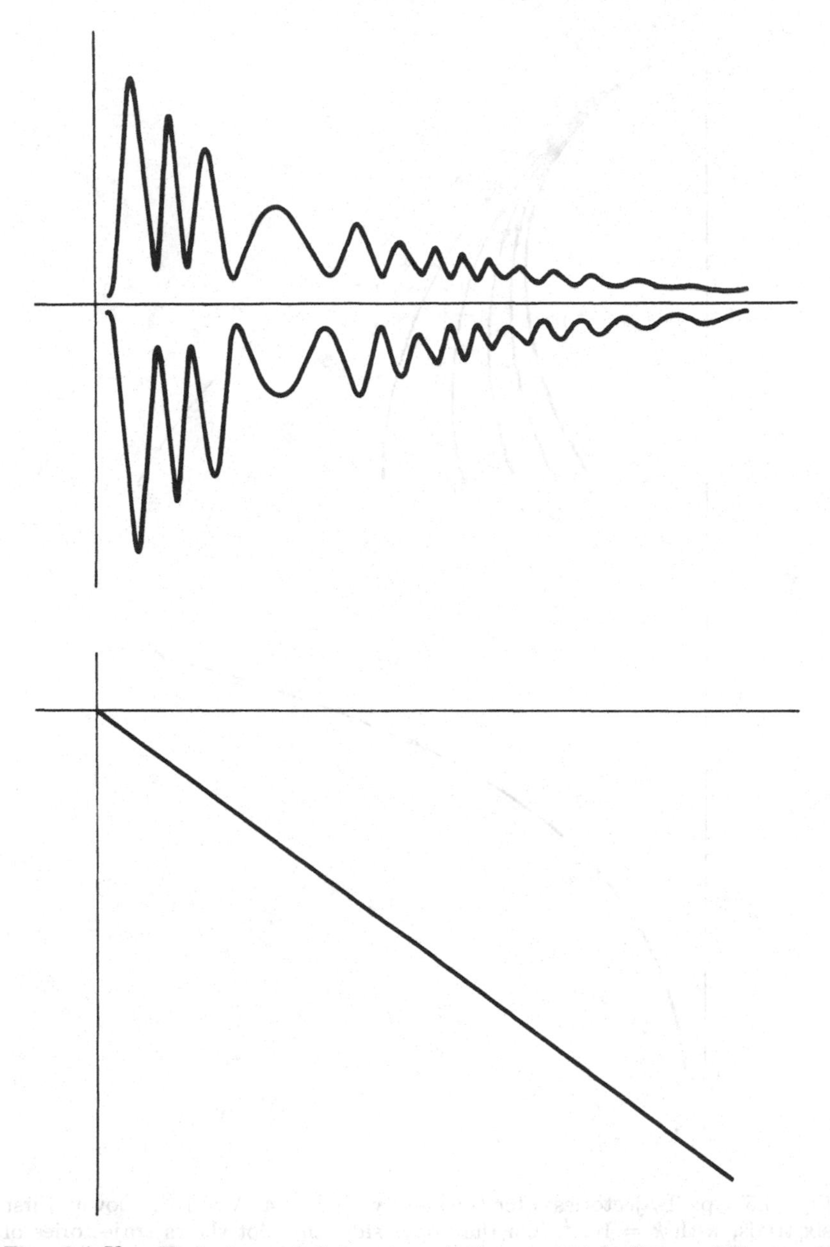

Fig. 6.4 Up: Trajectory of the system during one trial. Down: Velocity of the system during the same. On the upper side, we have plotted velocity v and w versus time. One can check that the system reach the target with zero velocities. On the bottom, the plot shows the target (centered ball of radius 10^{-1}), and a trajectory: The network control easily this system (These plots represent the first trials).

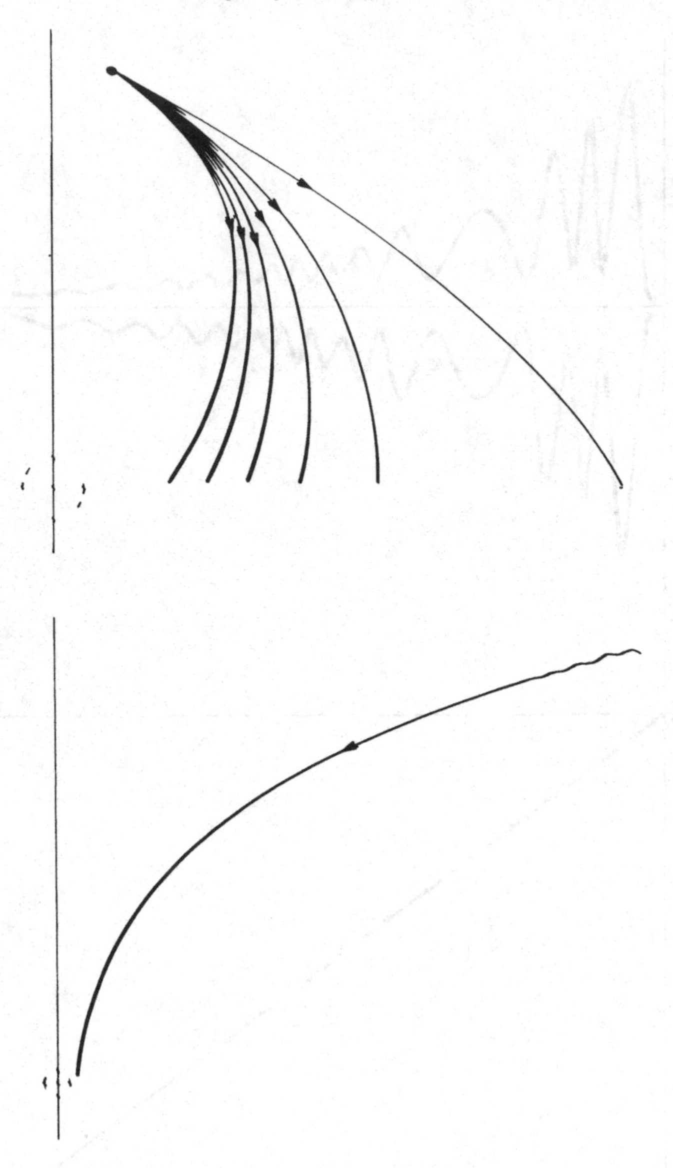

Fig. 6.5 Up: Trajectories after learning, with $k = 4$, $\lambda = 10^2$. Down: First six trials, with $k = 10^{-2}$. On the upper side, the plot shows trajectories of the system for the first six learning trials: One can see that learning seems very long, but remember that in this case a very low temperature was used. Trajectories bends in the direction of the target. On the bottom, the plot shows the target (centered ball of radius 10^{-1}), and the trajectory during the first trial. One can check that the neural network learns faster with high temperature.

those six trials, the sigmoid temperature was very low ($k = 10^{-2}$). In that case, the learning time was very long. Afterward, we set a high temperature ($k = 4$) to make the learning time shorter (i.e., so that the network would find the equilibrium state more rapidly). Figure 6.5 shows that that was effectively the case: After one trial, the control took the system to the target.[7]

In Figure 6.6 we set a higher temperature ($k = 5$), with the same criterion J. We obtained shorter learning times. After the first trial, the trajectory was closer to the target than in Figure 6.5.

6.4 Uniform External Algorithm

Let us consider the following discrete control problem:

$$x_{n+1} = f(x_n, u_n), \qquad u_n \in U$$

where f is a C^1 map from the state space X to the control space Z, $x_n \in X$ denotes the state of the system, and $u_n \in U$ is the control input, where $U \subset Z$ denotes the set of feasible controls. The problem we want to solve consists in finding a neural-network controller able to keep the state of the system inside a given domain of the state space, denoted by K:

$$\forall n \geq 0, \qquad x_n \in K$$

Let us note $d_K(x) = \inf_{y \in K} \| y - x \|$ the distance from x to K. Let us recall that $\Pi_K(\cdot)$ denotes the projector (of best approximation) onto K, which is a continuous single-valued map whenever K is closed and convex. We want the control to be viable, in the sense that for any initial state, at least one solution is viable in the sense that it remains in K.

Let us denote by $\mathcal{L}(X, U)$ the space of linear operators, mapping the state space X to the control space U. For the sake of simplicity, we are looking for a one-layer network (without so-called hidden layers) as a viable controller, whose propagation function is given by $u_n = \phi(W x_n)$, where ϕ is the C^1 threshold function, and W is a synaptic matrix belonging to $\mathcal{L}(X, U)$.

The approach described in the first section consists in updating the

[7] Let us recall that our method is a real-time learning method. Learning occurs during evolution and control of the system.

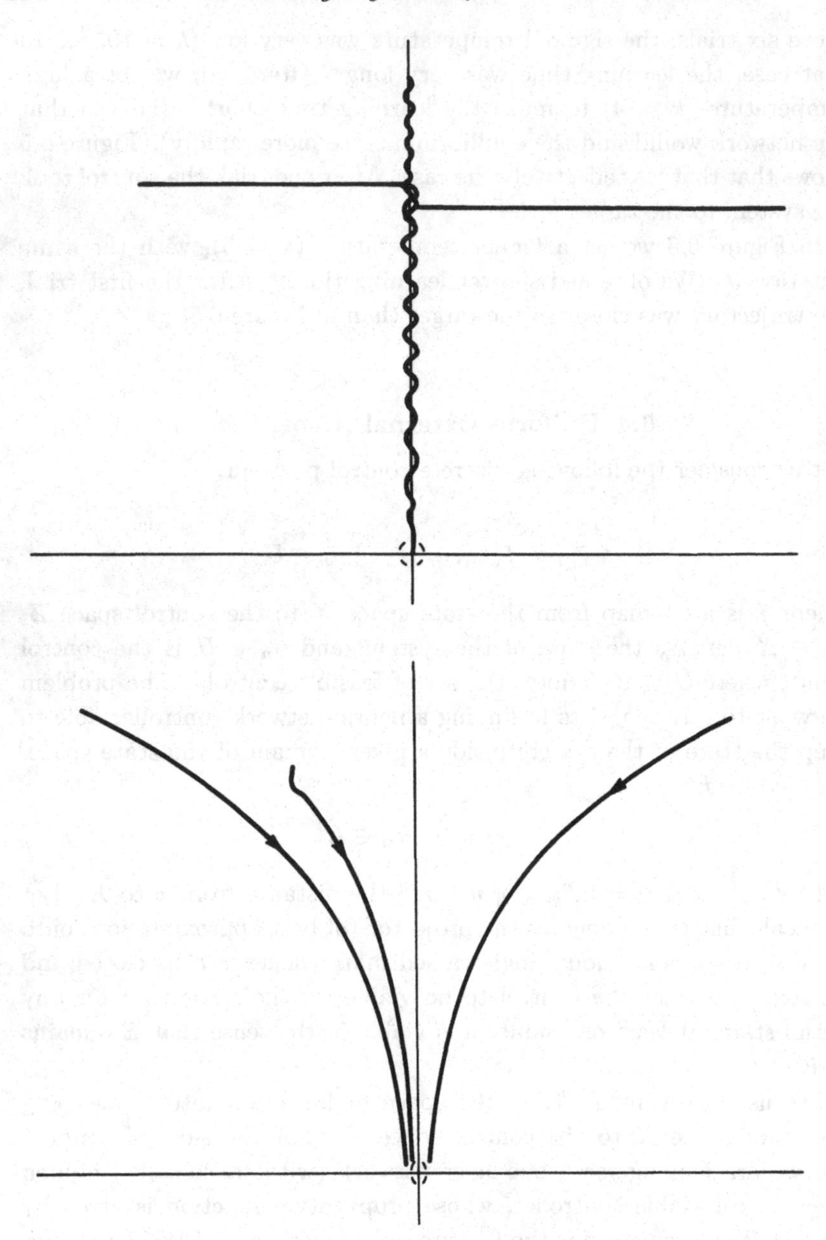

Fig. 6.6 Up: Trajectories after learning, with $k = 7$, $\lambda = 10^2$. Down: First six trials, with $k = 7$, $\lambda = 10^4$. On the upper side, the gradient J is stiff with respect to x. The network tends to make the system reach the vertical axis at first, and to reach the target afterwards. On the bottom, the gradient is the same than in figure 6.5, but the temperature is higher. One can check that the system learns faster.

network synaptic matrix in order to minimize the following local error function

$$\min_{W \in \mathcal{L}(X,U)} d_K^2(f(x_n, \phi(Wx_n)))$$

along particular trajectories of the system, called "learning trajectories," which correspond to trials of the learning algorithm.

Instead of that, one can try to minimize a *global error*, defined by

$$\min_{W \in \mathcal{L}(X,U)} \sup_{x \in K} (d_K^2(f(x, \phi(Wx))))$$

This means that we want the worst error that could occur in controlling the system with the current synaptic matrix W to be minimal. This is a nonsmooth optimization problem. In order to derive a subgradient algorithm, it is necessary to make use of nonsmooth calculus.

For practical implementation of the method, let us consider now a grid of the domain K, denoted as G. We replace the previous optimization problem by the finite-dimensional problem

$$\min_{W \in \mathcal{L}(X,U)} \sup_{x_i \in G} d_K^2(f(x_i, \phi(Wx_i)))$$

so that now the problem becomes to minimize the supremum of a finite number of functions

$$g(W) = \sup_{x_i \in G} d_K^2(f(x_i, \phi(Wx_i)))$$

Let us also recall the definition of the generalized gradient of a locally Lipschitz function and the result allowing us to compute the subdifferential of $g(W)$.

Definition 6.4.1 *Let $f : X \mapsto \mathbf{R} \cup \{+\infty\}$ be a locally Lipschitz function defined on a nonempty open subset. Its Clarke directional derivative is defined by*

$$C_\uparrow f(x)(v) := \limsup_{h \to 0+, y \to x} \frac{f(y+hv) - f(y)}{h}$$

The set $\partial f(x) := \{p \in X^\star \mid \forall v \in X, \quad \langle p, v \rangle \leq C_\uparrow f(x)(v)\}$ is called the Clarke generalized gradient of f at point x.

We recall the following statement allowing us to compute the directional derivative and the generalized gradient of the supremum of a finite number of functions.

Proposition 6.4.2 *Let us consider l locally Lipschitz functions $h_i : X \mapsto \mathbf{R} \cup \{+\infty\}$ and $g(x) = \sup_{1 \le i \le l} h_i(x)$. Let $I(x) = \{1 \le i \le l \mid g(x) = h_i(x)\}$, and $x \in \cap_{1 \le i \le l} \mathrm{Int}(\mathrm{Dom}(h_i))$. Then the Clarke directional derivative of g at x is defined by*

$$C_{\uparrow} g(x)(v) = \max_{i \in I(x)} \langle h'_i(x), v \rangle$$

and its generalized gradient is

$$\partial g(x) = \overline{\mathrm{co}} \left(\bigcup_{i \in I(x)} h'_i(x) \right)$$

where $\overline{\mathrm{co}}(A)$ denotes the closed convex envelope of the set A.

This allows us to compute the generalized gradient of the function g we want to minimize:

Theorem 6.4.3 *Let $f : X \times U \mapsto X$ and $\phi : U \mapsto U$ be two C^1 single-valued maps. Let W be a linear operator from X to U, and let $K \subset X$ be a closed convex subset of X. We denote by*

$$I(W) \ := \ \bar{x} \in G \mid d_K^2(f(\bar{x}, \phi(W\bar{x}))) = \sup_{x_i \in G} d_K^2(f(x_i, \phi(Wx_i)))$$

and we set $u(\bar{x}) = \phi(W\bar{x})$. Then the generalized gradient of the function

$$g(W) \ := \ \sup_{x_i \in G} d_K^2(f(x_i, \phi(Wx_i)))$$

is the subset of $\mathcal{L}(X^\star, U^\star)$ defined by

$$\partial g(W) \ = \ \overline{\mathrm{co}} \left(\bigcup_{\bar{x} \in I(W)} \bar{x} \otimes \phi'(W\bar{x})^\star \frac{\partial f(\bar{x}, u(\bar{x}))}{\partial u}^\star (I - \Pi_K) f(\bar{x}, u(\bar{x})) \right)$$

which is the convex hull of the matrices $V = (v_{ij})$, the entries of which are given by the formula

$$v_{ij} \ = \ \bar{x}_i \left(\phi'(W\bar{x})^\star \frac{\partial f(\bar{x}, u(\bar{x}))}{\partial u}^\star (I - \Pi_K) f(\bar{x}, u(\bar{x})) \right)_j$$

This theorem enables us to build the *uniform external algorithm*, based on a generalized gradient method. The uniform external algorithm is then the following:

1. For each point x_i of K, evaluate the value of the errors

$$d_K^2(f(x_i, \phi(Wx_i)))$$

and select one point $\bar{x} \in G$ that achieves the maximal error.

2. Perform a gradient descent in the direction of the following synaptic matrix: $W^{t+1} - W^t = -\varepsilon_n \bar{V}$, the entries v_{ij} of which are given by

$$\bar{v}_{ij} = (\bar{x})_i \left(\phi'(W\bar{x})^\star \frac{\partial f(\bar{x}, \phi(W\bar{x}))^\star}{\partial u} (I - \Pi_K) f(\bar{x}, \phi(W\bar{x})) \right)_j$$

3. Go back to step 1, and repeat until the error is greater than is desirable.

7
Internal Learning Algorithm for Feedback Controls

Introduction

This chapter is devoted to *internal learning algorithms* for learning feedback controls for solutions to continuous control systems

$$\begin{cases} \text{(i)} & x'(t) = f(x(t), u(t)) \\ \text{(ii)} & u(t) \in U(x(t)) \end{cases}$$

that are viable in a closed subset K, in the sense that

$$\forall t \geq 0, \quad x(t) \in K$$

Viability theory[1] provides, under adequate assumptions, a set-valued *regulation map* $R_K \subset U$ associating with each state $x \in K$ the subset $R_K(x) \subset U(x)$ of *viable controls* (i.e., controls regulating viable solutions). The viability subset is viable under the control system if and only if $R_K(x) \neq \emptyset$ for every $x \in K$, and in this case the viable solutions are regulated by controls obeying the *regulation law*:

$$\forall t \geq 0, \quad u(t) \in R_K(x(t))$$

We shall find neural networks with two layers X_1 and X_2, defined by an input map h from the state space X to the first layer and by a propagation map g from the second layer X_2 to the control space Z, that *learn this regulation law* in the sense that

$$u(t) = g(W(t)h(x(t)))$$

This amounts to replacing the usual controls $u \in Z$ by synaptic matrices $W \in \mathcal{L}(X_1, X_2)$ to control the system by a regulation law given through the neural network according to the foregoing law.

[1] We refer to the book *Viability Theory* (Aubin 1991) for an exhaustive exposition of this theory.

To illustrate the point, we can regard the control system as describing a car, and the neural network as its driver. The car-driver system is then described by the new control system

$$\begin{cases} \text{(i)} & x'(t) = f(x(t), g(W(t)h(x(t)))) \\ \text{(ii)} & g(W(t)h(x(t))) \in U(x(t)) \end{cases}$$

subject to the viability constraints

$$\forall\, t \geq 0, \quad x(t) \in K$$

We shall compute a differential equation that governs the evolution of synaptic matrices $W(t)$ yielding solutions to the control system viable in K. We therefore show that learning, in this way, an a priori feedback control law with a neural network requires only the knowledge of its derivative. The learning scheme we design is an adaptive one. We can use this learning scheme for tracking problems. As a matter of fact, the evolution law of the synaptic matrix enables the network controller to adapt at each point in time to state constraints of the kind

$$\forall t \geq 0, \quad x(t) = y(t)$$

where $y(t)$ is a reference trajectory to track. Numerical results concerning the tracking control of autonomous underwater vehicles (AUVs) will be presented and compared with other control modes.

7.1 Viability Theory at a Glance

Let us describe the (nondeterministic) dynamics of a system by a set-valued map[2] F from the state space X to itself. We consider initial-value problems (or Cauchy problems) associated with differential inclusion,

$$\text{for almost all } t \in [0,T], \quad x'(t) \in F(x(t)) \tag{7.1}$$

satisfying the initial condition $x(0) = x_0$. We shall look for solutions among *absolutely continuous functions.*

Definition 7.1.1 (Viability Properties) *Let K be a subset of the domain of F. A function $x(\cdot) : I \mapsto X$ is said to be* viable in K on the interval I *if and only if*

$$\forall\, t \in I, \quad x(t) \in K$$

[2]See Appendix A, Section A.9, for definitions related to set-valued maps.

We shall say that K is locally viable under F *(or enjoys the* local viability property for the set-valued map F*) if for any initial state x_0 in K* there exist $T > 0$ *and a solution on $[0, T]$ to differential inclusion (7.1) starting at x_0 that is* viable *in K. It is said to be* (globally) viable under F *[or to enjoy the* (global) viability property*] if we can take $T = \infty$.*

When K is a subset of X, and x belongs to K, we introduce the *contingent cone* $T_K(x)$ to K at x, which is the closed cone of elements v such that

$$\liminf_{h \to 0+} \frac{d(x + hv, K)}{h} = 0$$

defined in Appendix A, Section A.8.

Definition 7.1.2 (Viability Domain) *Let $F : X \rightsquigarrow X$ be a nontrivial set-valued map. We shall say that a subset $K \subset \mathrm{Dom}(F)$ is a viability domain of F if and only if*

$$\forall x \in K, \quad F(x) \cap T_K(x) \neq \emptyset$$

Because the contingent cone to a singleton obviously is reduced to zero, we observe that a singleton $\{\overline{x}\}$ is a viability domain if and only if $\overline{x}$ is an equilibrium of F, that is, a stationary solution to the differential inclusion, which is a solution to the inclusion

$$0 \in F(\overline{x}) \tag{7.2}$$

In other words, the equilibria of a set-valued map provide the first examples of viability domains - actually, of *minimal viability domains.*

7.1.1 The Viability Theorem

The main viability theorems hold true for the class of *Marchaud maps*, that is, the nontrivial closed set-valued maps with closed domain and convex values and linear growth. We observe that the only truly restrictive condition is the convexity of the images of these set-valued maps, since the continuity requirements (closed graph) are kept minimal. But we cannot dispense with the convexity condition, as the following counter example shows.

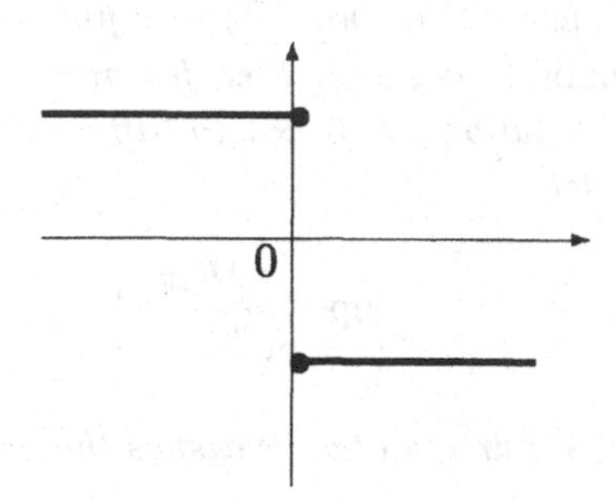

Fig. 7.1 Example of a Map without Convex Values. There is no solution starting at 0

Example Let us consider $X := \mathbf{R}$, $K := [-1, +1]$, and the set-valued map $F : K \rightsquigarrow \mathbf{R}$ defined by (Figure 7.1)

$$F(x) \quad := \quad \begin{cases} -1 & \text{if} \quad x > 0 \\ \{-1, +1\} & \text{if} \quad x = 0 \\ +1 & \text{if} \quad x < 0 \end{cases}$$

Obviously, no solution to the differential inclusion $x'(t) \in F(x(t))$ can start from from zero because zero is not an equilibrium of this set-valued map! We note, however, that

- The graph of F is closed
- F is bounded
- K is convex and compact
- K is a viability domain of F

But the value $F(0)$ of F at zero is not convex. Observe that if we had set $F(0) := [-1, +1]$, then zero would have been an equilibrium.

This example shows that closedness of the graph is not strong enough to compensate for the lack of convexity of the images.[3] We consider the weighted Sobolev space $W^{1,1}(0, \infty; X; e^{-bt}dt)$, which we supply with the topology for which a sequence $x_n(\cdot)$ converges to $x(\cdot)$ if and only if

$$\begin{cases} \text{(i)} & x_n(\cdot) \text{ converges uniformly to } x(\cdot) \text{ on compact sets} \\ \text{(ii)} & x_n'(\cdot) \text{ converges weakly to } x'(\cdot) \text{ in } L^1(0, T; X, e^{-bt}dt) \end{cases}$$

[3]Stronger continuity or differentiability requirements allow us to relax this assumption.

Theorem 7.1.3 (Viability Theorem) *Consider a* Marchaud map $F : X \rightsquigarrow X$ *and a* closed *subset* $K \subset \mathrm{Dom}(F)$ *of a finite-dimensional vector space* X. *If* K *is a* viability domain, *then for any initial state* $x_0 \in K$, *there exists a* viable *solution on* $[0, \infty[$ *to differential inclusion (7.1). More precisely, if we set*

$$c := \sup_{x \in \mathrm{Dom}(F)} \frac{\|F(x)\|}{\|x\| + 1}$$

then every solution $x(\cdot)$ *starting at* x_0 *satisfies the estimates*

$$\forall\, t \geq 0, \quad \|x(t)\| \;\leq\; (\|x_0\| + 1)e^{ct}$$

and

$$\text{for almost all } t \geq 0, \quad \|x'(t)\| \;\leq\; c(\|x_0\| + 1)e^{ct}$$

and thus belongs to the weighted Sobolev space $W^{1,1}(0, \infty; X; e^{-bt}dt)$ *for* $b > c$.

7.1.2 Solution Map

We denote by $\mathcal{S}(x_0)$ or by $\mathcal{S}_F(x_0)$ the (possibly empty) set of solutions to differential inclusion (7.1).

Definition 7.1.4 (Solution Map) *We shall say that the set-valued map* $\mathcal{S}$ *defined by* $\mathrm{Dom}(F) \ni x \mapsto \mathcal{S}(x)$ *is the* solution map *of* F *[or of differential inclusion (7.1)].*

The solution map depends continuously on the initial state, in the following sense:

Theorem 7.1.5 (Continuity of the Solution Map) *Let us consider a finite-dimensional vector space* X *and a* Marchaud map $F : X \rightsquigarrow X$. *We set*

$$c := \sup_{x \in \mathrm{Dom}(F)} \frac{\|F(x)\|}{\|x\| + 1}$$

Then for $b > c$, *the solution map* $\mathcal{S}$ *has a closed graph and compact images in the space* $W^{1,1}(0, \infty; X; e^{-bt}dt)$.

7.1.3 Limit Sets of Solutions

The limit sets of the solutions to the differential inclusion

$$\text{for almost all } t \geq 0, \quad x'(t) \in F(x(t)) \tag{7.3}$$

provide examples of closed viability domains:

Definition 7.1.6 (Limit Set) *Let $x(\cdot)$ be a solution to differential inclusion (7.3). We say that the subset*

$$L(x(\cdot)) \ := \ \bigcap_{T>0} \mathrm{cl}(x([T,\infty[))$$

of its cluster points when $t \to \infty$ is the limit set of $x(\cdot)$.

Theorem 7.1.7 (Limit Sets Are Viability Domains) *Let us consider a* Marchaud map $F : X \rightsquigarrow X$. *Then the limit sets of the solutions to differential inclusion (7.3) are closed viability domains. In particular, the limits of solutions to differential inclusion (7.3), when they exist, are equilibria of F, and the trajectories of periodic solutions to differential inclusion (7.3) are also closed viability domains.*

We can derive the existence of an equilibrium when ergodic averages of the velocities converge to zero:

Theorem 7.1.8 *Let us assume that F is upper-hemicontinuous, with closed convex images, and that $K \subset$ Dom(F) is compact. If there exists a solution $x(\cdot)$ viable in K such that*

$$\inf_{t>0} \frac{1}{t} \int_0^t \|x'(\tau)\| d\tau \ = \ 0$$

then there exists a viable equilibrium $\bar{x}$, i.e., a state $\bar{x} \in K$ that is a solution to the inclusion $0 \in F(\bar{x})$.

7.1.4 Viability Kernels

If a closed subset K is not a viability domain, the question arises whether or not there are closed viability subsets of K that are viable under F, as well as whether or not there exists a largest closed subset of K viable under F. The answer is affirmative for Marchaud maps, and we call the *viability kernel* of a closed subset K the largest closed subset viable under F contained in K.

Consider a closed subset K of the domain of F. We shall prove the existence of the largest closed subset of K viable under F.

Definition 7.1.9 (Viability Kernel) *Let K be a subset of the domain of a set-valued map $F : X \rightsquigarrow X$. We shall say that the largest closed subset of K viable under F (which may be empty) is the viability kernel of K for F and denote it by* $\mathrm{Viab}_F(K)$, *or simply* $\mathrm{Viab}(K)$. *If the viability kernel of K is empty, we say that K is a repeller.*

Theorem 7.1.10 *Let us consider a Marchaud map $F : X \rightsquigarrow X$. Let $K \subset \mathrm{Dom}(F)$ be closed. Then the viability kernel of K exists (possibly empty) and is equal to the subset of initial states such that at least one solution starting from them is viable in K.*

It may be useful to state the following consequence:

Proposition 7.1.11 *Let F be a Marchaud map and K be a closed subset of the domain of F. If x_0 belongs to $K \backslash \mathrm{Viab}_F(K)$, then every solution $x(\cdot) \in \mathcal{S}(x_0)$ must eventually leave K in finite time [in the sense that for some $T > 0, \quad x(T) \notin K$] and never meets* $\mathrm{Viab}_F(K)$ *as long as it remains in K. In particular, if K is a repeller, every solution starting from K leaves it in finite time.*

Following the terminology coined by Poincaré in the case of differential equations, we could call the complement of the viability kernel of K its *shadow (ombre)*: It is the subset of K from which every solution leaves K in finite time.

It is not only an attractive concept in the framework of viability theory but also an efficient mathematical tool we shall use quite often. We illustrate this fact by characterizing the *permanence property* introduced by J. Hofbauer and K. Sigmund and the *fluctuation property* introduced by V. Krivan in terms of viability kernels.

Theorem 7.1.12 *Let $F : X \rightsquigarrow X$ be a Marchaud map and K be a closed subset of its domain with a nonempty interior. Assume that there exists a nonempty open subset Ω such that $\overline{\Omega} \subset \mathrm{Int}(K)$ and*

$$\partial K = \mathrm{Viab}_F(K \backslash \Omega)$$

Then

1. *The boundary ∂K of K is semipermeable under F in the sense that*

for every $x_0 \in \partial K$, any solution $x(\cdot) \in \mathcal{S}(x_0)$ viable in $K \backslash \Omega$ is actually viable in ∂K.

2. *The interior* $\text{Int}(K)$ *of* K *is invariant under* F.

Actually, starting from $x_0 \in \text{Int}(K)\backslash\Omega$, *we can associate with any solution* $x(\cdot) \in \mathcal{S}(x_0)$ *an increasing sequence of instants* t_n *satisfying*

$$x(t_{2k}) \in \text{Int}(K)\backslash\Omega \quad \text{and} \quad x(t_{2k+1}) \in \Omega$$

such that $t_{2k+1} > t_{2k}$ *is finite whenever* t_{2k} *is finite. If this sequence is finite, it stops at an odd instant* $t_{2k_\infty - 1}$ *for the first index* k_∞ *such that*

$$\inf\{s \geq t_{2k_\infty - 1} \mid x(s) \notin \overline{\Omega}\} = \infty$$

It stops at t_1 *whenever* $\overline{\Omega}$ *is invariant under* F (the permanence property *introduced by J. Hofbauer and K. Sigmund*). *It never stops if the viability kernel* $\text{Viab}_F(\overline{\Omega})$ *of* $\overline{\Omega}$ *is empty* (the fluctuation property *introduced by V. Krivan*).

7.2 Regulation of Control Systems

7.2.1 Viable Control Systems

We now translate the viability theorems in the language of control theory. From now on, we introduce the state space X, the constraint space Y, the control space Z, and a *feedback set-valued map* $U : X \rightsquigarrow Z$ associating with any state x the (possibly empty) subset $U(x)$ of feasible controls when the state of the system is x. In other words, we assume that the available controls of the system are required to obey constraints that may depend upon the state.

The dynamics of the system are further described by a (single-valued) map $f : \text{Graph}(U) \mapsto X$ that assigns to each state-control pair $(x, u) \in \text{Graph}(U)$ the *velocity* $f(x, u)$ of the state. Hence the set $F(x) := \{f(x, u)\}_{u \in U(x)}$ is the set of available velocities for the system when its state is x. We shall assume from now on that $f(x, u) := c(x) + g(x)u$, where $g(x) \in \mathcal{L}(Z, X)$ are linear operators (i.e., the system is affine with respect to the control).

We also consider the case in which the viability domain $K := h^{-1}(M)$ is defined by more explicit constraints through a map h from X to the constraint space Y: The evolution of the system (U, f) is governed by the differential inclusion

$$\begin{cases} \text{(i)} & \text{for almost all } t, \quad x'(t) \;=\; f(x(t),u(t)) \\ \text{(ii)} & \text{where } u(t) \;\in\; U(x(t)) \end{cases} \tag{7.4}$$

When $U(x) = Z$ for all $x \in K$ and when $M = \{0\}$, we recognize the traditional control systems.

The regulation map R_K is the set-valued map defined by

$$R_K(x) \;:=\; \{u \in U(x) \mid h'(x)g(x)u \in T_M(h(x)) - h'(x)c(x)\}$$

Theorem 7.2.1 *Let us assume that the dynamics satisfy*

$$\begin{cases} \text{(i)} & \forall (x,u) \in \text{Graph}(U), \quad f(x,u) := c(x) + g(x)u \\ \text{(ii)} & \text{Graph}(U) \textit{ is closed, and the images of } U \textit{ are convex} \\ \text{(iii)} & c : \text{Dom}(U) \mapsto X \textit{ is continuous} \\ \text{(iv)} & g : \text{Dom}(U) \mapsto \mathcal{L}(Z,X) \textit{ is continuous and bounded} \\ \text{(v)} & c \textit{ and } U \textit{ have linear growth} \end{cases} \tag{7.5}$$

and that the constraints verify

$$\begin{cases} \text{(i)} & M \textit{ is a closed sleek subset of } Y \\ \text{(ii)} & h \textit{ is a } C^1 \textit{ map from } X \textit{ to } Y \\ \text{(iii)} & \forall x \in K := h^{-1}(M), \quad Y := \text{Im}(h'(x)) - T_M(h(x)) \\ \text{(iv)} & \forall x \in h^{-1}(M), \quad \exists u \in U(x) \\ & \textit{such that } h'(x)g(x)u \in T_M(h(x)) - h'(x)c(x) \end{cases}$$

Then K is a viability domain of the control system, and the viable solutions are regulated through the regulation law

$$\text{for almost all } t, \quad u(t) \;\in\; R_K(x(t)) \tag{7.6}$$

The regulation map R_K has compact nonempty convex values.

7.2.2 Closed-Loop Controls and Slow Solutions

Viable solutions to the control system (7.4) are regulated by viable controls whose evolution is governed by the regulation law (7.6). Continuous single-valued selections r_K of the regulation map R_K are *viable closed-loop controls*, because the viability theorem states that the differential equation

$$x'(t) \;=\; f(x(t), r_K(x(t)))$$

enjoys the viability property. Indeed, by construction, K is a viability domain of the single-valued map $x \in K \mapsto f(x, r_K(x))$.

So we have to investigate under which assumptions there exists a continuous selection of the regulation map. An answer can be provided by Michael's theorem, but that is not useful in practice, because Michael's selection theorem does not provide constructive ways to find those continuous closed-loop controls. Therefore, we are tempted to use explicit selections of the regulation map R_K, such as the minimal selection R_K° defined by

$$R_K^\circ(x) := \{u \in R_K(x) \mid \|u\| = \min_{y \in R_K(x)} \|y\|\} \tag{7.7}$$

It is continuous only when R is continuous with closed convex images. Unfortunately, there is no hope of having, in general, continuous regulation maps R_K (as soon as we have inequality constraints). Hence this minimal selection is not necessarily continuous when the regulation map is only lower-semicontinuous. But we can still prove that by taking the minimal selection R_K°, the differential equation

$$x'(t) = f(x(t), R_K^\circ(x(t))) \tag{7.8}$$

does enjoy the viability property.

Definition 7.2.2 *The solutions to the differential equation (7.8) are called slow viable solutions to the control system (7.4).*

Theorem 7.2.3 *If K is a closed viability domain, and if the regulation R_K map is lower-semicontinuous,*[4] *then the control system has slow viable solutions.*

The reason why this theorem holds true is that the minimal selection is obtained through a "selection procedure" defined in the following way:

Definition 7.2.4 (Selection Procedure) *A selection procedure of a set-valued map $R : X \rightsquigarrow Y$ is a set-valued map $S_R : X \rightsquigarrow Y$*

$$\begin{cases} \text{(i)} & \forall x \in \mathrm{Dom}(R), \quad S(R(x)) := S_R(x) \cap R(x) \neq \emptyset \\ \text{(ii)} & \textit{the graph of } S_R \textit{ is closed} \end{cases}$$

and the set-valued map $S(R) : x \rightsquigarrow S(R(x))$ is called the selection of R.

[4]See Chapter 6 of *Viability Theory* (Aubin 1991) for conditions under which the regulation map is lower-semicontinuous.

The selection procedure providing the minimal selection is given by

$$S_R(x) := \{u \in X \quad \text{such that } \|u\| \leq d(0, R(x))\}$$

which is closed with convex values.

We can easily provide other examples of selection procedures through optimization and/or game theory. The foregoing theorem follows from the following general principle:

Theorem 7.2.5 *We suppose that K is a viability domain. Let S_{R_K} be a selection of the regulation map R_K with convex values. Then, for any initial state $x_0 \in K$, there exist a viable solution starting at x_0 and a viable control to the control system (7.4) that are regulated by the selection $S(R_K)$ of the regulation map R_K, in the sense that*

$$\text{for almost all } t \geq 0, \quad u(t) \in S(R_K)(x(t)) := R_K(x(t)) \cap S_{R_K}(x(t))$$

7.2.3 Smooth Solutions

There are many reasons for looking for "smooth viable controls," which are absolutely continuous instead of being measurable. Too much oscillation of the controls can damage them, for instance. We can obtain smooth viable solutions by setting a bound to the growth of the evolution of controls. For that purpose, we shall associate with this control system and with any nonnegative continuous function $u \to \varphi(x, u)$ with linear growth[5] the system of differential inclusions

$$\begin{cases} \text{(i)} & x'(t) = f(x(t), u(t)) \\ \text{(ii)} & u'(t) \in \varphi(x(t), u(t))B \end{cases} \tag{7.9}$$

We are looking for closed set-valued feedback maps R contained in U in which we can find the initial state controls yielding smooth viable solutions to the control system. We shall characterize them as (set-valued) solutions to systems of first-order partial differential inclusions. For that purpose, we need the concept of *contingent derivatives* of set-valued maps defined in Appendix A, Section A.10. Indeed, the viability theorem implies the following:

Theorem 7.2.6 *Let us assume that the control system (7.4) satisfies*

$$\begin{cases} \text{(i)} & \text{Graph}(U) \textit{ is closed} \\ \text{(ii)} & f \textit{ is continuous and has linear growth} \end{cases} \tag{7.10}$$

[5]This can be a constant c, or the function $c\|u\|$, or the function $(x, u) \to c(\|u\| + \|x\| + 1)$.

Let $(x,u) \to \varphi(x,u)$ *be a nonnegative continuous function with linear growth and* $R : Z \rightsquigarrow X$ *a closed set-valued map contained in* U*. Then the two following conditions are equivalent:*

1. R *regulates* φ*-smooth viable solutions in the sense that for any initial state* $x_0 \in \text{Dom}(R)$ *and any initial control* $u_0 \in R(x_0)$*, there exists a* φ*-smooth state-control solution* $(x(\cdot), u(\cdot))$ *to the control system (7.4) starting at* (x_0, u_0) *and viable in the graph of* R*.*
2. R *is a solution to the partial differential inclusion*

$$\forall\, (x,u) \in \text{Graph}(R), \quad 0 \in DR(x,u)(f(x,u)) - \varphi(x,u)B \quad (7.11)$$

satisfying this constraint: $\forall\, x \in K, \;\; R(x) \subset U(x)$*.*

In this case, such a map R *is contained in the regulation map* R_K *and is thus called a* φ*-subregulation map of* U *or simply a subregulation map. The metaregulation law regulating the evolution of state-control solutions viable in the graph of* R *takes the form of the system of differential inclusions*

$$\begin{cases} \text{(i)} & x'(t) \;=\; f(x(t), u(t)) \\ \text{(ii)} & u'(t) \;\in\; G_R(x(t), u(t)) \end{cases} \quad (7.12)$$

where the set-valued map G_R *defined by*

$$G_R(x,u) \;:=\; DR(x,u)(f(x,u)) \cap \varphi(x,u)B$$

is called the metaregulation map associated with R*.*

Furthermore, there exists a largest φ*-subregulation map denoted* R^φ *contained in* U*.*

Proof Indeed, to say that R is a regulation map regulating φ- smooth solutions amounts to saying that its graph is viable under the system (7.9).

In this case, we deduce that for any $(x_0, u_0) \in \text{Graph}(R)$, there exists a solution $(x(\cdot), u(\cdot))$ viable in the graph of U, so that $x(\cdot)$ is, in particular, viable in K. Since $x'(t) = f(x(t), u(t))$ is absolutely continuous, we infer that $f(x_0, u_0)$ is contingent to K at x_0, that is, that u_0 belongs to $R_K(x_0)$.

The regulation map for the system (7.9) associates with any $(x,u) \in \text{Graph}(R)$ the set of pairs $(x', u') \in \{f(x,u)\} \times \varphi(x,u)B$ such that (x', u') belongs to the contingent cone to the graph of R at (x,u), that is, such that

$$u' \;\in\; DR(x,u)(f(x,u)) \cap \varphi(x,u)B \;=:\; G_R(x,u)$$

The graph of R^{φ} is the viability kernel of Graph(U) for the system of differential inclusions (7.9). □

7.2.4 Dynamical Closed Loops

Let us consider a control system (U, f), a regulation map $R \subset U$ that is a solution to the partial differential inclusion (7.11), and the metaregulation map

$$(x,u) \rightsquigarrow G_R(x,u) \;:=\; DR(x,u)(f(x,u)) \cap \varphi(x,u)B$$

regulating smooth state-control solutions viable in the graph of R through the system (7.12) of differential inclusions. The question arises whether or not we can construct selection procedures for the control component of this system of differential inclusions. It is convenient for this purpose to introduce the following definition.

Definition 7.2.7 (Dynamical Closed Loops) *Let R be a φ-growth subregulation map of U. We shall say that a selection g of the metaregulation map G_R associated with R mapping every $(x,u) \in$ Graph(R) to*

$$g(x,u) \in G_R(x,u) \;:=\; DR(x,u)(f(x,u)) \cap \varphi(x,u)B \qquad (7.13)$$

is a dynamical closed loop of R. The system of differential equations

$$\begin{cases} \text{(i)} & x'(t) \;=\; f(x(t),u(t)) \\ \text{(ii)} & u'(t) \;=\; g(x(t),u(t)) \end{cases} \qquad (7.14)$$

is called the associated closed loop differential system.

Clearly, every solution to (7.14) is also a solution to (7.12). Therefore, a dynamical closed loop being given, solutions to the system of ordinary differential equations (7.14) (if any) are smooth state-control solutions of the initial control problem (7.4).

Such solutions do exist when g is continuous (and if such is the case, they will be continuously differentiable). But they also may exist when g is no longer continuous, as we saw when we built closed-loop controls. This is the case, for instance, when $g(x,u)$ is the element of minimal norm in $G_R(x,u)$.

7.2.5 Heavy Viable Solutions

The simplest example of dynamical closed-loop control is the minimal selection of the metaregulation map G_R, which in this case is equal to the map g_R° associating with each state-control pair (x,u) the element $g_R^\circ(x,u)$ of minimal norm of $DR(x,u)(f(x,u))$, because for all (x,u), $\|g_R^\circ(x,u)\| \leq \varphi(x,u)$ whenever $G_R(x,u) \neq \emptyset$.

Definition 7.2.8 (Heavy Viable Solutions) *Denote by $g_R^\circ(x,u)$ the element of minimal norm of $DR(x,u)(f(x,u))$. We shall say that the solutions to the associated closed-loop differential system*

$$\begin{cases} \text{(i)} \quad x'(t) = f(x(t),u(t)) \\ \text{(ii)} \quad u'(t) = g_R^\circ(x(t),u(t)) \end{cases}$$

are heavy viable solutions to the control system (U,f) associated with R.

Theorem 7.2.9 (Heavy Viable Solutions) *Let us assume that U is closed and that f and φ are continuous and have linear growth. Assume that the metaregulation $G_R(\cdot,\cdot)$ is lower-semicontinuous with closed convex images. Then for any initial state-control pair (x_0,u_0) in* Graph*(R), there exists a heavy viable solution to the control system (7.4).*

Remark *Any heavy viable solution $(x(\cdot),u(\cdot))$ to the control system (7.4) satisfies the inertia principle*: Indeed, we observe that if for some t_1, the solution enters the subset $C_R(u(t_1))$, where we set

$$C_R(u) := \{x \in K \mid 0 \in DR(x,u)(f(x,u))\}$$

the control $u(t)$ remains equal to $u(t_1)$ as long as $x(t)$ remains in $C_R(u(t_1))$. Because such a subset is not necessarily a viability domain, the solution may leave it.

If for some $t_f > 0$, $u(t_f)$ is a punctuated equilibrium, then $u(t) = u_{t_f}$ for all $t \geq t_f$, and thus $x(t)$ *remains in the viability cell $N_1^0(u(t_f))$ for all $t \geq t_f$.* □

The reason why this theorem holds true is that the minimal selection is obtained through the selection procedure defined by

$$S_{G_R}^\circ(x,u) := \|g_R^\circ(x,u)\| \, B \tag{7.15}$$

It is this fact that matters. So Theorem 7.2.9 can be extended to any selection procedure of the metaregulation map $G_R(x,u)$ defined in Chapter 6 (see Definition 7.2.4).

Theorem 7.2.10 *Let us assume that the control system (7.4) satisfies*

$$\begin{cases} \text{(i)} & \text{Graph}(U) \textit{ is closed} \\ \text{(ii)} & f \textit{ is continuous and has linear growth} \end{cases} \tag{7.16}$$

Let $(x,u) \to \varphi(x,u)$ *be a nonnegative continuous function with linear growth and* $R : Z \rightsquigarrow X$ *a closed set-valued map contained in* U. *Let* $S_{G_R} : \text{Graph}(R) \rightsquigarrow X$ *be a selection procedure with convex values of the metaregulation map* G_R. *Then, for any initial state* $(x_0, u_0) \in \text{Graph}(R)$, *there exists a state-control solution to the associated closed loop system*

$$x' = f(x,u), \qquad u' \in G_R(x,u) \cap S_{G_R}(x,u) \tag{7.17}$$

defined on $[0, \infty[$ *and starting at* (x_0, u_0). *In particular, if for any* $(x,u) \in \text{Graph}(R)$, *the intersection*

$$G_R(x,u) \cap S_{G_R}(x,u) \ = \ \{s\,(G_R(x,u))\}$$

is a singleton, then there exists a state-control solution, defined on $[0, \infty[$ *and starting at* (x_0, u_0), *to the associated closed loop system*

$$x'(t) \ = \ f(x(t), u(t)), \qquad u'(t) \ = \ s\,(G_R(x(t), u(t)))$$

Proof Consider the system of differential inclusions

$$x' \ = f(x,u), \qquad u' \ \in \ S_{G_R}(x,u) \cap \varphi(x,u)B \tag{7.18}$$

subject to the constraints

$$\forall\, t \geq 0, \quad (x(t), u(t)) \ \in \ \text{Graph}(R)$$

Since the selection procedure S_{G_R} has a closed graph and convex values, the right-hand side is an upper-semicontinuous set-valued map with nonempty compact convex images and with linear growth. On the other hand, Graph(R) is a viability domain of the map $\{f(x,u)\} \times (S_{G_R}(x,u) \times \varphi(x,u)B)$. Therefore, the viability theorem can be applied. For any initial state control $(x_0, u_0) \in \text{Graph}(R)$, there exists a solution $(x(\cdot), u(\cdot))$ to (7.18) that is viable in Graph(R). Consequently, for almost all $t \geq 0$, the pair $(x'(t), u'(t))$ belongs to the contingent cone to the graph of R at $(x(t), u(t))$, which is the graph of the contingent derivative $DR(x(t), u(t))$. In other words, for almost all $t \geq 0$, $u'(t) \in DR(x(t), u(t))(f(x(t), u(t)))$. Since $\|u'(t)\| \leq \varphi(x(t), u(t))$, we deduce that $u'(t) \in G_R(x(t), u(t))$ for almost all $t \geq 0$. Hence, the state-control pair $(x(\cdot), u(\cdot))$ is a solution to (7.17). □

Proof of Theorem 7.2.9 By the maximum theorem, the map $(x, u) \mapsto \|g_R^\circ(x, u)\|$ is upper-semicontinuous. It has linear growth on Graph(R). Thus the set-valued map $(x, u) \rightsquigarrow \|g_R^\circ(x, u)\| B$ is a selection procedure satisfying the assumptions of Theorem 7.2.10. □

Because we know many examples of selection procedures, it is possible to multiply examples of dynamical closed loops as we did for the usual closed loops.

7.3 Internal Learning Algorithm

7.3.1 Control Systems

Naturally, we can consider other evolution laws for open-loop controls associated with the control system (U, f) that provide smooth open-loop controls yielding viable solutions. First, we can introduce an observation space Y, replace the initial control space Z by another finite-dimensional space Z_1, introduce an observation map $\beta : X \mapsto Y$, and relate the new controls $v \in Z_1$ and the observation y to the former controls $u \in Z$ by a single-valued map of the form

$$u = \alpha(\beta(x), v)$$

where

$$\alpha : Y \times Z_1 \mapsto Z$$

We then define a new control system (g, V) defined by

$$\begin{cases} \text{(i)} & g(x, v) := f(x, \alpha(\beta(x), v)) \\ \text{(ii)} & V(x) := \{v \in Z_1 \mid \alpha(\beta(x), v) \in U(x)\} \end{cases}$$

Therefore, the new control system governed by

$$\begin{cases} \text{(i)} & x'(t) = g(x(t), v(t)) \\ \text{(ii)} & v(t) \in V(x(t)) \end{cases} \tag{7.19}$$

provides the same dynamics of the state, although through another parameterization. That being done, we can propose any evolution laws for the open-loop controls as long as they are compatible with the constraints $v(t) \in V(x(t))$ [or $u(t) \in U(x(t))$].

For instance, if $A \in \mathcal{L}(Z_1, Z_1)$ and $\Phi : X \times Z_1 \rightsquigarrow Z_1$ and $\varphi : X \times Z_1 \rightsquigarrow Z_1$ is a Marchaud map, we can replace system (7.9) by the system of differential inclusions

$$\begin{cases} \text{(i)} & x'(t) = g(x(t), v(t)) \\ \text{(ii)} & v'(t) \in Av(t) + \Phi(x(t), v(t)) \end{cases} \tag{7.20}$$

Then the regularity theorem becomes the following:

Theorem 7.3.1 *Assume that U is closed and sleek, that f and φ are continuous with linear growth, that the maps α and β are continuously differentiable with linear growth, and that*

$$\forall\, (x,v) \in \text{Graph}(V), \quad \alpha'_v(\beta(x),v) \text{ is surjective}$$

Then the following two statements are equivalent:

1. *For any initial state $x_0 \in \text{Dom}(V)$ and control $v_0 \in V(x_0)$, there exists a solution $(x(\cdot), v(\cdot))$ to the control system (7.20) starting at (x_0, v_0) [so that $x(\cdot)$ is still a solution to the control system (7.4)].*
2. *The set-valued map V satisfies the following: for every $(x,v) \in$ Graph(V),*

$$\begin{aligned}&Av \in -\Phi(x,v)+\\&\alpha'_v(\beta(x),v)^{-1}\left[DU(x,\alpha(\beta(x),v))(g(x,v)) - \alpha'_y(\beta(x),v)\beta'(x)g(x,v)\right]\end{aligned}$$

Proof By the viability theorem, Theorem 7.1.3, we have to check that the graph of V is a viability domain for the set-valued map

$$(x,v) \rightsquigarrow \{g(x,v)\} \times (Av + \Phi(x,v))$$

Since the graph of V is the inverse image of the graph of U under the differentiable map $h : X \times Z_1 \mapsto X \in Z$ defined by

$$h(x,v) \;=\; (x, \alpha(\beta(x),v))$$

we can derive a formula to compute its contingent cone whenever U is sleek and the following transversality condition holds true:

$$\text{Im}(h'(x,v)) - T_{\text{Graph}(U)}(h(x,v)) \;=\; X \times Z$$

But the surjectivity of $\alpha'_v(\beta(x),v)$ obviously implies the surjectivity of $h'(x,v)$, so that this condition is satisfied. Hence, the contingent derivative of V is given by the formula

$$\begin{aligned}&DV(x,v)(x')\\&= \alpha'_v(\beta(x),v)^{-1}\left[DU(x,\alpha(\beta(x),v))(x') - \alpha'_y(\beta(x),v)\beta'(x)x'\right]\end{aligned}$$

Therefore, we observe that the second statement of the theorem states that the graph of V is a viability domain. □

7.3.2 Learning Feedback Rules by Neural Networks

Consider a control problem

$$\begin{cases} \text{(i)} & x'(t) = f(x(t), u(t)) \\ \text{(ii)} & u(t) \in U(x(t)) \end{cases} \tag{7.21}$$

Let us introduce a neural network with two layers X_1 and X_2, an input map h from the state space X to the first layer, and a propagation map g from the second layer X_2 to the control space Z. This amounts to replacing the usual controls $u \in Z$ by synaptic matrices $W \in \mathcal{L}(X_1, X_2)$ to control the system by a regulation law given through the neural network according to the law

$$u(t) = g(W(t)h(x(t))) \tag{7.22}$$

Therefore, we want to solve the system of differential inclusions that governs the initial control system together with the neural network that guides it:

$$\begin{cases} \text{(i)} & x'(t) = f(x(t), g(W(t)h(x(t)))) \\ \text{(ii)} & W'(t) \in \varphi(x(t), W(t))B \end{cases}$$

subject to the viability constraints

$$\forall\, t \geq 0, \quad (x(t), g(W(t)h(x(t)))) \in \text{Graph}(U)$$

Theorem 7.3.2 *Suppose that the graph of the set-valued map U is closed and that $f : \text{Graph}(U) \mapsto X$ is continuous with linear growth. Let us assume that the neural network defined by the input map h and the propagation rule g satisfies the following:*

$$\forall\, x \in \text{Dom}(U),\ \forall\, W \in \mathcal{L}(X_1, X_2), \quad g'(Wh(x)) \text{ is surjective,}$$

and $h(x) \neq 0$. Let us assume also that for every $x \in \text{Dom}(U)$ and any synaptic matrix W such that $g(Wh(x)) \in U(x)$, there exists a synaptic matrix $V \in \mathcal{L}(X_1, X_2)$ satisfying

$$\begin{cases} \text{(i)} & g'(Wh(x))Vh(x) \in DU(x, g(Wh(x)))(f(x, g(Wh(x)))) \\ & -g'(x)Wh'(x)f(x, g(Wh(x))) \\ & \\ \text{(ii)} & \|V\| \leq \varphi(x, W) \end{cases}$$

Then for any initial state $x_0 \in \text{Dom}(U)$ and any initial synaptic matrix W_0 such that $g(Wh(x_0)) \in U(x_0)$, there exist a solution $x(\cdot)$ to the *control system (7.21) starting at* x_0 *and a solution* $W(\cdot)$ *to the implicit differential inclusion*

$$g'(p(t))\left(\frac{d}{dt}W(t)\right)h(x(t)) \in DU(x(t),g(p(t)))(f'(x(t),g(p(t))))$$
$$- g'(p(t))W(t)h'(x(t))f'(x(t),g(p(t)))$$

[where we set $p(t) := W(t)h(x(t))$*] starting at* W_0 *regulating the system in the sense that*

$$\text{for almost all } t \geq 0, \quad u(t) = g(W(t)h(x(t)))$$

Proof This is a consequence of Theorem 7.3.2, with $Z_1 := \mathcal{L}(X_1, X_2)$, $A := 0$, $\Phi(x,v) := \varphi(x,W)B$ [where B denotes the unit ball of $\mathcal{L}(X_1,X_2)$], and where

$$\alpha(\beta(x),v) := g(Wh(x))$$

We observe that

$$\alpha_v'(\beta(x),v) = h(x) \otimes g'(Wh(x))$$

is surjective whenever $g'(x)$ is surjective and $h(x) \neq 0$. □

In the case when U is a single-valued map $\tilde{u}$, the foregoing system of differential equations becomes the following system of differential equations:

$$x'(t) = f(x(t),g(W(t)h(x(t))))g'(W(t)h(x(t)))\left(\tfrac{d}{dt}W(t)\right)h(x(t))$$
$$= (\tilde{u}'(x(t)) - g'(W(t)h(x(t)))W(t)h'(x(t)))\,(f'(x(t),g(W(t)h(x(t)))))$$

Since the derivatives $g'(y)$ of the propagation function g of the neural network are surjective and $h(x) \neq 0$, we infer from Corollary 2.5.2 that the map $h(x) \otimes g'(x)$ is surjective from the space $\mathcal{L}(X_1, X_2)$ of synaptic matrices to the control space Z and that its right-inverse is given by the formula

$$(h(x) \otimes g'(x))^+ z = \frac{Lh(x)}{\lambda(h(x))^2} \otimes g'(x)^+ z$$

Then the heavy solutions for the system for differential equations are the solutions to the system

$$\begin{cases} \text{(i)} \quad x'(t) = f(x(t),g(W(t)h(x(t)))) \\ \text{(ii)} \quad W'(t) = \frac{h(x(t))}{\|h(x(t))\|^2} \otimes g'(W(t)h(x(t)))^+\,(\tilde{u}'(x(t)) \\ \qquad - g'(W(t)h(x(t)))W(t)h'(x(t))\,(f'(x(t),g(W(t)h(x(t)))))) \end{cases}$$

We observe that the differential equation governing the evolution of synaptic matrices $W(t)$ is of a Hebbian type: The velocity at a synapse (i,j) is proportional to the product of the ith component of $\frac{h(x(t))}{\|h(x(t))\|^2}$ and the jth component of

$$g'(W(t)h(x(t)))^+(\tilde{u}'(x(t)) \\ -g'(W(t)h(x(t)))W(t)h'(x(t))(f'(x(t),g(W(t)h(x(t))))))$$

7.4 Stabilization Problems

Let us consider a twice differentiable function $V : X \to \mathbf{R}_+$ and a positive real number a. If we want to stabilize the system around an equilibrium, we must check this Lyapunov stability condition

$$\forall t \geq 0, \quad V(x(t)) \leq V(x_0)e^{-at} \tag{7.23}$$

for a solution $x(\cdot)$ to the control problem regulated by a control $u(t) = \phi(W(t)h(x(t)))$ provided by a neural network.

This problem is a particular case of the general problem where the set-valued map U is defined by

$$U(x) \ := \ \{u \mid \langle V'(x), f(x,u)\rangle + aV(x) \leq 0\}$$

Applying the previous theorem, we derive the adaptive learning rule that is able to adapt the neural network such that condition (7.23) is satisfied:

$$\left\langle W'(t), \left[(\phi'(p(t))^\star \frac{\partial f}{\partial u}^\star (x(t),u(t))V'(x(t)))_i(h(x(t)))_j\right]_{i,j}\right\rangle \leq \mu(t) \tag{7.24}$$

where $\mu(t) = -\langle \Pi(x(t),W(t)), f(x(t),u(t))\rangle$ and

$$\begin{aligned}\Pi(x(t),W(t)) \ = \ & h'(x(t))^\star W(t)^\star \phi'(p(t))^\star \tfrac{\partial f}{\partial u}^\star (x(t),u(t))V'(x(t)) - \\ & V''(x(t))f(x(t),u(t)) + \tfrac{\partial f}{\partial x}^\star (x(t),u(t))V'(x(t)) + \\ & aV'(x(t))\end{aligned}$$

In order to select a differential equation from this differential inclusion, we can choose to minimize the norm of the synaptic matrix velocity. This yields the following result:

Corollary *Let $V : X \to \mathbf{R}_+$ be a C^2 function. Condition (7.23) is verified if the network synaptic matrix evolution is governed by the following* *differential equation:*

$$w'_{ij}(t) = \lambda(t)(\phi'(p(t))^\star \frac{\partial f}{\partial u}^\star (x(t), u(t))V'(x(t)))_i(h(x(t)))_j$$

where

$$\lambda(t) = \frac{-\langle \Pi(x(t), W(t)), f(x(t), \phi(p(t)))\rangle}{\| h(x(t)) \|^2 \| \phi'(p(t))^\star \frac{\partial f}{\partial u}^\star (x(t), \phi(p(t)))V'(x(t)) \|^2}$$

8

Learning Processes of Cognitive Systems

Introduction

We propose in this chapter a speculative dynamical description of an abstract cognitive system that goes beyond neural networks to attempt to take into account some features of nervous systems and, in particular, adaptations to environmental constraints. This personal viewpoint of the author is but one of the several attempts to model cognitive processes mathematically. It is presented primarily for the purpose of stirring up reaction and prompting further research involving other techniques and other approaches to this wide field.

Before we look at the evolution of nervous systems for useful suggestions regarding the means they have used to master more and more complex cognitive faculties,[1] we shall start from the fact that an organism must adapt to environmental constraints by perceiving them and recognizing them through "metaphors" with what we shall call "conceptual controls."[2] This problem of adaptation is not dealt with explicitly in most studies of neural networks. This chapter is devoted to highlighting the roles of cognitive systems in this process.[3]

[1] Even if we assume that the basic principle is simple (biological communication), its structure has become more and more complex during the course of evolution. Indeed, the improvement consists in adding new structures to old ones, without always destroying them.
An understanding and mastering of the basic structures would allow a possible computer translation by economizing the useless part of complexity that results from (nonteleological) biological evolution, retaining only what is relevant to allow adaptation.

[2] These may perhaps be the "synaptic weight matrices," as in the case of neural networks.

[3] waiting the possibility of positioning oneself in the intersection of behavioral sciences (ethology, psychology, sociology), as far as the description of the survival objectives of organisms are concerned, and the neurosciences, for the means to achieve them.

The variables of the cognitive system are described by its state and a regulatory control (conceptual control[4]). The state of the system (henceforth called the *sensorimotor state*) is described by

- the state and the variations of the environment on which the cognitive system acts,
- the state of cerebral motor activity of the cognitive system, which guides an individual's action on the environment.

The regulatory control of the cognitive system is described by

- an endogenous cerebral activity that is not genetically programmed, but is acquired by learning and is recorded in the memory.

The purpose of this activity is to "interpret" (or "illuminate") the sensory perception of the environment, and we shall call it the *conceptual control*.

We shall study both the evolution of the state of the cognitive system and the evolution of its regulatory control. For that purpose, we must identify the laws that constrain and govern the evolution of the system. They are as follows:

- *a recognition mechanism*, with genetically programmed evolution, which selects metaphors among the state, the variation of the environment, and the conceptual controls by matching at each moment the sensory perception of the environment and its variations with the conceptual control to be chosen,
- *viability constraints*, which restrict the potential action of the cognitive system on the environment and constrain it to "adapt" at each instant, transforming it and consuming scarce resources,
- *an action law*, which governs the endogenous evolution of the environment and the evolution that results from the action of the cognitive system, the acceleration of this evolution depending on both the environment and the cerebral motor activity,
- *a perception law*, which governs the evolution of the cerebral motor activity from the sensory perception of the environment and its variations and the conceptual control.

[4]This "duality" between the sensorimotor state and the conceptual control is of the same nature as those encountered in the regulation of many "macrosystems," such as the duality between phenotypes and genotypes, heredity and development, nature and nurture in biology, members of a society and cultural codes in sociology, commodities and prices in economy, in which the "controls" regulate the states and obey an inertia principle.

Instead of imposing a priori learning rules,[5] we shall select them, according to a given principle,[6] from among *the largest learning rule consistent with the cognitive system.* Namely, we shall consider a *learning process* as a nondeterministic[7] feedback law, which is a *feedback set-valued map associating a set of "learnable" conceptual controls with each sensorimotor state,* which is consistent with the viability constraints and the recognition mechanism.

The purpose of this chapter is to characterize the learning processes that are consistent with the action law and the perception law, in a sense that we shall make precise later, but which states that, using the conceptual controls provided by the learning process, the cognitive system evolves and remains viable. We shall prove that, given the recognition mechanism, the viability constraints, and the action and perception laws, there exists a *largest learning process.*

We shall postulate also that the evolution of conceptual controls obeys an *inertia principle,* which states that *whenever a conceptual control "works"* (i.e., keeps the evolution of the cognitive system viable), *then it is kept.*

Actually, we shall implement this principle by selecting *heavy viable solutions,* which minimize at each instant (the norm of) the velocity of the conceptual controls. Hence we derive from any viable learning process a deterministic law that governs the evolution of the cognitive system.

This will lead us to associate with any conceptual control a (possibly empty) "sensorimotor cell" that enjoys the following property: When the state of the system reaches such a sensorimotor cell, it can evolve inside it while being regulated by its conceptual control, which remains constant.[8]

In their continuous versions, most models of neural networks belong to the following class: If W denotes the synaptic matrix (the entries of which are the synaptic weights between two neurons) and x the state (the intensity of the signal at each neuron), then the model's evolution is governed by a differential equation of the form $x'(t) = f(x(t), W(t)$.

[5] as in neural networks, by choosing a rule from among the ones suggested by scientists from Hebb to Widrow and Hoff.

[6] such as the inertia principle.

[7] which, then, is only remotely analogous to the learning rules usually employed in artificial intelligence.

[8] It is a property of conceptual controls that makes precise, in this framework, the concept of "punctuated equilibrium" proposed by Eldredge and Gould in paleontology.

The learned states are the (stable or asymptotically stable) equilibria of the system, that is, the solutions to the equation $f(x, W) = 0$. To ask the neural network to learn $|k|$ states x^k amounts to finding a synaptic matrix $\overline{W}$ such that $f(x^k, \overline{W}) = 0$ for $k = 1, \ldots, |k|$, with possibly other conditions to guarantee such or such stability property. The other equilibria, solutions to the equation $f(x, \overline{W}) = 0$, are the states "discovered" by the neural network. Most of them are using constant synaptic matrices, which are obtained by adequate algorithms, regarded as learning rules.

One can consider that the recognition mechanism between signals and synaptic matrices amounts to saying that x is recognized by W if and only if $f(x, W) = 0$.

Since environments do not appear explicitly in most models of neural networks, there is no requirement in these models to adapt to viability constraints by acting on the environment. It is for taking into account this phenomenon of adaptation as a central theme that I suggest a mathematical metaphor for a cognitive system such as the one studied in this chapter. I shall also provide reasons not to choose, necessarily, signals induced by nervous influx as the state of the system and synaptic matrices as support for the conceptual controls.

The proofs in this chapter rely on the viability theorems[9] and the differential calculus of set-valued maps, which may seem rather "heavy tools" to provide only qualitative conclusions (which may look reasonable, if not obvious, to specialists in neurosciences or cognitive psychology[10]).

We cannot avoid the use of set-valued maps for defining the recognition mechanism, the viability constraints, and the learning processes. But

[9] which have been proved to take into account Darwinian evolution by a mathematical metaphor of nondeterministic systems whose solutions are selected by viability constraints. See Chapter 7 for a summary of viability theory.

[10] This situation is analogous to the one found in economics, where we had to wait a century after Adam Smith for Léon Walras to formulate the concept of economic equilibrium, and another century for Arrow and Debreu to prove the existence of such an equilibrium with up-to-date mathematical tools, thus making precise the concept of price decentralization and giving a kind of mathematical warrant to the relevance of such a concept. It is precisely because natural and social sciences require a *modern* mathematical arsenal that mathematical techniques have not often been used. The fact that the useful mathematical tools are quite sophisticated impairs their acceptance by specialists in neurosciences and cognitive psychology.

for the sake of simplicity, we do not use differential inclusions[11] nor differential inclusions with memory to model the action of the cognitive system on the environment. We present the main ideas of this chapter in the framework of differential equations.

We gather, in the last section, several comments that justify the mathematical metaphor that we suggest, and we present in the first section a mathematical description of what we call a cognitive system, the statements of the theorems and their proofs.

8.1 Viable Learning Processes

8.1.1 Sensorimotor States and Conceptual Controls

We represent the environment by a finite-dimensional vector space X. We emphasize the phenomenon of *chemical communication.*[12] For simplicity, we shall neglect the endocrine system.[13] We then assume that *the state of cerebral activity is described by the evolution of neurotransmitters in each synapse.*

[11] The set-valued character of the differential inclusion takes into account the uncertainties of the events of the environment and the actions of the other cognitive systems, as well as the lack of knowledge of the consequences of the perception. It also allows one to take into account the diffusion of neurotransmitters by a given neuron in many other neurons, i.e., the "parallelism" and "distributed" character of information processing by nervous systems.
We also note that differential inclusions appear whenever we take into account in the recognition mechanism the perceptions of variations in the environment.

[12] This chemical communication process, through the sending of a macromolecule from an emitter cell to a receptor cell, the emission and the reception of such a molecule modifying the chemical properties of these cells, could constitute a relevant framework for many biological phenomena at several levels, which could be approached mathematically. One can consider "adaptive systems" as systems formed of receptors (of the environment), effectors (on the environment), and intermediate (or processing) cells. Their behavior consists in acting at each instant to bring an unfavorable state of the environment to a more favorable one.
The perception of the environment by the emitters provokes the transmission of molecular messages toward the cells of the system. These messages go from one cell to the others, the reception by an intermediate cell inducing the emission of another message. The action starts after the reception of the message by the effectors, modifying the state of the environment, which is again perceived by the system and leads to a new action by the system.

[13] At this level, the endocrine system and the nervous system may be distinguished by the *mode of transport* of the chemical messenger. This transport is slow and nonspecific in the case of the endocrine system: Hormones are carried by the blood. It is fast and specific in the case of the nervous system: Neurotransmitters have to cross a synapse (the place where the axon of a "presynaptic" neuron meets the dendrite of a "postsynaptic" one) that is only 0.02 μm wide. When the pulse-coded information sent by the postsynaptic neuron reaches a certain threshold value, it releases neurotransmitters in the synapse, inducing an electrical response on the postsynaptic membrane after about 10^{-3} seconds.

We denote by S the set of synapses (about one hundred thousand billion of them) and by $C := \mathbf{R}^S$ the *cerebral space*, which is the finite-dimensional vector space of cerebral activity describing the number of neurotransmitters in each synapse. Hence the component y_s of an element $y = (y_s)_{s \in S}$ of this cerebral space C denotes the number of neurotransmitter molecules passing through synapse s (with a minus sign if the role of the neurotransmitter is inhibitor). We shall describe the temporal cerebral activity by several functions of time into the cerebral space $C := \mathbf{R}^S$.

The complex mechanism describing the processing of the presynaptic signals by each neuron will not be taken explicitly into account.[14]

We distinguish two classes of time-dependent functions:

1. the function $t \mapsto a(t) \in C$, which describes the evolution of *cerebral motor activity*, that is, the knowledge, in each synapse $s \in S$, at each time t, of the number $a_s(t)$ of neurotransmitters involved in the perception process,
2. the function $t \mapsto c(t) \in C$, which describes the endogenous evolution of the *conceptual controls*, that is, the knowledge, in each synapse $s \in S$, at each time t, of the number $c_s(t)$ of neurotransmitters involved in the conceptual activity (which has to be compared with the other cerebral functions in the recognition mechanism).

With each function $t \mapsto z(t) \in C$ we can associate its "neuronal trace"[15] $N(z(t))$ defined by

$$N(z(t)) \;:=\; \{\, s \in S \mid z_s(t) \neq 0 \,\}$$

which specifies the set of active synapses.

Hence these two functions $a(\cdot)$ and $c(\cdot)$ determine at each time t the active "neuronal networks" and their evolution.

The *states* of the systems are therefore sensorimotor states (x, v, a) ranging over the vector space $X \times X \times C$.

The *controls* of the system are elements c of C.

[14] either by threshold functions involved in synaptic matrices, as in neural networks, or, in a more sophisticated way, by parabolic equations of the Hodgkin-Huxley type.

[15] We do not choose to emphasize the topological nature of the neural network, because it can be described in terms of the "traces" left by the circulation of neurotransmitters; a given synapse is "weighted" by the total number of neurotransmitters crossing it during each period.
There also is little hope of obtaining through "functional" logical reasoning an explanation of the topological structure of the nervous system in large units without a thorough study of the apparitions of these structures in phylogenesis.

8.1.2 The Recognition Mechanism

The recognition mechanism compares the sensory perception of the environment, the sensory perception of its variation, and the state of the conceptual control at each instant. We shall describe it by a family of subsets $\mathcal{R}(t)$ of the space $C \times X \times X$. There is recognition at instant t of the state of the environment x and its variation v by the conceptual control c if we have

$$(c, x, v) \in \mathcal{R}(t) \tag{8.1}$$

In other words, the subsets $\mathcal{R}(t)$ describe the set of possible *metaphors* (c, x, v) between a conceptual control and the sensory perception of the environment that can be recognized at time t.

We can regard $\mathcal{R}(t)$ as the graph of the set-valued map $R(t) : X \times X \rightsquigarrow C$ associating with the state of the environment x and its variation[16] v at time t the subset $R(t; x, v)$ of conceptual controls c that can recognize the pair (x, v). This is the "data driven" version of the recognition mechanism. The inverse $R^{-1}(t; \cdot)$ associates with any conceptual control c the set of pairs (x, v) that can be recognized by c. This is the "conceptually driven" version of the recognition mechanism.[17]

8.1.3 The Viability Constraints

The viability constraints translate the fact that the cognitive system consumes scarce resources of the environment in order to maintain and improve through a recognition mechanism an internal state (which tends

[16]This translates the frequently observed fact that we are sensitive to both the state and the variations of the environment, and we may be more sensitive to the variations.

[17]By introducing the set-valued map F associated with the recognition mechanism by the formula

$$v \in F(t, x, c) \iff (c, x, v) \in \mathcal{R}(t)$$

we see that recognition of the evolution of the sensorimotor state can be translated by the "controlled" differential inclusion

$$x'(t) \in F(t, x(t), c(t))$$

to deteriorate) with respect to a reference[18] state (which may evolve with time).

We translate into a quite general form these constraints to which the cognitive system must adapt by requiring that

$$\forall\, t \geq 0, \quad a(t) \;\in\; K(t, x(t)) \tag{8.2}$$

where K is a set-valued map from $[0, \infty[\times X$ to C. In other words, this set-valued map K specifies the states of the environment that allow the cognitive system to survive and the constraints that are set on its motor activity.

8.1.4 The Action Law

We assume that the evolution of the environment due to the action of the cognitive system depends on both the environment and the cerebral motor activity of the cognitive system. At this level of generality, we can assume that it is governed by a second-order differential equation of the form[19]

$$x''(t) \;=\; f(t, x(t), x'(t), a(t)) \tag{8.3}$$

where f is a map from $[0, \infty[\times X \,\times\, X \times C$ to X.

8.1.5 The Perception Law

The evolution of the cerebral motor activity induced by the perception of the environment and regulated by conceptual controls is governed by a controlled differential equation: The velocity of the flux of neurotransmitters involved in the motor activity depends at each time t not only upon the perception of the environment and the cerebral motor activity

[18] A. Danchin has proposed an interesting theory to explain the maintenance of internal reference states (self-referent controls, where the set-valued character plays a crucial role). The inputs received by a receptor give birth to several (two, for instance) evolutions processed (with different delays) in different networks of intermediate cells, which are compared by the same effector, which then, on the one hand, feeds back on one part of the input (by acting on the environment and perceiving it) and, on the other hand, feeds back on some consequences of the input.

[19] To take into account the laws of mechanics. In full generality, we should use here a second-order differential inclusion with memory. Differential inclusions with memory and functional viability have been studied in *Viability Theory* (Aubin 1991). For simplicity, we restrict ourselves in this book to the simple case of second-order differential equations.

but also upon the conceptual control that is currently active. Hence, we can write that

$$a'(t) \quad = \quad g(t, x(t), x'(t), a(t), c(t)) \tag{8.4}$$

where g is a map from $[0, \infty[\times X \times X \times C \times C$ to C.

The cerebral activity induced by sensory perception of the environment (both external and internal) is not included explicitly, but implicitly in the perception law, which also takes into account the propagation of synaptic excitation along the set S of synapses.[20]

8.1.6 The Learning Process

Once the viability constraints and the recognition mechanism are given, we can formulate their compatibility by introducing the set-valued map P defined by

$$P(t, x, v, a) := \begin{cases} R(t, x, v) & \text{if } a \in K(t, x) \\ \emptyset & \text{if } a \notin K(t, x) \end{cases} \tag{8.5}$$

and require that the domain of this set-valued map P not be empty.

Definition 8.1.1 (Learning Processes) *A learning process is a set-valued map L with closed graph[21] that at each time[22] t associates with every sensorimotor state (x, v, a) a subset $L(t, x, v, a)$ of conceptual controls that is consistent with the recognition mechanism (8.1) and the viability constraints (8.2) in the sense that*

$$\forall\, (t, x, v, a) \in \mathrm{Dom}(P), \quad L(t, x, v, a) \subset P(t, x, v, a) \tag{8.6}$$

We do not exclude the case in which L has empty values.

We have now to distinguish among learning processes the ones that are consistent with the evolution laws of the cognitive process. From now on, we regard the "growth rate" $\rho \geq 0$ of a learning process as a parameter.

[20] At this time, these propagation laws are too complex to be described explicitly. However, it may be possible to use the laws derived from networks of automata (or neurons) that are currently under investigation.

[21] This is the weakest requirement for continuity we can think of for set-valued maps.

[22] We shall regard L either as the set-valued map $(t, x, v, a) \rightsquigarrow L(t, x, v, a)$ or, for each time t, as the set-valued map $(x, v, a) \rightsquigarrow L(t; x, v, a)$. The context will indicate which interpretation is currently being used.

Definition 8.1.2 (Viable Learning Processes) *We shall say that a closed ρ learning process L_ρ enjoys the viability property [with respect to the evolution laws (8.3) and (8.4)] if and only if for any initial time t_0 and any initial state $(t_0, x_0, v_0, a_0, c_0)$ in the graph of the learning process there is a solution*

$$t \in [t_0, \infty[\mapsto (x(t), a(t), c(t))$$

to the system of differential inclusions

$$\begin{cases} \text{(i)} & x''(t) = f(t, x(t), x'(t), a(t)) \\ \text{(ii)} & a'(t) = g(t, x(t), x'(t), a(t), c(t)) \\ \text{(iii)} & \|c'(t)\| \leq \rho(\|c(t)\| + 1) \end{cases} \tag{8.7}$$

satisfying

$$\forall\, t \geq t_0, \quad c(t) \in L_\rho(t, x(t), x'(t), a(t)) \tag{8.8}$$

The case of a zero growth rate deserves some comment. Indeed, when $\rho = 0$, differential inclusion (8.7iii) implies that the conceptual control must remain constant. Furthermore, the zero learning process must obey equation (8.8), which specifies that for any $(t_0, x_0, v_0, a_0, c_0)$ of the graph of L_0, one must have

$$\forall\, t \geq t_0, \quad c_0 \in L_\rho(t, x(t), x'(t), a(t))$$

which can be rewritten in the form

$$\forall\, t \geq t_0, \quad (x(t), x'(t), a(t)) \in L_0^{-1}(t; c_0)$$

Hence the set-valued map $c \rightsquigarrow L_0^{-1}(t; c)$ yields at each time t_0 the subset of sensorimotor states (x_0, v_0, a_0) that are "learnable" by the conceptual control c_0 at time t, in the sense that there exists a solution to the system of differential equations

$$\begin{cases} \text{(i)} & x''(t) = f(t, x(t), x'(t), a(t)) \\ \text{(ii)} & a'(t) = g(t, x(t), x'(t), a(t), c_0) \end{cases}$$

starting at (x_0, v_0, a_0) and satisfying

$$\forall\, t \geq 0, \quad (x(t), x'(t), a(t)) \in L_0^{-1}(t; c_0)$$

The subsets $L_0^{-1}(t; c)$ are called the "sensorimotor cells" of the conceptual control c at time t, and the set-valued maps $t \to L_0^{-1}(t; c)$ are the "sensorimotor tubes."

We shall characterize viable ρ learning processes and prove that there exits a largest viable ρ learning process. For that purpose, we need the concept of contingent derivatives of set-valued maps.[23]

Let us denote by $\mathcal{L}_\rho$ the graph of the set-valued map L_ρ, and by

$$DL_\rho(t,x,v,a,c)(1,x',v',a')$$

the contingent derivative of L_ρ at (t,x,v,a,c) in the direction $(1,x',v',a')$. We shall set, for simplicity,

$$M_\rho(t,x,v,a,c) \;:=\; DL_\rho(t,x,v,a,c)(1,v,f(t,x,v,a),g(t,x,v,a,c))$$

Hence, M_ρ is the set-valued map from $\mathcal{L}_\rho$ to X that specifies the "directions" in which the conceptual controls can evolve according to the learning process and the evolution laws.

Theorem 8.1.3 (Learning Process) *Let us assume that the functions f and g governing the action law and the perception law of the cognitive system are continuous and have linear growth. Assume also that the graphs of the set-valued maps $(t,x,v) \rightsquigarrow R(t,x,v)$ and $(t,x) \rightsquigarrow K(t,x)$ are closed. Then a ρ learning process L_ρ enjoys the viability property if and only if it is* viable *in the sense that*

$$\forall\,(t,x,v,a,c) \in \mathcal{L}_\rho, \quad M_\rho(t,x,v,a,c) \cap \rho(\|c\|+1)B \;\neq\; \emptyset \tag{8.9}$$

where B denotes the unit ball of C. Furthermore, there exists a largest viable ρ learning process, and thus there are largest sensorimotor cells.

Proof We can apply the viability theorem, for example, to the following system of differential inclusions defined on the closed subset [which is

[23]We recall (e.g., Appendix A, Sections A.8 and A.10) that the *contingent cone* $T_K(x)$ to a subset K at $x \in K$ is the closed cone of elements v satisfying

$$\liminf_{h\to 0+} d(x+hv,K)/h \;=\; 0$$

The *contingent derivative* of the set-valued map F from X to Y at a point (x,y) of its graph is the closed, positively homogeneous set-valued map $DF(x,y)$ from X to Y defined by

$$\mathrm{Graph}(DF(x,y)) \;:=\; T_{\mathrm{Graph}(F)}(x,y)$$

or, equivalently, by

$$v \in DF(x,y)(v) \iff \liminf_{h\to 0+,u'\to u} d\left(v, \frac{F(x+hu')-y}{h}\right) \;=\; 0$$

Naturally, if F is single-valued and Gâteaux-differentiable, $DF(x,y)$ coincides with the directional derivative $F'(x)$.

the graph of the set-valued map P defined by (8.5)]

$$\mathcal{K} := \{ (t, x, v, a, c) \mid a \in K(t, x) \quad \text{and} \quad c \in R(t, x, v) \}$$

by

$$\begin{cases} \text{(i)} & \tau'(t) = 1 \\ \text{(ii)} & x'(t) = v(t) \\ \text{(iii)} & v'(t) = f(t, x(t), v(t), a(t)) \\ \text{(iv)} & a'(t) = g(t, x(t), v(t), a(t), c(t)) \\ \text{(v)} & c'(t) \in \rho(\|c(t)\| + 1)B \end{cases}$$

We observe at once that a set-valued map L_ρ is a ρ learning process if and only if its graph $\mathcal{L}_\rho$ is contained in $\mathcal{K}$ and that the ρ learning process L_ρ enjoys the viability property if and only if its graph $\mathcal{L}_\rho$ enjoys the viability property for the foregoing system of differential inclusions. The viability theorem states that $\mathcal{L}_\rho$ enjoys the viability property if and only if it is a viability domain in the sense that

$$\begin{array}{l} \forall\, (\tau, x, v, a, c) \in \mathcal{L}_\rho, \quad \exists\, w \in \rho(\|c\| + 1)B \quad \text{such that} \\ (1, v, f(t, x, v, a), g(t, x, v, a, c), w) \in T_{\mathcal{L}_\rho}(t, x, v, a, c) \end{array}$$

Taking into account the definition of the contingent derivative of L_ρ, this is the definition of a viable ρ learning process. Consequently, the learning process L_ρ enjoys the viability property if and only if it is viable.

We also know that there exists a largest closed viability domain (possibly empty) $\widetilde{\mathcal{L}_\rho}$ contained in $\mathcal{K}$ that is then the graph of the largest learning process $\widetilde{L_\rho}$. □

Remark Naturally, if the set-valued map P defined by (8.5) is a ρ learning process, it will be the largest one.

For that purpose, we need to compute the contingent derivative of P in terms of the contingent derivatives of K and R and check whether or not they satisfy condition (8.9). One can prove under adequate technical conditions, which are too complicated to explain here (we have to assume that the set-valued maps R and K are "sleek" and "transversal") that

$$DP(t, x, v, a, c)(1, \xi, \nu, \alpha) = \begin{cases} DR(t, x, v)(1, \xi, \nu) & \text{if } \alpha \in DK(t, x)(1, \xi) \\ \emptyset & \text{if } \alpha \notin DK(t, x)(1, \xi) \end{cases}$$

With this formula at hand, one can check that P is the largest viable learning process if and only if

$$\forall\, (t, x, v, a, c) \in \mathcal{K}, \quad g(t, x, v, a, c) \in DK(t, x)(1, v)$$

and

$$DR(t,x,v)(1,v,f(t,x,v,c)) \cap \rho(\|c\|+1)B \neq \emptyset \quad \square$$

Remark Although time-independent cognitive systems do not make much sense for modeling "real" cognitive systems, one can still ask the traditional question: Is the limit when $t \to +\infty$ of the learning process a time-independent learning process? The answer is affirmative when we give a precise meaning to the foregoing statement.

A time-independent cognitive system is defined by set-valued maps R and K and single-valued maps f and g that do not depend on the time. Hence a time independent ρ learning process L_ρ^∞ is a set-valued map from $X \times X \times C$ to C that is contained in the set-valued map P^∞ defined by

$$P^\infty(x,v,a) := \begin{cases} R(x,v) & \text{if } a \in K(x) \\ \emptyset & \text{if } a \notin K(x) \end{cases}$$

It is viable if and only if $\forall\, (x,v,a,c) \in \mathcal{L}_\rho^\infty$,

$$DL_\rho^\infty(x,v,a,c)(v,f(x,v,a),g(x,v,a,c)) \cap \rho(\|c\|+1)B \neq \emptyset \qquad (8.10)$$

Next, we consider a time-dependent learning process L, and we associate with it its "upper graphical limit" L_ρ^∞, whose graph $\mathcal{L}_\rho^\infty$ is defined by

$$\mathcal{L}_\rho^\infty := \bigcap_{T>0} \bigcup_{t \geq T} \overline{\mathcal{L}_\rho(t)}$$

where $\mathcal{L}_\rho(t)$ denotes the graph of $(x,v,a) \rightsquigarrow L_\rho(t,x,v,a)$. Therefore, one can prove that *if the time-dependent ρ learning process L_ρ is viable, so is its upper graphical limit L_ρ^∞.* $\square$

A fundamental question arises: How can we select a specific evolution in a nondeterministic viable ρ learning process L_ρ? We shall present a selection mechanism obeying the "inertia principle," which states that *whenever a conceptual control "works"* (i.e., keeps the evolution of the cognitive system viable), *we keep it.*

We can translate this principle by saying that we choose to keep the velocity $c'(t)$ of the conceptual control equal to zero whenever that is possible.

One way (among others that we do not describe here) to implement this principle is to select "heavy viable solutions," which *minimize at each instant the norm $\|c'(t)\|$ of the velocity of the conceptual controls.*

This is possible because we know how the velocity of the conceptual controls evolve, thanks to the differential calculus of set-valued maps: We recall that we set

$$M_\rho(t,x,v,a,c) := DL_\rho(t,x,v,a,c)(1,v,f(t,x,v,a),g(t,x,v,a,c))$$

Hence the conceptual controls evolve according to the differential inclusion

$$\text{for almost all } t \geq 0, \quad c'(t) \in M_\rho(t,x(t),x'(t),a(t),c(t)) \qquad (8.11)$$

We define now the selection m_ρ of M_ρ defined by

$$m_\rho(t,x,v,a,c) \in M_\rho(t,x,v,a,c)$$

such that

$$\|m_\rho(t,x,v,a,c)\| = \inf_{m \in M_\rho(t,x,v,a,c)} \|m\|$$

We deduce the following result:

Theorem 8.1.4 (Heavy Conceptual Controls) *We posit the assumptions of Theorem 8.1.3, and we assume that the set-valued map M_ρ is lower semicontinuous with closed convex images. Then for every initial time t_0 and initial state (x_0, v_0, a_0, c_0), there exists a solution to the system of differential equations*

$$\begin{cases} \text{(i)} & x''(t) = f(t,x(t),x'(t),a(t)) \\ \text{(ii)} & a'(t) = g(t,x(t),x'(t),a(t),c(t)) \\ \text{(iii)} & c'(t) = m_\rho(t,x(t),x'(t),a(t),c(t)) \end{cases} \qquad (8.12)$$

that satisfies

$$\forall\, t \geq 0, \quad c(t) \in L_\rho(t,x(t),x'(t),a(t)) \qquad (8.13)$$

Definition 8.1.5 *The solutions to the system (8.12) of differential equations are called ρ heavy solutions to the cognitive system.*

Observe that if at time t_1 a heavy solution satisfies

$$(x(t_1),x'(t_1),a(t_1)) \in L_0^{-1}(t_1;c(t_1))$$

then we know that the conceptual control $c(t_1)$ can remain constant after t_1 and that the state of the cognitive system will evolve for $t \geq t_1$ according to the law

$$\begin{cases} \text{(i)} & x''(t) = f(t,x(t),x'(t),a(t)) \\ \text{(ii)} & a'(t) = g(t,x(t),x'(t),a(t),c(t_1)) \end{cases}$$

while remaining in the sensorimotor cell

$$\forall\, t \geq t_1, \quad (x(t), x'(t), a(t)) \in L_0^{-1}(t; c(t_1))$$

If this happens, one can say that the conceptual control c_1 is a "punctuated equilibrium" (in the sense of Eldredge and Gould) because it remains constant, whereas the state may evolve.

Naturally, we should have the same qualitative result with any other selection mechanism for velocities of conceptual controls that assigns the value 0 whenever 0 belongs to $M_\rho(t, x, v, a, c)$.

8.2 Motivations and Comments

There should be no difficulty in accepting the idea of an environment[24] on which the cognitive systems act.

One of the universal properties of living matter is to consume resources provided by the environment and thus to transform, create, or destroy it[25] in order to keep alive their organisms and to transfer their genes to their offspring.

For that purpose, they must perceive in the environment sources of nutrients and must have access to them, in one way or another. From perception to action, most organisms process a very complex sequence of chemical communications, one molecule leaving an emitter cell and arriving to a receptor cell, modifying its chemical composition.

There should also be no problem in accepting the existence of cerebral activity that operates the internal organs of the body and the muscular activity by which interaction with the environment is possible.

The existence of conceptual controls and their use in a recognition mechanism are more questionable assumptions.

1. First, nothing forbids, in principle, the synaptic weights playing the

[24] Further studies should make precise the structure of the space describing the environment. It should be at least represented as the product of a physical environment, a biological one, formed of other species (which thus becomes a dynamical environment by involving the relations with other organisms), and a conceptual environment, formed of cultural artifacts available to human brains (Popper's third world). In the case of pluricellular organisms, one can consider that the internal organization is a part of the environment of the system, whose objective is the survival of the organism (the receptor cells of the internal states are called the proprioreceptor cells). For the sake of simplicity, we represent this too complex a structure by a finite-dimensional vector space, whose dimension is the number of characteristic features that can be measured with respect to the unit.

[25] Some 4 billion years ago the photosynthesis of the first organisms transformed the existing atmosphere of methane and ammonia to the oxygenated one we know today, and that probably was the first example of pollution.

role devoted to the conceptual controls, but that probably would require fairly large variability margins (if evolution of the learning is required). Second, this representation isolates nervous systems from the endocrine systems, for which the concept of "weight" between an emitter cell and a receptor cell is difficult to conceive. Perhaps we should be cautious and leave open the problem of the precise nature of those conceptual controls, even if we agree to postulate their existence.

However, we shall attempt to justify the introduction of conceptual controls at several levels.

2. The ambiguous concept of *perception* includes both an "objective" component and a "subjective" component. The objective component, which we call sensory perception, is provided by the neuronal circuit activated by the sensory receptors. But everyone knows that there is also a subjective component by which this sensory perception is interpreted. This interpretation may depend on many factors (previous experiences, emotional state, attention level, etc.), that is, on a state of cerebral activity independent of the sensory inputs. This independent activity represents part of the regulatory control that we called conceptual control.
3. If we accept the existence of endogenous cerebral activity that "interprets" the sensory perception of the environment, we must postulate the existence of a recognition mechanism that tells us whether or not a conceptual control and the sensory perception of the environment and its variations are consistent.

   It seems that brains have evolved, during phylogenesis, systems that transform information on bodily needs and environmental events into cerebral activity producing either pleasure (comfort) or pain (discomfort). These systems are known by psychologists as *motivational systems*, and are naturally more sophisticated than strictly pleasure-seeking or pain-avoiding systems. They include the emotional system and the homeostatic drive systems, which basically keep the organism functioning (e.g., the hunger drive). They are often regulated by hormones.

   These systems reveal the relation between the perception of the environment and the conceptual controls; if these are not consistent, the situation can be remedied by

   (a) acting on the environment or

   (b) changing the conceptual control when action on the environment is not consistent with the existing conceptual control.

The latter strategy (change of conceptual controls) appears to be less frequent than the former, and for many subsystems (such as the homeostatic systems) it is quite impossible. This is probably due to an *inertia principle*, which we have postulated, which states that whenever a conceptual control "works" (i.e., keeps the evolution of the cognitive system viable), we keep it.

4. The idea of a recognition mechanism based on conceptual controls is consistent with the concept of *epigenesis*. The recognition mechanism outlined earlier is basically a selection mechanism with a definite Darwinian flavor, choosing conceptual controls as a function of the environment and changes in the environment. By representing the cerebral activity as the flux of neurotransmitters in individual synapses, as discussed later, one could suppose that the synapses used most frequently would be stabilized, while those used less frequently would deteriorate. But the mere description of the synapses that are stabilized after a period of activity is capable of explaining epigenesis only to the extent that a road network can determine the routes taken by cars - in this case, one can agree to observe that if an existing network constrains the traffic without determining it, this is the existing travel pattern that influences the evolution of the network by requiring the maintenance of commonly used routes, while the others can be neglected.[26]

5. We also postulated that the evolution of the recognition mechanism is programmed genetically. This recognition mechanism probably is rather simple in its principle: It may simply open or close (activate or deactivate) a number of neuronal circuits during one or several specified periods of time, allowing both the neurotransmitters released by the perception of the environment and the conceptual controls to pass through circuits allowing their comparison.

It seems likely that some components of this mechanism (which should obey the laws of biochemistry[27]) are *periodic*, with overlapping periodicity, as specialists in chronobiology propose. These components are the many biological clocks involved in maintain-

[26] This is the reason we have chosen to study the evolution of "cars" at the crossroads of the network to infer its evolution, instead of studying this network per se.

[27] One can find these "hypercycles" in chemistry, where the cycles are maintained by a sequence of catalysts, each one acting on the next one. Enzymes are biological catalysts. It has been suggested that such hypercycles of catalysts gave birth to life in the celebrated primordial soup.

ing the homeostatic equilibrium of the organism.[28] These periodic components of the recognition mechanism probably lie at the heart of the ability to learn rules for recognizing regularities supplied by these cyclic processes on the basis of "analogous" endogenous cyclic processes, as well as the ability to "memorize" them by "autosustaining" these cyclic processes, to extrapolate them, and to attempt to look for causal relations.

Comparison of trajectories with different periods may provide (through differentiated processing) reading mechanisms that are either retrograde (recall processes) or anterograde (anticipation processes). Moreover, this principle allows for noncoded memorization. Memorized phenomena can be reconstructed from initial conditions, which can be pieces of functions from the time line to the set of neurotransmitters (and/or hormones) by dynamical processes, the nature of which could be set-valued. *The set-valued character of the consequences of initial data could be as important as the support of the initial function is small.* The richer is the input, the more reduced is the uncertainty or the availability of the outputs. This cyclic character also allows recovery of the same sequence (piece of trajectory) from several different pieces of a periodic initial datum, which can range over different geographic areas of the nervous system.

One can also take into account the mechanism of *generalization*: If two time-dependent functions coincide on a given time interval, one can say that the common piece of these functions is an abstraction of each of them. The difficulty in tracing two functions backward in time to discover their common piece may explain the difficulties of abstraction (or induction), as compared with deduction, which amounts to letting the dynamical process (whose set-valued character opposes the deterministic rigor to which the deduction professionals aspire) go forward.

One can also revisit, in this framework, the problems studied through "semantic networks," by replacing the representation of concepts as the nodes of the semantic network by the representation of concepts as (multivalued) concatenations of elementary functions from the time line to the set of synapses.

[28]It may be postulated that the recognition of the periodicity of the trajectories of the sun and the moon by periodic components of the recognition mechanism, in combination with suitable conceptual controls, leads to the perception and concept of time.

6. Other components of this mechanism are not periodic, but are active only during certain periods. This may be illustrated by the phenomenon of "imprinting" in ethology: Among animal species in which the young are able to walk almost immediately after birth, the newborn animals follow the first moving object that they perceive, whatever it may be. (In practice, it is usually a parent.)

   However, this susceptibility does not last indefinitely,[29] and it involves the perception of movement.
7. The assumption of a recognition mechanism using conceptual controls allows us to explain the *adaptability* and *redundancy* of cerebral activities. A cognitive system can recognize the same sensory perception using different conceptual controls at different times - this is redundancy. On the other hand, thanks to the periodic nature of many components of cerebral activities, this sensory perception can be "interpreted" in several ways, provoking different actions (since we have assumed that the action taken depends upon the conceptual controls) - and this is adaptability.
8. The components of the recognition mechanism based on one or a small number of conceptual controls operate the automatic biological systems, since in this case the subsystem inherits the genetic program of the component of the recognition mechanism. Stimulus-response (SR) processes belong to this class.
9. The concept of a recognition mechanism reflects the dichotomy between "conceptually driven processes" and "data-driven processes" introduced by specialists in cognitive psychology and pattern recognition. In this case the data-driven process is the cerebral activity provoked by the sensory perception of the environment, and the conceptually driven process takes the form of conceptual controls (this is the origin of our terminology).
10. The idea of a recognition mechanism is also consistent with the concept of *metaphor*, regarded as a combination of a sensory perception of the environment and conceptual control *recognized* by the recognition mechanism. A feeling of understanding, which amounts to a feeling of pleasure, occurs when a metaphor is recognized by the recognition process.

The set-valued character of the recognition mechanism is not for-

[29] For example, ducklings can be imprinted only during the first 24 hours of their life, with sensitivity at a maximum between the 14th and 17th hours. The crucial factor in imprinting is the mobility of the object to be imprinted, and this reveals the importance of the perception of variations in the environment.

eign to the mechanism of discovery allowed by the possibility of exploring new conceptual controls to interpret the sensory perception.

Perhaps thought processes can also fit into this representation, since they involve setting up conceptual controls in the form of assumptions and then comparing them with the perception of the environment (including the cultural environment). This dynamical process of *making* and *matching* seems to be quite universal.

11. The foregoing mathematical metaphor describes a *learning process* as a feedback relation that associates a set of (learnable) conceptual controls with each sensorimotor state. The larger the set of conceptual controls associated with a sensorimotor state, the less deterministic the learning process.

    This role of the sensorimotor state in learning has been observed and emphasized by Piaget and others when they have described the learning processes of children. Here, we characterize the viable learning processes, and we deduce the existence of a largest learning process.

    This role is consistent with several observed facts. For instance, studies of the imprinting phenomenon have shown that the greater the effort made by the young animal to follow the moving object, the stronger is the imprint. When one of the components of the sensorimotor state is suppressed, the learning mechanism does not work normally. For instance, if kittens are raised in a visual environment composed of black and white vertical lines, they are unable to "see" horizontal stripes later in life. In another experiment, two kittens from the same litter spent several hours each day in a contraption that allowed one kitten almost complete freedom to explore and perceive its environment, while the other was suspended passively in a "gondola" whose motion was controlled by the first kitten. The two animals received the same visual stimulation, but the active kitten learned to interpret those signals to give it an accurate picture of its environment, whereas the passive kitten learned nothing and was, in practical terms, "blind" to the real world.

8.3 Recognition Mechanisms

It is possible to conceive, in an abstract way, recognition mechanisms between two circuits of neurotransmitters. Let us consider, for example, two subnetworks with the same numbers of synapses, through which circulate two circuits of neurotransmitters. For comparing these two circuits, we introduce a third subnetwork with the same number of synapses. From each synapse of the same rank in each of the two original subnetworks there are branch neurons - one excitatory, and the other inhibitory - connected to the synapse of the same rank in the third subnetwork.

Consider now two circuits of neurotransmitters circulating in each subnetwork. They influence, through the branching neurons, the synapse of the third subnetwork, with antagonistic effects. It is sufficient that the synapses of the third circuit be silent - excitation produced by one circuit is inhibited by the other circuit - for these two circuits of neurotransmitters to be recognized. Furthermore, when these two circuits are not recognized, the synapses of the third circuit become excited. The farther apart are the two circuits, the more active is the third circuit. One can imagine that this third circuit, when it is active, triggers activity in other circuits that eventually induce some action on the environment in order that one of the circuits can become recognized by the other.

The same type of mechanism can be used to recognize time-dependent functions of neurotransmitters journeying through a given synapse by another time dependent function of neurotransmitters going through a second synapse. Assume that each is a discrete function of n periods of time. We introduce a subnetwork of n neurons, and we link each of the two neurons to the synapses of the subnetwork with antagonistic effects (if one neuron arriving at a given synapse is excitatory, the other one is inhibitory, and vice versa). Assume, furthermore, that the duration of transit of the influx from each of the two neurons to the jth neuron of the subnetwork is equal to j units of time. In this case, again, if each synapse of the subnetwork is silent, one can say that the two functions passing through each of the neurons are recognized. The activity in the subnetwork is a measure of the lack of recognition of these two functions. This activity can then trigger other subnetworks and, finally, the action of the system on the environment.

9
Qualitative Analysis of Static Problems

Introduction

This chapter deals with problems motivated by economics in the 1960s (comparative statics) and, recently, by a domain of artificial intelligence known by the name of "qualitative simulation" or "qualitative physics," to which special volumes of *Artificial Intelligence* (1984) and of *IEEE Transactions on Systems, Man and Cybernetics* (1987) have been devoted. It concerns "sign solutions" (or confluence solutions, to adopt the terminology used in artificial intelligence) of nonlinear equations and inclusions. We shall provide a criterion involving adequate linearizations of the sign problem, which then will provide sign solutions for both the original problem and its linearization.

The purpose of this chapter is then to offer several theorems on the qualitative solutions to a linear or nonlinear equation,

$$\text{given } y \in \mathbf{R}^m, \quad \text{find } x \in \mathbf{R}^n \quad \text{such that } f(x) = y$$

in terms of "confluences" (i.e., of the signs of the components of the vector y).

Qualitative Frames For defining in a rigorous way qualitative solutions to an equation, we introduce the definition of qualitative frames: A *qualitative frame* $(\mathcal{X}, Q)$ of a set X is defined by

1. a set $\mathcal{X}$, called the *qualitative set*[1]
2. a set-valued map[2] $Q : \mathcal{X} \rightsquigarrow X$, called the *value map*

[1] which is generally assumed to be a finite set.
[2] A set-valued map Q from $\mathcal{X}$ to X maps each $a \in \mathcal{X}$ to a subset $Q(a) \subset X$, possibly empty.

where we assume that

$$\begin{cases} \text{(i)} & \forall\, a \in \mathcal{X},\ \ Q(a) \neq \emptyset \\ \text{(ii)} & Q \text{ is surjective } (\forall\, x \in X,\ \exists\, a \in \mathcal{X} \mid x \in Q(a)) \\ \text{(iii)} & Q \text{ is injective } (\forall\, a, b \in \mathcal{X},\ a \neq b,\ Q(a) \neq Q(b)) \end{cases} \tag{9.1}$$

We shall say that the qualitative frame is *strict* if we assume further that

$$\forall\, a, b \in \mathcal{X},\ a \neq b,\ Q(a) \cap Q(b) \ = \emptyset \tag{9.2}$$

In this chapter, we shall use only the confluence frames of finite-dimensional vector spaces, defined as follows:

Strict Confluence Frame We associate with $X := \mathbf{R}^n$ the n-dimensional *confluence space* $\mathcal{R}^n$ defined by

$$\mathcal{R}^n \ := \ \{\, -, 0, +\, \}^n$$

The *strict confluence frame* is defined by $(\mathcal{R}^n, Q_n)$ where the value map Q_n maps each qualitative value $a \in \mathcal{R}^n$ to the convex cone

$$Q_n(a) \ := \ \mathbf{R}^n_a \ := \ \{\, v \in \mathbf{R}^n \mid \text{sign of } (v_i) = a_i \,\}$$

For $n = 1$, we have

$$Q_1(-) \ =]-\infty, 0[, \quad Q_1(0) \ = \ \{0\}, \quad \text{and} \quad Q_1(+) \ =]0, +\infty[$$

Large Confluence Frame We still associate with $\mathbf{R}^n$ the n-dimensional confluence space $\mathcal{R}^n$. The *large confluence frame* is then defined by the set-valued map $\overline{Q}_n$ associating with every $a \in \mathcal{R}^n$ the convex cone

$$\overline{Q}_n(a) \ := \ a\mathbf{R}^n_+ \ := \ \{\, v \in \mathbf{R}^n \mid \text{ sign of } (v_i) = a_i \text{ or } \ 0\,\}$$

which is the closure of $Q_n(a) = \mathbf{R}^n_a$.

Dual Confluence Frame Let $\mathbf{R}^{n^\star}$ denote the dual of $X := \mathbf{R}^n$. The *dual confluence frame* $(\mathcal{R}^n, Q_n^\star)$ is made of the n-dimensional confluence space $\mathcal{R}^n := \{-, 0, +\}^n$ and the set-valued map $Q_n^\star$ from $\mathcal{R}^n$ to $\mathbf{R}^{n^\star}$ that maps every $a \in \mathcal{R}^n$ to the closed cone $Q^\star(a)$ of elements $p := (p^1, \ldots, p^n)$ defined by

$$\begin{cases} p_i \geq 0 & \text{if } a_i \ = \ + \\ p_i \in \mathbf{R} & \text{if } a_i \ = \ 0 \\ p_i \leq 0 & \text{if } a_i \ = \ - \end{cases}$$

Qualitative Solutions Let us consider a map $f : X \mapsto Y$ from a set X to a set Y. Let $(\mathcal{X}, Q_X)$ and $(\mathcal{Y}, Q_Y)$ be qualitative frames of the two sets X and Y, and let $b \in \mathcal{Y}$ be a qualitative right-hand side. We define the *standard qualitative solution* $a \in \mathcal{X}$ satisfying

$$f(Q_X(a)) \cap Q_Y(b) \neq \emptyset$$

which depends naturally on the choice of the qualitative frames and, in particular, on the choice of the value maps Q_X and Q_Y.

For example, let the two quantitative spaces $X := \mathbf{R}^n$ and $Y := \mathbf{R}^m$ be finite-dimensional vector spaces. It is usually more difficult to prove the existence of strict confluence solutions than that of large ones. We shall provide a criterion for the existence of strict standard solutions.

Let us begin with the case when $f := A \in \mathcal{L}(X, Y)$ is a linear operator. In this case, there exist always large qualitative solutions, because $A(0) = 0$! But if we assume that the *dual condition*

$$\begin{cases} 0 \text{ is the only solution } q \text{ to} \\ A^\star(q) \in Q_n^\star(a) \quad \text{and} \quad q \in -Q_m^\star(b) \end{cases} \tag{9.3}$$

is satisfied, then there exists a standard strict qualitative solution $a \in \mathcal{R}^n$:

$$A(Q_n(a)) \cap Q_m(b) \neq \emptyset$$

This theorem can be extended to the nonlinear case through linearization and duality.

Let $x_0 \in \overline{Q}_n(a)$ [where $y_0 := f(x_0) \in \overline{Q}_n(b)$] be a representative of a solution a to the large qualitative equation

$$f(\overline{Q}_n(a)) \cap \overline{Q}_m(b) \neq \emptyset$$

Assume that f is continuous and continuously differentiable at x_0. If

$$\begin{cases} 0 \text{ is the only solution } q \text{ to} \\ f'(x_0)^\star(q) \in Q_n^\star(a) \quad \text{and} \quad q \in -Q_m^\star(b) \end{cases} \tag{9.4}$$

then a is a solution to the qualitative equation

$$f(Q_n(a)) \cap Q_m(b) \neq \emptyset$$

There are other ways to define qualitative solutions to an equation. For instance, we can propose

(a) *the upper qualitative solution* $a \in \mathcal{X}$ satisfying

$$f(Q_X(a)) \subset Q_Y(b)$$

(b) *the lower qualitative solution* $a \in \mathcal{X}$ satisfying

$$f^{-1}(Q_Y(b)) \subset Q_X(a)$$

(c) *the anti-upper qualitative solution* $a \in \mathcal{X}$ satisfying

$$Q_Y(b) \subset f(Q_X(a))$$

and we see at once that any upper qualitative or lower qualitative solution is a standard qualitative solution.

For upper qualitative solutions, we shall prove in the linear case the following duality principle: Let $A \in \mathcal{L}(X, Y)$ be linear. Then the two following conditions are equivalent:

$$A^{-1}(\overline{Q}_m(b)) \subset \overline{Q}_n(a)$$

and

$$Q_m^\star(a) \subset \overline{A^\star(Q_m^\star(b))}$$

There is an equivalent statement for lower qualitative solutions: The two conditions

$$A(\overline{Q}_n(a)) \subset \overline{Q}_m(b)$$

and

$$A^\star(Q_m^\star(b)) \subset Q_n^\star(a)$$

are equivalent.

The Need for Set-valued Analysis As can be seen from the definitions of qualitative solutions, our theorems rely on "set-valued analysis," which was developed for various purposes ranging from Painlevé's early works to recent results on graphical convergence and the differential calculus of set-valued maps, including an inverse-function theorem that we shall use. [See, e.g., *Set-Valued Analysis* (Aubin and Frankowska 1990).] Hence, we can now study qualitative analysis of set-valued maps $F : X \rightsquigarrow Y$. The mathematical cost will be approximately the same, and furthermore there is an important motivation to do that at the onset in the framework of qualitative analysis.

Indeed, when the problem is the resolution of an equation $f(x) = y$, the single-valued map f is not exactly known, even when physicists and other scientists model such maps by classical and familiar "special functions" through their behavior; it is sufficient to mention the favorite use of the exponential, logarithmic, logistic, trigonometric, and other functions in many models.

Actually, the choices of these functions often are made because there are "representatives" of a class of functions defined by a list of properties, whether the list is exhaustive or not, conscious or not.

Keeping in mind the philosophy of qualitative reasoning, it is more to the point to start with such a list of requirements on maps from X to Y and build from it the "largest" set-valued map F from X to Y that will satisfy them. Hence we are led to propose to solve right away the qualitative solutions to "inclusions"

$$F(x) \ni y$$

We shall prove the set-valued versions of the results described at the beginning of the introduction in the rest of this chapter, after adapting to the set-valued case the foregoing definitions, summarized in the following table:

Table 9.1. *The ménagerie of qualitative solutions*

Name	Definition of the qualitative solution
Standard	$F(Q_X(a)) \cap Q_Y(b) \neq \emptyset$
Upper	$F(Q_X(a)) \subset Q_Y(b)$
Anti-lower	$Q_X(a) \subset F^{-1}(Q_Y(b))$
Lower	$F^{-1}(Q_Y(b)) \subset Q_X(a)$
Anti-upper	$Q_Y(b) \subset F(Q_X(a))$

9.1 Qualitative Frames

The general features of qualitative analysis, and, more particularly, of confluences, can be captured in the mathematical framework we propose here. Let X, regarded as and called the *quantitative space*, denote the set of elements on which qualitative reasoning operates.

Definition 9.1.1 *A qualitative frame* $(\mathcal{X}, Q)$ *of a set* X *is defined by*

- *a set* $\mathcal{X}$, *called the qualitative set*
- *a set-valued map*[3] $Q : \mathcal{X} \rightsquigarrow X$, *called the value map satisfying*

[3]See Section A.9 for further definitions concerning set-valued maps.

$$\begin{cases} \text{(i)} & \forall\, a \in \mathcal{X},\ Q(a) \neq \emptyset \\ \text{(ii)} & Q \text{ is surjective } (\forall\, x \in X,\ \exists\, a \in \mathcal{X} \mid x \in Q(a)) \\ \text{(iii)} & Q \text{ is injective } (\forall\, a, b \in \mathcal{X},\ a \neq b,\ Q(a) \neq Q(b)) \end{cases} \tag{9.5}$$

We shall say that the qualitative frame is strict if we assume further that

$$\forall\, a, b \in \mathcal{X},\ a \neq b,\ Q(a) \cap Q(b) = \emptyset \tag{9.6}$$

We denote by $P := Q^{-1}$ the inverse[4] *of Q, called the qualitative map. The values $Q(a)$ are called the qualitative cells.*

Remark: *Strict Qualitative Frames* When the qualitative frame is strict, the qualitative map P is single-valued and is then denoted by p. In this case, the subsets $Q(a)$ form a partition of X when a ranges over the qualitative set $\mathcal{X}$, so that they constitute the equivalence classes of the the binary relation $\mathcal{R}$ defined on X by

$$x \,\mathcal{R}\, y \iff p(x) = p(y)$$

which is an equivalence relation. Hence, in this case we can regard the qualitative set $\mathcal{X}$ as the factor space $\mathcal{X} := X/\mathcal{R}$, and p as the canonical surjection.[5] □

Let M be a subset of X. Usually, the inclusions

$$M \subset Q \circ Q^{-1}(M)$$

always hold true. We shall say that a subset $M \subset X$ is *saturated* if $M = Q \circ Q^{-1}(M)$.

We observe that when the qualitative frame is strict, the subsets $Q(a)$ are saturated.

[4]This means that $a \in P(x)$ if and only if $x \in Q(a)$. In particular,

$$\forall\, x \in X,\ x \in Q(P(x))$$

[5]Conversely, we can associate with any equivalence relation $\mathcal{R}$ a strict qualitative frame, where the factor space $\mathcal{X} := X/\mathcal{R}$ is the qualitative set, and the canonical surjection

$$p : X \mapsto \mathcal{X} := X/\mathcal{R}$$

associating with each element $x \in X$ its equivalence class $Q(x) \in \mathcal{X}$ is the qualitative map.

Remark: *Closure of a Strict Qualitative Frame* When X is a topological space, it is convenient to associate with a strict qualitative frame $(\mathcal{X}, Q)$ its closure $(\mathcal{X}, \overline{Q})$, where

$$\forall\, a \in \mathcal{X},\ \ \overline{Q}(a) \ :=\ \mathrm{cl}(Q(a)) \ =:\ \overline{Q(a)} \quad \square \tag{9.7}$$

Example: Strict Confluence Frame We consider the usual finite-dimensional vector space $X := \mathbf{R}^n$ as a quantitative space, and we associate with it the n-dimensional *confluence space* $\mathcal{R}^n$ defined by

$$\mathcal{R}^n \ :=\ \{\,-,\ 0,\ +\,\}^n$$

whose elements are denoted by $a := (a_1, \ldots, a_n)$.

The *strict confluence frame* is defined by $(\mathcal{R}^n, \mathcal{Q}_n)$, where the value map $\mathcal{Q}_n$ maps each qualitative value $a \in \mathcal{R}^n$ to the convex cone

$$\mathcal{Q}_n(a) \ :=\ \mathbf{R}^n_a \ :=\ \{\, v \in \mathbf{R}^n \ \mid \ \text{sign of } (v_i) = a_i \,\}$$

It is obviously a strict qualitative frame, so that the inverse of $\mathcal{Q}_n$, denoted by s_n, is the single-valued map from $\mathbf{R}^n$ to $\mathcal{R}^n$ defined by

$$\forall\, i \in \{1, \ldots, n\},\ \ s_n(x)_i \ :=\ a_i \quad \square$$

Example: Large Confluence Frame We still consider the finite-dimensional vector space $X := \mathbf{R}^n$ as a quantitative space, and we associate with it the n-dimensional confluence space $\mathcal{R}^n$. The *large confluence frame* is then defined by the set-valued map $\overline{\mathcal{Q}}_n$ associating with every $a \in \mathcal{R}^n$ the convex cone

$$\overline{\mathcal{Q}}_n(a) \ :=\ a\mathbf{R}^n_+ \ :=\ \{\, v \in \mathbf{R}^n \mid \ \text{sign of } (v_i) = a_i \text{ or } 0 \,\}$$

which is the closure of $\mathcal{Q}_n(a) = \mathbf{R}^n_a$, as well as the image of $\mathbf{R}^n_+$ by the map $a : x \mapsto ax = (a_1 x_1, \ldots, a_n x_n)$.

Its inverse is the set-valued map from $\mathbf{R}^n$ to $\mathcal{R}^n$ denoted by S_n, which is defined by

$$\forall\, i \in \{1, \ldots, n\},\ \ S_n(x)_i \ :=\ \begin{cases} - & \text{if } x_i \leq 0 \\ \{\,-, 0, +\,\} & \text{if } x_i = 0 \\ + & \text{if } x_i \geq 0 \quad \square \end{cases}$$

Example: Dual Confluence Frame Let $\mathbf{R}^{n^\star}$ denote the dual of $X := \mathbf{R}^n$. The *dual confluence frame* $(\mathcal{R}^n, \mathcal{Q}^\star_n)$ is made of the n-dimensional confluence space $\mathcal{R}^n := \{-, 0, +\}^n$ and the set-valued map $\mathcal{Q}^\star_n$ from $\mathcal{R}^n$

to $\mathbf{R}^{n^\star}$ that maps every $a \in \mathcal{R}^n$ to the closed cone $Q_n^\star(a)$ of elements $p := (p^1, \ldots, p^n)$ defined by

$$\begin{array}{lll} p^i \leq 0 & \text{if } a_i & = - \\ p^i \in \mathbf{R} & \text{if } a_i & = 0 \\ p^i \geq 0 & \text{if } a_i & = + \end{array}$$

Its inverse is the set-valued map from $\mathbf{R}^{n^\star}$ to $\mathcal{R}^n$ denoted by $S_n^\star$ and defined by

$$S_n^\star(p)_i = \begin{cases} \{+, 0\} & \text{if } p^i > 0 \\ \{-, 0\} & \text{if } p^i < 0 \\ \{-, 0, +\} & \text{if } p^i = 0 \end{cases}$$

The reason that this qualitative frame is called the dual confluence frame is given by the following lemma:

Lemma 9.1.2 *The positive polar cone to the convex cone* $\overline{Q}_n(a) := a\mathbf{R}_+^n$ *is the cone* $Q^\star(a) := a^{-1}(\mathcal{R}_+^n)$.

Proof We associate with any $a \in \mathcal{R}^n$ the subsets

$$\begin{array}{ll} I_0(a) & := \{ i \mid a_i = 0 \} \\ I_+(a) & := \{ i \mid a_i > 0 \} \\ I_-(a) & := \{ i \mid a_i < 0 \} \end{array}$$

of $I := \{1, \ldots, n\}$. We observe that

$$p \in (a\mathcal{R}_+^n)^+ \quad \text{if and only if } p^i \geq 0 \text{ if } i \in I_+(a) \text{ and } p^i \leq 0 \text{ if } i \in I_-(a)$$

because this is equivalent to

$$\forall\, y \in \mathcal{R}_+^n, \ \sum_{i \in I} a_i p^i y_i = \sum_{i \in I_+(a)} p^i y_i - \sum_{i \in I_-(a)} p^i y_i \geq 0$$

□

9.2 Qualitative Equations and Inclusions

Let us consider two quantitative spaces X and Y, a single-valued map $f : X \mapsto Y$, and the equation

$$\text{find } x \in X \text{ such that } f(x) = y \tag{9.8}$$

which we shall call the "quantitative equation."

More generally, we can also start with a set-valued map $F : X \rightsquigarrow Y$ (see appendix Section A.9 for definitions) and the "quantitative inclusion"

$$\text{find } x \in X \quad \text{such that } F(x) \ni y \tag{9.9}$$

In order to make a qualitative analysis of such an equation or an inclusion, we introduce two qualitative frames $(\mathcal{X}, Q_X)$ and $(\mathcal{Y}, Q_Y)$ and their qualitative maps Q_X and Q_Y:

$$\begin{array}{ccccc} & X & \overset{F}{\mapsto} & Y & \\ Q_X^{-1} & \downarrow\uparrow \quad Q_X & & Q_Y^{-1} & \downarrow\uparrow \quad Q_Y \\ & \mathcal{X} & \underset{\mathcal{F}}{\mapsto} & \mathcal{Y} & \end{array}$$

There are many ways to associate with a set-valued map F "projections" that map $\mathcal{X}$ to $\mathcal{Y}$. We shall mention only three of them:

The Standard Projection It is the set-valued map $\pi_0(F) : \mathcal{X} \rightsquigarrow \mathcal{Y}$ defined by

$$\pi_0(F) \ := \ Q_Y^{-1} \circ F \circ Q_X \tag{9.10}$$

where $\circ$ denotes the usual product of set-valued maps. Hence, we can associate with the quantitative inclusion the standard qualitative inclusion:

$$\text{find } a \in \mathcal{X} \quad \text{such that } \pi_0(F)(a) \ni b \tag{9.11}$$

which is equivalent to any of the formulations[6]

$$\left\{ \begin{array}{ll} \text{(i)} & F(Q_X(a)) \cap Q_Y(b) \neq \emptyset \\ \text{(ii)} & Q_X(a) \cap F^{-1}(Q_Y(b)) \neq \emptyset \\ \text{(iii)} & \text{Graph}(F) \cap (Q_X(a) \times Q_Y(b)) \neq \emptyset \\ \text{(iv)} & a \ \in \ Q_X^{-1}(F^{-1}(Q_Y(b))) \end{array} \right. \tag{9.12}$$

The latter property follows from the observation that

[6]When the quantitative spaces X and Y are vector spaces, we can associate with F the set-valued map $\Phi : X \times Y \rightsquigarrow Y$ defined by

$$\Phi(x, y) \ := \ F(x) - y$$

Hence $\pi_0(F)(a) \ni b$ if and only if

$$\exists\, (x, y) \in Q_X(a) \times Q_Y(b) \quad \text{such that } \Phi(x, y) \ = \ 0$$

$$(\pi_0(F))^{-1} \;=\; \pi_0(F^{-1}) \tag{9.13}$$

which is obvious when we remark that

$$\text{Graph}(\pi_0(F)) \;=\; (Q_X^{-1} \times Q_Y^{-1})\text{Graph}(F)$$

Solving the standard qualitative equation means that, given a "qualitative right-hand side" $b \in \mathcal{Y}$, there exist a quantitative right-hand side $y \in Q_Y(b)$ and a solution $x \in F^{-1}(y)$ that belongs to $Q_X(a)$.

Let us single out the following observation:

$$\forall\, \mathcal{A} \subset \mathcal{X}, \quad \pi_0(F)(\mathcal{A}) = Q_Y^{-1}(F(Q_X(\mathcal{A})))$$

so that we always have $F(Q_X(\mathcal{A}) \subset Q_Y(\pi_0(F)(\mathcal{A}))$, equality holding true whenever $\pi_0(F)$ is saturated.

The Upper Projection This is the set-valued map $\pi_+(F) : \mathcal{X} \rightsquigarrow \mathcal{Y}$ defined by

$$\pi_+(F) \;:=\; Q_Y^{-1} \square (F \circ Q_X) \tag{9.14}$$

Hence, we can also associate with the quantitative inclusion the upper qualitative inclusion:

$$\text{find } a \in X \quad \text{such that } \pi_+(F)(a) \ni b \tag{9.15}$$

which is equivalent to either property

$$\left\{ \begin{array}{lrcl} \text{(i)} & F(Q_X(a)) & \subset & Q_Y(b) \\ \text{(ii)} & Q_X(a) & \subset & F^{+1}(Q_Y(b)) \end{array} \right. \tag{9.16}$$

In other words, solving the upper qualitative equation means that, given a "qualitative right-hand side" $b \in \mathcal{Y}$, a is an upper qualitative solution if for all representative $x \in Q_X(a)$, every $y \in F(x)$ is a representative of b.

The Anti-Lower Projection This is the set-valued map $\varpi_-(F)$: $\mathcal{X} \rightsquigarrow \mathcal{Y}$ defined by

$$\varpi_-(F) \;:=\; (Q_Y^{-1} \circ F) \square Q_X \tag{9.17}$$

Hence, we can also associate with the quantitative inclusion the anti-lower qualitative inclusion:

$$\text{find } a \in X \quad \text{such that } \varpi_-(F)(a) \ni b \tag{9.18}$$

which can be written in the following equivalent form:

$$Q_X(a) \subset F^{-1}(Q_Y(b)) \tag{9.19}$$

Therefore, solving the anti-lower qualitative equation means that, given a "qualitative right-hand side" $b \in \mathcal{Y}$, a is an anti-lower qualitative solution if for each representative $x \in Q_X(a)$, there exists a representative $y \in F(x)$ of b.

We can also invert the set-valued map F and "project" the set-valued map F^{-1}. We then obtain the two following concepts of projections and qualitative solutions:

The Lower Projection This is the set-valued map $\pi_-(F) : \mathcal{Y} \leadsto \mathcal{X}$ defined by

$$\pi_-(F) \ := \ Q_X^{-1} \square (F^{-1} \circ Q_Y) \tag{9.20}$$

Hence, we can also associate with the quantitative inclusion the lower qualitative inclusion:

$$\text{find } a \ \in \ \pi_-(F)(b) \tag{9.21}$$

which is equivalent to the property

$$F^{-1}(Q_Y(b)) \ \subset \ Q_X(a) \tag{9.22}$$

Therefore, to say that a is a lower qualitative solution amounts to saying that for all representatives $y \in Q_Y(b)$ of the qualitative right-hand side b, all solutions to the inclusion $F(x) \ni y$ are representatives of a.

The Anti-Upper Projection This is the set-valued map $\varpi_+(F)$: $\mathcal{Y} \leadsto \mathcal{X}$ defined by

$$\varpi_+(F) \ := \ (Q_X^{-1} \circ F^{-1}) \square Q_Y \tag{9.23}$$

Hence, we can also associate with the quantitative inclusion the anti-upper qualitative inclusion:

$$\text{find } a \ \in \ \varpi_+(F)(b) \tag{9.24}$$

which can be written in the following equivalent form:

$$Q_Y(b) \subset F(Q_X(a)) \tag{9.25}$$

Therefore, solving the anti-upper qualitative equation means that, given a "qualitative right-hand side" $b \in \mathcal{Y}$, for each representative $y \in Q_Y(b)$, there exists a solution x to the inclusion $F(x) \ni y$ that is a representative of a.

We observe at once the following:

Lemma 9.2.1 *Let F be a set-valued map from X to Y. The following hold true:*

1. *Any upper qualitative solution is an anti-lower qualitative solution,*
2. *Any lower qualitative solution is an anti-upper qualitative solution.*
3. *Both any anti-lower and anti-upper qualitative solutions are standard qualitative solutions.*

We shall say that a is a "strong lower qualitative solution" if it is both a lower qualitative solution and an anti-lower qualitative solution, that is, if

$$F^{-1}(Q_Y(b)) = Q_X(a) \tag{9.26}$$

and that a is a "strong upper qualitative solution" if it is both an upper qualitative solution and an anti-upper qualitative solution, that is, if

$$Q_Y(b) = F(Q_X(a)) \tag{9.27}$$

Remark Naturally, when F is a single-valued map f, the inverse image $f^{-1}(M)$ and the core $f^{+1}(M)$ coincide, so that the upper and anti-lower projections coincide:

$$\pi_+(f) = \varpi_-(f) \quad \square \tag{9.28}$$

Remark We could also have considered the projection of a set-valued map F to the set-valued map $Q_Y^{-1} \square (F \square Q_X)$, but its inverse has no interesting properties for our concern: Indeed, a belongs to $(Q_Y^{-1} \square (F \square Q_X))^{-1}(b)$ if and only if

$$\text{whenever } Q_X(a) \subset F^{-1}(y), \text{ then } y \in Q_Y(b) \quad \square$$

9.3 Qualitative Operations

Let us consider a composition law on a quantitative space X, denoted by $+$, that is, a single-valued map $+ : X \times X \mapsto X$ and a qualitative frame $(X, \mathcal{X}, Q)$ of X. Because its standard projection may be a set-valued map, we define a "qualitative composition law" on a qualitative space

Table 9.2. *Definitions of the various concepts of qualitative solutions*

Name	Definition	Definition of $b \in \Pi_{(F)}(a)$
Standard	$Q_Y^{-1} \circ F \circ Q_X$	$\exists\, y \in Q_Y(b), \quad \exists\, x \in F^{-1}(y) \mid x \in Q_X(a)$
Upper	$Q_Y^{-1} \square (F \circ Q_X)$	$\forall\, x \in Q_X(a), \quad \forall\, y \in F(x) \Longrightarrow y \in Q_Y(b)$
Anti-lower	$(Q_Y^{-1} \circ F) \square Q_X$	$\forall\, x \in Q_X(a), \quad \exists\, y \in F(x) \mid y \in Q_Y(b)$
Lower	$Q_X^{-1} \square (F^{-1}) \circ Q_Y$	$\forall\, y \in Q_X(b), \quad \forall\, x \in F^{-1}(y) \Longrightarrow x \in Q_X(a)$
Anti-upper	$(Q_X^{-1} \circ F^{-1}) \square Q_Y$	$\forall\, y \in Q_X(b), \quad \exists\, x \in F^{-1}(y) \mid x \in Q_X(a)$

$\mathcal{X}$ as a set-valued map from $\mathcal{X} \times \mathcal{X}$ to $\mathcal{X}$. Hence, we shall denote by $\oplus : \mathcal{X} \times \mathcal{X} \rightsquigarrow \mathcal{X}$ the standard projection of $+$, and we observe that

$$a \oplus b \;=\; Q^{-1}(Q(a) + Q(b))$$

In other words, $c \in a \oplus b$ if and only if $Q(c) \cap (Q(a) + Q(b)) \neq \emptyset$.

We observe that the qualitative composition law $\oplus$ is commutative whenever the composition law $+$ is commutative. From now on, we shall assume that the composition laws are commutative.

Assume that 0 is the neutral element of $+$. If 0 is saturated [this means that $Q(Q^{-1}(0)) = 0$] and if $Q^{-1}(0)$ is a singleton $\{e\}$, we shall say that $e = Q^{-1}(0)$ is the "qualitative neutral element." Hence, if the qualitative frame has a qualitative neutral element, it satisfies the property $a = a \oplus e = e \oplus a$.

We could say that $b \in \mathcal{X}$ is a *weak inverse* of a if and only if $e \in a \oplus b$, that is, if and only if $0 \in Q(a) + Q(b) \neq \emptyset$. If each element $x \in Q(a)$ has a unique inverse denoted by $-x$, the foregoing formula becomes $Q(b) \cap -Q(a) \neq \emptyset$.

Actually, we shall introduce the following:

Definition 9.3.1 *We shall say that a qualitative element $a \in \mathcal{X}$ has an inverse denoted by $-a$ if $Q(-a) = -Q(a)$.*

Therefore, if the qualitative frame is strict and if every element has a weak inverse, an element $a \in \mathcal{X}$ has at most a unique inverse $-a$, which is defined by $Q(-a) = -Q(a)$.

Assume now that each element $x \in X$ has an inverse $-x$. We define the composition law $-$ by $z = x - y$ if and only if $x = z + y$. Let $\ominus$

denote the standard projection of $-$. It is easy to observe that

$$c \in a \ominus b \text{ if and only if } a \in c \oplus b$$

since

$$Q(c) \cap (Q(a) - Q(b)) \neq \emptyset \text{ if and only if } Q(a) \cap (Q(b) + Q(c)) \neq \emptyset$$

If any qualitative element has an inverse, we observe that $a \ominus b = a \oplus (-b)$. If the qualitative frame is strict and if $e \in a \ominus b$, then $a = b$.

Let $\mathcal{A}$ and $\mathcal{B}$ be subsets of the qualitative space $\mathcal{X}$. We shall denote by $\mathcal{A} \oplus \mathcal{B}$ the subset

$$\mathcal{A} \oplus \mathcal{B} := Q^{-1}(Q(\mathcal{A}) + Q(\mathcal{B})) = \bigcup_{a \in \mathcal{A}, b \in \mathcal{B}} a \oplus b$$

Let us consider now three elements a, b, c of the qualitative space $\mathcal{X}$. We see at once that

$$Q^{-1}(Q(a) + Q(b) + Q(c)) \subset ((a \oplus b) \oplus c) \cap (a \oplus (b \oplus c))$$

since $Q(a) + Q(b) \subset Q(Q^{-1}(Q(a) + Q(b)))$.

We shall say that a $\oplus$ is *saturated* if for any finite sequence of elements $a_i \in \mathcal{X}$, the subset $\sum_i Q(a_i)$ is saturated. This happens whenever the sum of two saturated subsets is saturated. In this case, the standard projection of an associative composition law is associative in the sense that

$$a \oplus (b \oplus c) = (a \oplus b) \oplus c = Q^{-1}(Q(a) + Q(b) + Q(c))$$

Finally, let us point out that $\oplus$ is idempotent whenever for any $a \in \mathcal{X}$, $Q(a) + Q(a) = Q(a)$.

Example: Strict Confluence Frame Let $\mathcal{R} := \{-, 0, +\}$ be the confluence frame of $\mathbf{R}$. We denote the standard qualitative addition by $\oplus$ and the standard qualitative product by $\otimes$. Let us give the Pythagorean table of $\otimes$ and $\oplus$ derived immediately from the definition of standard projection.

Remark Using the upper projection, we obtain the following table:

Remark We obtain for $\oplus$ a "set-valued" composition law,[7] and $\otimes$ is a simple-valued composition law (of $\mathcal{R}$ into itself). □

[7] In previous works, a new element "?" was introduced in the set $\mathcal{R}$ to obtain a composition law with $\oplus$.

Table 9.3. *Addition table*

$\oplus$	$-$	0	$+$
$-$	$-$	$-$	$\{-, 0, +\}$
0	$-$	0	$+$
$+$	$\{-, 0, +\}$	$+$	$+$

Table 9.4. *Multiplication table*

$\otimes$	$-$	0	$+$
$-$	$+$	0	$-$
0	0	0	0
$+$	$-$	0	$+$

We note that
$\oplus$ is associative and commutative, 0 is the neutral element, and every $x \in \mathcal{R}$ has an inverse element.
$\otimes$ is associative and commutative, $+$ is the neutral element, and every $x \in \mathcal{R}$ has an inverse element (except 0).
We denote the lower inverse element x for $\oplus$ by $(-x)$.

We can also use the standard projection to define qualitative subtraction, obtaining the following table, where we denote the standard qualitative subtraction by $\ominus$:
Thus we remark that $x \oplus (-y) \;=\; x \ominus y$

Because we have a strict confluence frame, we know that $0 \in a \ominus b$ if and only if $a = b$. We also observe that $a = a \oplus a$ and that for any $b \in \mathcal{R}$, $a \in a \oplus b$, because $Q(a) \subset Q(a) + Q(b)$.

Qualitative addition enjoys the following property: If $a_1 \in a \oplus c$ and $b_1 \in b \ominus c$, then $(a_1 \oplus b_1) \cap (a \oplus b) \neq \emptyset$. In particular,

$$0 \in a \oplus c \quad \text{and} \quad 0 \in b \ominus c \implies 0 \in a \oplus b$$

This corresponds to the "Gauss rule" or the "resolution law" proposed by J. L. Dormoy.

Indeed, there exist $x \in Q(a)$, $x_1 \in Q(a_1)$, $y \in Q(b)$, and $y_1 \in Q(b_1)$ such that $x_1 = x + \varepsilon c$ and $y_1 = y - \eta c$, where ε and η are positive. By dividing the foregoing inequalities by ε and η, we infer that $z :=$

Table 9.5. *Upper-addition table*

$\uplus$	$-$	0	$+$
$-$	$-$	$-$	$\emptyset$
0	$-$	0	$+$
$+$	$\emptyset$	$+$	$+$

Table 9.6. *Subtraction table*

$\ominus$	$-$	0	$+$
$-$	$\{-,0,+\}$	$-$	$-$
0	$+$	0	$-$
$+$	$+$	$+$	$\{-,0,+\}$

$x_1\varepsilon + y_1/\eta = x/\varepsilon + y/\eta$ belongs to the intersection of $Q(a_1) + Q(b_1)$ and $Q(a) + Q(b)$.

We can define on $\mathcal{R}^n$ the qualitative addition $\oplus$ defined by $a \oplus b := \prod_{i=1}^n a_i \oplus b_i$ and a qualitative external law defined by

$$\forall\, \lambda \in \mathcal{R},\ \forall\, a\ \in \mathcal{R}^n,\quad \lambda \cdot a\ =\ \text{sign}(\lambda) \otimes a$$

9.4 Qualitative Solutions to Linear Problems

In the linear case, some concepts of qualitative solutions enjoy duality properties. Let the two quantitative spaces $X := \mathbf{R}^n$ and $Y := \mathbf{R}^m$ be finite-dimensional vector spaces and let $A \in \mathcal{L}(X, Y)$ be a linear operator from X to Y. We introduce the strict confluence frames $(\mathcal{R}^n, Q_n)$ and large confluence frames $(\mathcal{R}^m, Q_m)$, their dual confluence frames, and the transpose $A^\star \in \mathcal{L}(X^\star, Y^\star)$ of A defined by

$$\forall x \in X,\ \forall p \in Y^\star,\quad \langle A^\star p, x\rangle = \langle p, Ax\rangle$$

We consider the upper large projection of A and the anti-lower projection of the transpose $A^\star$ defined by

$$\left\{\begin{array}{lll} \text{(i)} & \overline{\pi_+}(A) & := S_m \,\square\, (A \circ \overline{Q}_n) \\ \text{(ii)} & \varpi_-(A^\star) & := (S_n^\star \circ A^\star) \,\square\, Q_m^\star \end{array}\right. \tag{9.29}$$

and the lower and anti-upper projections

$$\begin{cases} \text{(i)} & \overline{\pi_-}(A) := S_n \square (A^{-1} \circ \overline{Q}_m) \\ \text{(ii)} & \varpi_+(A^\star) := (S_m^\star \circ {A^\star}^{-1}) \square Q_n^\star \end{cases} \tag{9.30}$$

Theorem 9.4.1 (Qualitative Duality) *Let us consider a linear operator $A \in \mathcal{L}(X, Y)$ and its transpose. The following conditions are equivalent:*

$$\begin{cases} \text{(i)} & a \in \mathcal{R}^n \text{ solves } \overline{\pi_+}(A)(a) = b, \quad \text{where } b \in \mathcal{R}^m \\ \text{(ii)} & b \in \mathcal{R}^m \text{ solves } \varpi_-(A^\star)(b) \ni a, \quad \text{where } a \in \mathcal{R}^n \end{cases} \tag{9.31}$$

as are the two conditions

$$\begin{cases} \text{(i)} & a \in \mathcal{R}^n \text{ is equal to } \overline{\pi_-}(A)(b), \quad \text{where } b \in \mathcal{R}^m \\ \text{(ii)} & b \in \mathcal{R}^m \text{ belongs to } \varpi_+(A^\star)(a), \quad \text{where } a \in \mathcal{R}^n \end{cases} \tag{9.32}$$

This result follows from the bipolar theorem for continuous linear operators:

The Bipolar Theorem We recall the definition of the *(negative) polar cone*

$$K^- := \{p \in X^\star \mid \forall x \in K, \ <p, x> \leq 0\}$$

of a subset introduced in Section 4.4 (see Definition 4.4.1).

Theorem 9.4.2 (Bipolar Theorem) *Let X, Y be Banach spaces, and let $K \subset X$. The bipolar cone K^{--} is the closed convex cone spanned by K. If $A \in \mathcal{L}(X, Y)$ is a continuous linear operator from X to Y and K is a subset of X, then*

$$(A(K))^- = {A^\star}^{-1}(K^-)$$

where $A^\star$ denotes the transpose of A.

Thus the closed cone spanned by $A(K)$ is equal to $\left({A^\star}^{-1}(K^-)\right)^-$

[See, e.g., *Set-Valued Analysis* (Aubin and Frankowska 1990) for a proof.

Proof of Theorem 9.4.1 Indeed, to say that $b \in \overline{\pi_+}(A)(a)$ amounts to saying that

$$A(a\mathbf{R}_+^n) \subset b\mathbf{R}_+^m$$

or, by polarity, that

$$(b\mathbf{R}_+^m)^+ \subset (A(a\mathbf{R}_+^n))^+$$

By the "bipolar theorem," this is equivalent to saying that

$$b^{-1}(\mathbf{R}_+^m) \subset A^{\star^{-1}}(a^{-1}(\mathbf{R}_+^n))$$

which we can write in the form

$$Q_m^\star(b) \subset A^{\star^{-1}}(Q_n^\star(a))$$

by definition of the dual confluence frame. This means that $a \in \varpi_-(A^\star)(b)$. The proof of the second statement is analogous. □

We can extend these theorems to the set-valued analogues of continuous linear operators, which are the closed convex processes.[8]

Definition 9.4.3 (Closed Convex Process) *Let $F : X \rightsquigarrow Y$ be a set-valued map from a Banach space X to a Banach space Y. We shall say that F is*

- *convex if its graph is convex,*[9]
- *closed if its graph is closed,*
- *a process (or positively homogeneous) if its graph is a cone.*[10] *Hence a closed convex process is a set-valued map whose graph is a closed convex cone.*

(See, for instance Chapter 2 of *Set-Valued Analysis* (Aubin and Frankowska 1990).] Because the graphs of continuous linear operators from a Banach space to another are closed vector subspaces, we justify our statement that closed convex processes are their set-valued analogues. Actually, most of the properties of continuous linear operators are enjoyed by closed convex processes.

[8]Closed convex processes provide a way to represent uncertainty for linear operators analogous to "sign matrices." They possess better mathematical properties than sign matrices. In particular, since we cannot extend "sign matrix" to nonlinear maps, we cannot say that a sign matrix is the derivative of a "sign operator," whatever that may mean.

[9]This means that

$$\begin{array}{l}\forall\, x_1,\ x_2,\ \in \mathrm{Dom}(F), \quad \lambda \in [0,1], \\ \lambda F(x_1) + (1-\lambda)F(x_2) \subset F(\lambda x_1 + (1-\lambda)x_2)\end{array}$$

[10]This means that

$$\forall\, x \in X, \quad \lambda > 0, \quad \lambda F(x) = F(\lambda x) \quad \text{and} \quad 0 \in F(0)$$

Definition 9.4.4 (Transpose of a Process) *Let $F : X \rightsquigarrow Y$ be a process. Its left-transpose (in short, its transpose) $F^\star$ is the closed convex process from $Y^\star$ to $X^\star$ defined by*

$$\begin{cases} p \in F^\star(q) \text{ if and only if} \\ \forall\, x \in X, \quad \forall\, y \in F(x), \quad \langle p, x\rangle \leq \langle q, y\rangle \end{cases} \tag{9.33}$$

[See, e.g., Chapter 2 of *Set-Valued Analysis* (Aubin and Frankowska 1990).] The graph of the transpose $F^\star$ of F is related to the polar cone of the graph of F in the following way: The conditions

$$\begin{cases} (q, p) \in \text{Graph}(F^\star) \text{ if and only if} \\ (p, -q) \in (\text{Graph}(F))^- \end{cases} \tag{9.34}$$

are equivalent.

The qualitative duality theorem, Theorem 9.4.1, can be extended to closed convex processes because we can adapt the bipolar theorem:

Theorem 9.4.5 *Let $F : X \rightsquigarrow Y$ a closed convex process and $K \subset X$ a closed convex cone satisfying*

$$\text{Dom}(F) - K = X$$

Then

$$(F(K))^+ = F^{\star^{-1}}(K^+)$$

[See, e.g., Chapter 2 of *Set-Valued Analysis* (Aubin and Frankowska 1990).]

Therefore, it is easy to deduce the following duality characterization of qualitative solutions to closed convex processes:

Theorem 9.4.6 (Set-valued Qualitative Duality) *Let us consider a closed convex process F from X to Y and its transpose. Assume that*

$$\text{Dom}(F) - \overline{Q}_n(a) = X$$

Then the following conditions are equivalent:

$$\begin{cases} \text{(i)} \quad a \in \mathcal{R}^n \text{ solves } \overline{\pi_+}(F)(a) \ni b, \quad \text{where } b \in \mathcal{R}^m \\ \text{(ii)} \quad b \in \mathcal{R}^m \text{ solves } \varpi_-(F^\star)(b) \ni a, \quad \text{where } a \in \mathcal{R}^n \end{cases} \tag{9.35}$$

If we assume that

$$\text{Im}(F) - \overline{Q}_m(b) = Y$$

then the two following conditions are equivalent:

$$\begin{cases} \text{(i)} & a \in \mathcal{R}^n \text{ belongs to } \pi_-(F)(b), \quad \text{where } b \in \mathcal{R}^m \\ \text{(ii)} & b \in \mathcal{R}^m \text{ solves } \varpi_+(F^\star)(a), \quad \text{where } a \in \mathcal{R}^n \end{cases} \tag{9.36}$$

9.5 The Constrained Inverse-Function Theorem

In order to state the constrained inverse function Theorem, we need first to adapt to the set-valued case the concepts of Lipschitz maps:

Definition 9.5.1 *When X and Y are normed spaces, we shall say that $F : X \leadsto Y$ is Lipschitz around $x \in X$ if there exist a positive constant l and a neighborhood $\mathcal{U} \subset \mathrm{Dom}(F)$ of x such that*

$$\forall\, x_1, x_2 \in \mathcal{U}, \quad F(x_1) \subset F(x_2) + l\|x_1 - x_2\|B_Y$$

In this case, F is also called Lipschitz (or $l-$Lipschitz) on $\mathcal{U}$.

If y is given in $F(x)$, F is said to be pseudo-Lipschitz around $(x, y) \in \mathrm{Graph}(F)$ if there exist a positive constant l and neighborhoods $\mathcal{U} \subset \mathrm{Dom}(F)$ of x and $\mathcal{V}$ of y such that

$$\forall\, x_1, x_2 \in \mathcal{U}, \quad F(x_1) \cap \mathcal{V} \subset F(x_2) + l\|x_1 - x_2\|B_Y$$

We begin by recalling the statement of the usual inverse function theorem due to Graves:

Theorem 9.5.2 (Graves's Theorem) *Let X, Y be Banach spaces and $f : X \mapsto Y$ a continuous (single-valued) map. We assume that f is continuously differentiable on a neighborhood of x_0 and that*

$$f'(x_0) \text{ is surjective}$$

Then the set-valued map $y \leadsto f^{-1}(y)$ is pseudo-Lipschitz around $(f(x_0), x_0)$.

For constrained problems, that is, when the solution is required to belong to a closed subset K of X instead of [illegible]le space, we now need to introduce the concept of contingent cones and Clarke tangent cones, the exhaustive study of which is the purpose of Chapter 4 of the book *Set-Valued Analysis* (Aubin and Frankowska 1990).

Graves's inverse-function theorem can be extended to a "constrained problem" of the form

$$\text{find } x \in K \quad \text{such that } f(x) = y$$

for any right-hand side y in a neighborhood of some

$$y_0 := f(x_0) \in f(K)$$

The surjectivity assumption of Graves's theorem is replaced by an adequate transversality condition that involves Clarke tangent cones, as introduced and defined in Appendix A, Section A.7.

Theorem 9.5.3 (Pointwise Inverse-Mapping Theorem) *Let X be a Banach space, K be a closed subset of X, Y be a finite-dimensional vector space, and $f : X \mapsto Y$ be continuously differentiable around an element $x_0 \in K$. If*

$$f'(x_0)(C_K(x_0)) = Y$$

then the set-valued map $y \rightsquigarrow f^{-1}(y) \cap K$ is pseudo-Lipschitz around $(f(x_0), x_0)$.

9.6 Single-valued Nonlinear Equations

Let us consider two finite-dimensional vector spaces $X := \mathbf{R}^n$ and $Y := \mathbf{R}^m$, a single-valued map $f : X \mapsto Y$, and the equation

$$\text{find } x \in X \text{ such that } f(x) = y \tag{9.37}$$

We associate with $X := \mathbf{R}^n$ the n-dimensional *confluence space* $\mathcal{R}^n$ defined by

$$\mathcal{R}^n := \{-, 0, +\}^n$$

whose elements are denoted by $a := (a_1, \ldots, a_n)$.

Given a multisign $b \in \mathcal{R}^m$, does there exist a "standard qualitative solution" $a \in \mathcal{R}^n$, defined by

$$\exists\, x \text{ such that } \operatorname{sign}(x) = a \quad \text{and} \quad \operatorname{sign}(f(x)) = b \tag{9.38}$$

The purpose of this section is to provide a criterion[11] for answering this question (which answer is not obvious whenever components of a and b are different from 0) and to provide generalizations to the set-valued case. This criterion requires, so to speak, that we prove the uniqueness

[11] As in numerical analysis, which deals with approximating infinite-dimensional problems by finite-dimensional ones, problems in qualitative analysis arise at two levels: the passage from quantitative to qualitative, which amounts to associating with continuous problems discrete ones, and algorithms for solving the latter. This is uniquely the first aspect that is investigated here in the framework of solving this equation.

of a dual problem[12] of the linearized problem[13] and thus is the same for both the nonlinear problem and its linearization.

Before stating our first theorem, it is convenient to recall the notations[14]

$$\begin{array}{rcl} Q_n(a) & := & \mathbf{R}^n_a := \{ v \in \mathbf{R}^n \mid \text{sign of } (v_i) = a_i \} \\ \overline{Q}_n(a) & := & a\mathbf{R}^n_+ := \{ v \in \mathbf{R}^n \mid \text{sign of } (v_i) = a_i \text{ or } 0 \} \end{array}$$

and

$$Q^\star_n(a) := v \text{ such that } \begin{cases} v_i \geq 0 & \text{if } a_i = + \\ v_i \leq 0 & \text{if } a_i = - \\ v_i \in \mathbf{R} & \text{if } a_i = 0 \end{cases}$$

Theorem 9.6.1 *Let x_0 satisfy*

$$x_0 \in \overline{Q}_n(a) \quad \text{and} \quad f(x_0) \in \overline{Q}_m(b)$$

Assume that f is continuous and continuously differentiable at x_0. Then the criterion

$$\begin{cases} 0 \text{ is the only solution } q \text{ to} \\ q \in -Q^\star_m(b) \quad \text{and} \quad f'(x_0)^\star q \in Q^\star_n(a) \end{cases} \tag{9.39}$$

implies that a is a solution to the qualitative equation (9.38).

Remark Because $f'(x_0)$ is also the derivative of the linear operator $A := f'(x_0)$, we emphasize the fact that criterion (9.39) states also that there exists a standard qualitative solution to the linearized equation

$$f'(x_0)x = y$$

around x_0. For the linear case, we thus deduce the following:

Corollary 9.6.2 *Let $A \in \mathcal{L}(X,Y)$ be a linear operator. If*

$$\begin{cases} 0 \text{ is the only solution } q \text{ to} \\ q \in -Q^\star_m(b) \quad \text{and} \quad A^\star(q) \in Q^\star_n(a) \end{cases} \tag{9.40}$$

then there exists a solution a to the qualitative equation

$$\exists\, x \in Q_n(a) \mid Ax \in Q_m(b) \tag{9.41}$$

[12] for the resolution of which algorithms have to be designed.

[13] As for any duality result, existence is exchanged with uniqueness.

[14] Observe that $\overline{Q}_n(a)$ is the closure of $Q_n(a)$ and that the positive polar cone of $\overline{Q}_n(a)$ is the cone $Q^\star_n(a)$.

Proof By assumption, we know that (x_0, y_0) belongs to the intersection of the graph of f and the closed convex cone $\overline{Q}_n(a) \times \overline{Q}_m(b)$, so that $((x_0, y_0), (x_0, y_0))$ is a solution in $\mathrm{Graph}(f) \times (\overline{Q}_n(a) \times \overline{Q}_m(b))$ to the equation $(x_1, y_1) - (x_2, y_2) = 0$.

Let $\mathbf{1} \in \mathbf{R}^n$ denote the unit vector

$$\mathbf{1} := (1, \ldots, 1)$$

We shall prove that there exist a solution $(x_1, y_1) \in \mathrm{Graph}(f)$ and a solution $(x_2, y_2) \in \overline{Q}_n(a) \times \overline{Q}_m(b)$ to the equation

$$(x_1, y_1) - (x_2, y_2) = \epsilon(a\mathbf{1}, b\mathbf{1})$$

for some $\epsilon > 0$, so that $x_1 = x_2 + \epsilon a\mathbf{1}$ belongs to $Q_n(a)$, and $f(x_1) = y_1 = y_2 + \epsilon b\mathbf{1}$ belongs to $Q_m(b)$.

For that purpose, we can apply the pointwise inverse-function theorem, Theorem 9.5.3, which states that a solution to the foregoing equation does exist provided that the following assumption is satisfied:

$$C_{\mathrm{Graph}(f)}(x_0, y_0) - C_{\overline{Q}_n(a)}(x_0) \times C_{\overline{Q}_m(b)}(y_0) = X \times Y$$

where $C_K(z)$ denotes the Clarke tangent cone to a subset K at a point $z \in K$.

Because x_0 belongs to $\overline{Q}_n(a) := a\mathcal{R}^n_+$, the Clarke tangent cone $C_{\overline{Q}_n(a)}(x_0)$ coincides with the tangent cone, which contains[15] $Q_n(a)$. In the same way,

$$C_{\overline{Q}_m(b)}(y_0) \supset \overline{Q}_m(b)$$

On the other hand, f being continuously differentiable at x_0, the Clarke tangent cone to the graph of f is the graph of the derivative $f'(x_0)$. Hence the foregoing assumption can be rewritten in the form

$$\mathrm{Graph}(f'(x_0)) - (\overline{Q}_n(a) \times \overline{Q}_m(b)) = X \times Y$$

By polarity, this is equivalent to the condition

$$(\mathrm{Graph}(f'(x_0)))^- \cap (\overline{Q}_n(a) \times \overline{Q}_m(b))^+ = 0$$

which is nothing other than condition (9.39). □

We can adapt this theorem to the set-valued case, but this requires further definitions and results from set-valued analysis.

[15] If Q is a convex cone,

$$T_Q(x) = \mathrm{cl}(Q + x\mathbf{R}) \supset Q$$

9.7 Set-valued Problems

Let us consider a closed set-valued map $F : \mathbf{R}^n \leadsto \mathbf{R}^m$ and the problem

$$\text{find } x \in \mathbf{R}^n \quad \text{such that } F(x) \ni y$$

We are looking for an existence criterion for standard qualitative solutions to the inclusion $F(x) \ni y$. Given a multisign $b \in \mathcal{R}^m$, does there exist a "qualitative solution" $a \in \mathcal{R}^n$, defined by

$$\exists\, x,\ y \in F(x) \text{ such that } \operatorname{sign}(x) = a \quad \text{and} \quad \operatorname{sign}(y) = b \qquad (9.42)$$

Theorem 9.7.1 *Let $a \in \mathcal{R}^n$, $b \in \mathcal{R}^m$, $F : \mathbf{R}^n \leadsto \mathbf{R}^m$ be a closed set-valued map. Consider $x_0 \in \operatorname{Dom}(F) \cap \overline{Q}_n(a)$ satisfying $F(x_0) \cap \overline{Q}_m(b) \neq \emptyset$. If the criterion*

$$\begin{cases} (0,0) \text{ is the only solution } (p,q) \text{ to } p \in CF(x_0,y_0)^\star(q) \\ \text{satisfying } p \in Q_n^\star(a) \quad \textit{and} \quad q \in -Q_m^\star(b) \end{cases} \qquad (9.43)$$

holds true, then a solves the qualitative inclusion

$$\exists\, x \in Q_n(a) \text{ such that } F(x) \cap Q_m(b) \neq \emptyset \qquad (9.44)$$

Proof The proof is the same as that for the single-valued case. Pick $y_0 \in F(x_0) \cap \overline{Q}_m(b)$; then (x_0, y_0) belongs to the intersection of the graph of F and the closed convex cone $\overline{Q}_n(a) \times \overline{Q}_m(b)$, so that $((x_0,y_0),(x_0,y_0))$ is a solution in $\operatorname{Graph}(F) \times (\overline{Q}_n(a) \times \overline{Q}_m(b))$ to the equation

$$(x_1, y_1) - (x_2, y_2) = 0$$

Let $\mathbf{1} \in \mathbf{R}^n$ denote the unit vector $\mathbf{1} := (1, \ldots, 1)$, and set

$$a\mathbf{1} := (a_i 1)_{i=1,\ldots,n}$$

We shall prove that for some $\varepsilon > 0$ there exists a solution

$$(x_1, y_1) \in \operatorname{Graph}(F), \quad (x_2, y_2) \in \overline{Q}_n(a) \times \overline{Q}_m(b)$$

to the equation

$$(x_1, y_1) - (x_2, y_2) = \varepsilon(a\mathbf{1}, b\mathbf{1})$$

so that $x_1 = x_2 + \varepsilon a\mathbf{1}$ belongs to $Q_n(a)$, and $y_1 = y_2 + \varepsilon b\mathbf{1}$ belongs to $Q_m(b)$ and to $F(x_1)$.

For that purpose, we apply the constrained inverse-function theorem, which states that a solution to the foregoing equation does exist provided the assumption

$$C_{\mathrm{Graph}(F)}(x_0, y_0) - C_{\overline{Q}_n(a)}(x_0) \times C_{\overline{Q}_m(b)}(y_0) = \mathbf{R}^n \times \mathbf{R}^m$$

is satisfied.

Because x_0 belongs to $\overline{Q}_n(a) := a\mathcal{R}^n_+$, the cone $C_{\overline{Q}_n(a)}(x_0)$ coincides with the tangent cone, which contains $Q_n(a)$ [recall that when Q is a convex cone, $T_Q(x) = \mathrm{cl}(Q + x\mathbf{R}) \supset Q$]. In the same way, $C_{\overline{Q}_m(b)}(y_0) \supset \overline{Q}_m(b)$.

The foregoing assumption would follow from

$$\mathrm{Graph}(CF(x_0, y_0)) - (\overline{Q}_n(a) \times \overline{Q}_m(b)) = \mathbf{R}^n \times \mathbf{R}^m$$

By polarity, this is equivalent to the condition

$$(\mathrm{Graph}(CF(x_0, y_0)))^- \cap (\overline{Q}_n(a) \times \overline{Q}_m(b))^+ = \{0\}$$

which is nothing other than condition (9.43). □

Remark When A is a closed convex process, we can apply the foregoing theorem with

$$(x_0, y_0) = (0, 0) \in \mathrm{Graph}(A) \cap (\overline{Q}_n(a) \times \overline{Q}_m(b))$$

Since $CA(0,0) = A$ [because the tangent cone to the closed convex cone $\mathrm{Graph}(A)$ at the origin is equal to this closed convex cone], we deduce from the foregoing theorem the following consequence:

Corollary 9.7.2 *Let $a \in \mathcal{R}^n$, $b \in \mathcal{R}^m$, with A a closed convex process*[16] *from $\mathbf{R}^n$ to $\mathbf{R}^m$. If*

$$\begin{cases} (0,0) \text{ is the only solution } (p,q) \text{ to } p \in A^\star(q) \\ \text{such that } p \in Q_n^\star(a) \quad \textit{and} \quad q \in -Q_m^\star(b) \end{cases}$$

then a solves the qualitative inclusion

$$\exists\, x \in Q_n(a) \text{ such that } A(x) \cap Q_m(b) \neq \emptyset$$

Therefore, criterion (9.43) implies the existence of a qualitative solution to both problem (9.44) and the linearized problem

$$\exists\, u \in Q_n(a) \quad \text{such that } CF(x_0, y_0)(u) \cap Q_m(b) \neq \emptyset \;\; \square$$

[16] Observe that this corollary provides the same criterion for both the initial inclusion and its set-valued linearization.

Let us consider the case when $F : \mathbf{R}^n \leadsto \mathbf{R}^m$ is the set-valued map defined by

$$F(x) := \begin{cases} f(x) - M(x) & \text{when } x \in L \\ \emptyset & \text{when } x \notin L \end{cases}$$

where $f : \mathbf{R}^n \mapsto \mathbf{R}^m$ is continuous, L is a closed subset of $\mathbf{R}^n$, and $M : \mathbf{R}^n \leadsto \mathbf{R}^m$ is a closed set-valued map.

Corollary 9.7.3 *Let $a \in \mathcal{R}^n$, $b \in \mathcal{R}^m$, $M : \mathbf{R}^n \leadsto \mathbf{R}^m$ be a closed set-valued map and $L \subset \mathbf{R}^n$ be a closed set. Assume that x_0 and y_0 satisfy*

$$x_0 \in \overline{Q}_n(a) \cap L \quad \textit{and} \quad y_0 \in (f(x_0) - M(x_0)) \cap \overline{Q}_m(b)$$

that f is continuously differentiable at x_0, that M is Lipschitz at x_0, and that

$$C_L(x_0) - \mathrm{Dom}(CM(x_0, y_0 - f(x_0))) = \mathbf{R}^n$$

The criterion

$$\begin{cases} (0,0) \text{ is the only solution } (p,q) \text{ to} \\ p \in f'(x_0)^\star(q) + CM(x_0, y_0 - f(x_0))^\star(-q) + N_L(x_0) \\ \text{such that } p \in Q_n^\star(a) \quad \textit{and} \quad q \in -Q_m^\star(b) \end{cases} \tag{9.45}$$

implies that a solves the qualitative equation

$$\exists\, x \in Q_n(a) \cap L \quad \textit{and} \quad y \in Q_m(b) \text{ such that } y \in f(x) - M(x)$$

It follows from Theorem 9.7.1 and Proposition A1.10.3.

The particular case when M is constant yields the following:

Corollary 9.7.4 *Let $a \in \mathcal{R}^n$, $b \in \mathcal{R}^m$ and $L \subset \mathbf{R}^n$, $M \subset \mathbf{R}^m$ be closed sets. Assume that x_0 and y_0 satisfy*

$$x_0 \in \overline{Q}_n(a) \cap L \quad \textit{and} \quad y_0 \in (f(x_0) - M) \cap \overline{Q}_m(b)$$

Assume that f is continuously differentiable at x_0. Then the criterion

$$\begin{cases} 0 \text{ is the only solution } q \text{ to} \\ q \in -Q_m^\star(b) \cap N_M(f(x_0) - y_0) \quad \textit{and} \\ f'(x_0)^\star q \in Q_n^\star(a) - N_L(x_0) \end{cases}$$

implies that a solves the qualitative equation

$$\exists\, x \in Q_n(a) \cap L \quad \textit{and} \quad y \in Q_m(b) \text{ such that } f(x) \in y + M$$

Another example is provided by qualitative solutions to *variational inequalities.*

Example: Variational Inequalities Let us consider a closed convex subset $K \subset \mathbf{R}^n$ and a $\mathcal{C}^1$ map f from a neighborhood of K to $\mathbf{R}^n$. An element $y_0 \in \mathbf{R}^n$ being given, we recall that an element $x_0 \in K$ is a solution to the variational inequalities

$$\forall\, x \in K, \ < f(x_0) - y_0,\, x_0 - x > \leq 0$$

if and only if x_0 is a solution to the inclusion

$$y_0 \in f(x_0) + N_K(x_0)$$

where $N_K(x_0)$ is the normal cone to K at x_0.

Theorem 9.7.5 *Let $a \in \mathcal{R}^n$, $b \in \mathcal{R}^m$, and let $x_0 \in \overline{Q}_n(a) \cap K$, $y_0 \in \overline{Q}_m(b)$ satisfy*

$$\forall\, x \in K, \ < f(x_0) - y_0, x_0 - x > \leq 0$$

The criterion

$$\left\{ \begin{array}{l} (0,0) \ \text{is the only solution} \ (p,q) \ \text{to} \\ p \in f'(x_0)^\star(q) + CN_K(x_0, y_0 - f(x_0))^\star(-q) \\ \text{satisfying} \ \ p \in Q_n^\star(a) \quad \textit{and} \quad q \in -Q_m^\star(b) \end{array} \right. \tag{9.46}$$

implies that a solves the qualitative variational inequalities

$$\left\{ \begin{array}{l} \exists\, \overline{x} \in Q_n(a) \cap K \quad \textit{and} \quad \overline{y} \in Q_m(b) \ \text{such that} \\ \qquad \forall\, x \in K, \ < f(\overline{x}) - \overline{y}, \overline{x} - x > \leq 0 \end{array} \right.$$

Proof This a consequence of Proposition A1.10.3, with $M(x) := -N_K(x)$, and Theorem 9.7.1. $\square$

10

Dynamical Qualitative Simulation

Introduction

The purpose of this chapter is to revisit the QSIM algorithm introduced by Kuipers for studying the qualitative evolution of solutions to a differential equation $x' = f(x)$, where the state x ranges over a closed subset K of a finite-dimensional vector space $X := \mathbf{R}^n$.

The *qualitative state* of a solution to the differential equation at a given time t is given by the knowledge of the monotonicity property of each component $x_i(t)$ of a solution $x(\cdot)$ to this differential equation, that is, the knowledge of the sign of the derivatives $x_i'(t)$. Hence the *qualitative behavior* is the evolution of the qualitative states of the solution, that is, the evolution of the vector of signs of the components of $x'(t) = f(x(t))$, which must be determined without solving the differential equation.

We shall study the qualitative behavior of the differential equation, that is, the evolution of the functions $t \mapsto \text{sign}(x'(t))$ associated with solutions $x(\cdot)$ of the differential equation. Furthermore, we shall track down the *landmarks*, that is, the states at which the monotonic behavior of the solutions is modified. But instead of finding them a posteriori by following the qualitative behavior of a given solution, we shall find them a priori, without solving the dynamical system, neither qualitatively nor analytically.

In order to denote the qualitative states and track down their evolution, we introduce the n-dimensional *confluence space* $\mathcal{R}^n$ defined by

$$\mathcal{R}^n \ := \ \{ -, \ 0, \ + \}^n$$

the convex cones where

$$\mathbf{R}^n_a \ := \ \{ \ v \in \mathbf{R}^n \mid \text{sign of } (v_i) = a_i \ \}$$

and their closures

$$a\mathbf{R}^n_+ := \{ v \in \mathbf{R}^n \mid \text{ sign of } (v_i) = a_i \text{ or } 0 \}$$

For studying the qualitative behavior of the differential equation, we introduce the *monotonic cells* defined by

$$K_a := \{x \in K \mid f(x) \in \mathbf{R}^n_a\}$$

Indeed, the quantitative states $x(\cdot)$ evolving in a given monotonic cell K_a share the same monotonicity properties, because as long as $x(t)$ remains in K_a,

$$\forall\, i = 1, \ldots, n, \quad \text{sign of } \frac{dx_i(t)}{dt} = a_i$$

These monotonic cells are examples of what we can call *qualitative cells* of the subset K. In full generality, qualitative cells are subsets $K_a \subset K$ of a family of subsets covering K. The problem is then to check whether or not a family of qualitative cells is consistent with a differential equation $x' = f(x)$ in the sense that one can find a *discrete dynamical system* Φ mapping each cell to other cells such that every solution starting from one cell K_a arrives in one of the qualitative cells of the image $\Phi(K_a)$.

In other words, the problem arises whether or not we can map the differential equation $x' = f(x)$ to a discrete dynamical system $\Phi : \mathcal{R}^n \leadsto \mathcal{R}^n$ on the qualitative space $\mathcal{R}^n$ that prescribes the transitions between monotonicity cells.

This is not always possible, and we shall conclude this chapter with an extension of a result derived by D. Saari concerning chaos. "Chaos" here means the following property: Given any arbitrary infinite sequence of qualitative cells, there is always one solution that visits these cells in the prescribed order.

To the extent that qualitative cells describe phenomena in the framework of the model described by such a differential equation, this discrete dynamical system Φ provides *causality* relations, by specifying the phenomena caused by a given phenomenon. In this sense, we are able to deduce from the model "physical laws." This is one of the main motivations providing the name for this topic: *qualitative physics*.

But first we shall characterize the qualitative equilibria, which are the qualitative cells such that the solutions that arrive in a qualitative cell remain in that cell. We shall also single out the qualitative repellers, which are qualitative cells such that any solution that arrives in a qualitative cell must leave that cell in finite time. We shall then provide conditions to ensure that the qualitative cells are not empty.

The theoretical results concerning this version of the QSIM algorithm are illustrated by software due to Olivier Dordan. It operates on a class of differential equations called "replicator systems," which play an important role in biochemistry and biology. This software provides the monotonic cells and draws them on the screen of the computer for three-dimensional systems (the subset K being the probability simplex). It also supplies symbolically the transitions from one monotonic cell to the others. Finally, it provides a LaTeX report providing the list of qualitative cells, singling out qualitative equilibria and describing the discrete dynamical system Φ.

10.1 Monotonic Cells

We posit the assumptions of the viability theorem for differential equations (called the *Nagumo Theorem*):

$$\begin{cases} \text{(i)} & f \text{ is continuous with linear growth} \\ \text{(ii)} & K \text{ is a closed viability domain} \end{cases} \tag{10.1}$$

Therefore, from every initial state $x_0 \in K$ starts a solution to the differential equation

$$x'(t) \ = \ f(x(t)) \tag{10.2}$$

viable (remaining) in K.

10.1.1 Monotonic Behavior of the Components of the State

For studying the qualitative behavior of the differential equation (10.2), that is, the evolution of the functions $t \mapsto \text{sign}(x'(t))$ associated with solutions $x(\cdot)$ of the differential equation, we split the viability domain K of the differential equation into 3^n *monotonic cells* K_a and *large monotonic cells* $\overline{K}_a$, defined by

$$K_a \ := \ \{x \in K \mid f(x) \in \mathbf{R}^n_a\} \quad \text{and} \quad \overline{K}_a \ := \ \{x \in K \mid f(x) \in a\mathbf{R}^n_+\}$$

Indeed, the quantitative states $x(\cdot)$ evolving in a given monotonic cell K_a share the same monotonicity properties, because as long as $x(t)$ remains in K_a,

$$\forall\, i = 1, \ldots, n, \quad \text{sign of } \frac{dx_i(t)}{dt} \ = \ a_i$$

The monotonic cell K_0 is then the set of equilibria[1] of the system, because $K_0 = \{ x \in K \mid f(x) = 0 \}$.

These monotonic cells are examples of *qualitative cells* and for that reason often are called qualitative cells.

Studying the qualitative evolution of the differential equation amounts to determining the laws (if any) that govern the transition from one monotonic cell K_a to other cells without solving the differential equation.

In Kuipers's terminology, the boundaries of the monotonic cells are called the *landmarks*. They are naturally unknown and are derived through the formulas defining these monotonic cells. In the forthcoming algorithms we shall compute them, before studying the transition properties from one cell to another (or to others).

These laws thus *reveal causality relations* between qualitative phenomena concealed in the dynamical system, by specifying the successors of each monotonic cell, and constitue a major area of interest in physics for making some sense out of the maze of qualitative properties.

First, we mention the following result, due to O. Dordan, stating that starting from any monotonic cell, a solution either converges to an equilibrium or leaves the monotonic cell in finite time:

Theorem 10.1.1 *Assume that a monotonic cell K_a is not empty and is bounded. Then, for any initial state $x_0 \in K_a$, either the solution leaves K_a in finite time or it converges to an equilibrium.*

Proof Assume that a solution $x(\cdot)$ remains in K_a for all nonnegative $t \geq 0$.

Let any i be such that $a_i \neq 0$. Since

$$x_i(t) - x_i(0) = \int_0^t x_i'(\tau) d\tau$$

we deduce that $x_i(t)$ is monotonic and bounded. Therefore, it converges to some number x_i when $t \to +\infty$. Consequently, each component of the solution $x(\cdot)$ either is constant or converges, so that $x(t)$ converges to a limit, which is then an equilibrium of the dynamical system. □

[1]Such an equilibrium exists whenever the viability domain K is convex and compact, thanks to the Brouwer-Ky Fan theorem.

10.1.2 Monotonic Behavior of Observations of the State

But before proceeding further, we shall generalize our problem - free of any mathematical cost - to take care of physical considerations. Instead of studying the monotonicity properties of each component $x_i(\cdot)$ of the state of the system under investigation, which can be too numerous, we shall study only the monotonicity properties of m functionals $V_j(x(\cdot))$ on the state (e.g., energy or entropy functionals in physics, observations in control theory, various economic indexes in economics), which do matter. The previous case is the particular case in which we take the n functionals V_i defined by $V_i(x) := x_i$.

We shall assume for simplicity that these functionals V_j are continuously differentiable around the viability domain K. We denote by $\mathbf{V}$ the map from X to $Y := \mathbf{R}^m$, defined by

$$\mathbf{V}(x) \ := \ (V_1(x), \ldots, V_m(x))$$

Since the derivative of the observation $\mathbf{V}(x(\cdot))$ is equal to $\mathbf{V}'(x(\cdot))x'(\cdot) = \mathbf{V}'(x(\cdot))f(x(\cdot))$, it will be convenient to set

$$\forall\, x \in K, \ g(x) \ := \ \mathbf{V}'(x)f(x)$$

Hence, we associate with each qualitative state a the qualitative cells K_a and the large qualitative cells $\overline{K}_a$ defined by

$$K_a \ := \ \{x \in K \mid g(x) \in \mathbf{R}^m_a\} \quad \text{and} \quad \overline{K}_a \ := \ \{x \in K \mid g(x) \in a\mathbf{R}^m_+\}$$

In other words, the quantitative states $x(\cdot)$ evolving in a given monotonic cell K_a share the same monotonicity properties of their observations, because as long as $x(t)$ remains in K_a,

$$\forall\, j = 1, \ldots, m, \quad \text{sign of } \frac{d}{dt}V_j(x(t)) \ = \ a_j$$

In particular, the m functions $V_j(x(t))$ remain constant while they evolve in the qualitative cell K_0.

By using observation functionals chosen in such a way that many qualitative cells are empty, the study of transitions may be drastically simplified; this is a second reason to carry our study into this more general setting.

This is the case, for instance, when the observation functionals are *Lyapunov functions* $V_j : K \mapsto \mathbf{R}$. We recall that V is a Lyapunov function if $< V'(x), f(x) >\leq 0$ for all $x \in K$, so that $V(x(\cdot))$ decreases along the solutions to the differential equation. Hence, if the observation functionals are Lyapunov functions, the qualitative cells K_a are empty whenever a component a_i is positive. In this case, we have at most 2^m nonempty qualitative cells. (In some sense, one can say that Lyapunov was the originator of qualitative simulation a century ago.)

Naturally, we would like to know directly the laws that govern the transition from one qualitative cell K_a to other qualitative cells, without solving the differential equation, and therefore without knowing the state of the system, but only some of its properties.

10.2 Transitions between Qualitative Cells

We shall assume from now on that f is continuously differentiable and that the m functions V_j are twice continuously differentiable around the viability domain K. Let us denote by $s : K \mapsto \mathcal{C}^1(0,\infty;X)$ the "solution map" associating with each initial state $x_0 \in K$ the solution $s(\cdot)x_0$ to the differential equation (10.2) starting at x_0.

Definition 10.2.1 *Let us consider a map f from K to X and m observation functionals $V_j : K \mapsto \mathbf{R}$. We denote by $\mathcal{D}(f,\mathbf{V})$ the subset of qualitative states $a \in \mathcal{R}^n$ such that the associated qualitative cell K_a is not empty. We shall say that a qualitative state $c \in \mathcal{D}(f,\mathbf{V})$ is a "successor" of $b \in \mathcal{D}(f,\mathbf{V})$ if for all initial states $x_0 \in \overline{K}_b \cap \overline{K}_c$, there exists $\tau \in]0,+\infty]$ such that $s(t)x_0 \in K_c$ for all $t \in]0,\tau[$.*

A qualitative state $a \in \mathcal{D}(f,\mathbf{V})$ is said to be a "qualitative equilibrium" if it is its own successor. It is said to be a "qualitative repeller" if for any initial state $x_0 \in \overline{K}_a$, there exists $t > 0$ such that $s(t)x_0 \notin \overline{K}_a$.

Our first objective is to express the fact that c is a successor of b through a set-valued map Φ. For that purpose, we shall set

$$h(x) := g'(x)f(x) = V''(x)(f(x),f(x)) + V'(x)f'(x)f(x)$$

We introduce the notation

$$\overline{K}_a^i := \{ x \in \overline{K}_a \mid g(x)_i = 0 \}$$

(Naturally, $\overline{K}_a = \overline{K}_a^i$ whenever $a_i = 0$.)

We shall denote by Γ the set-valued map from $\mathcal{R}^m$ to itself defined by

$$\forall a \in \mathcal{R}^m, \quad (\Gamma(a))_i \text{ is the set of signs of } h_i(x) \text{ when } x \in \overline{K}_a^i$$

We also set $I_0(x) := \{ i = 1,\dots,m \mid g(x)_i = 0 \}$ and

$$\mathbf{R}_+^{I_0(x)} := \{ v \in \mathbf{R}^m \mid v_i \geq 0 \ \forall i \in I_0(x) \}$$

We introduce the operations $\wedge$ on $\mathcal{R}^m$, defined by

$$(b \wedge c)_i \; := \; \begin{cases} b_i & \text{if } b_i = c_i \\ 0 & \text{if } b_i \neq c_i \end{cases}$$

and the set-valued operation $\vee$, where $b \vee c$ is the subset of qualitative states a such that

$$a_i \; := \; b_i \quad \text{or} \quad c_i$$

We set

$$a \# b \iff \forall i = 1, \ldots, m, \qquad a_i \neq b_i$$

Proposition 10.2.2 *The set-valued map Γ satisfies the consistency property*

$$\Gamma(a \vee 0) \; \subset \; \Gamma(a)$$

and thus

$$\Gamma(b \wedge c) \; \subset \; \Gamma(b) \cap \Gamma(c)$$

Proof To say that $\overline{K}_b$ is contained in $\overline{K}_a$ amounts to saying that b belongs to $a \vee 0$. When this is the case, we deduce that for all $i = 1, \ldots, m$, $\overline{K}_b^i \subset \overline{K}_a^i$, so that the signs taken by $h(x)_i$ when x ranges over $\overline{K}_b^i$ belong to the set $\Gamma(a)_i$ of signs taken by the same function over $\overline{K}_a^i$. Therefore, $\Gamma(b)$ is contained in $\Gamma(a)$.

Since $b \wedge c$ belongs to both $b \vee 0$ and $c \vee 0$, we deduce that $\Gamma(b \wedge c)$ is contained in both $\Gamma(b)$ and $\Gamma(c)$. □

Definition 10.2.3 *We shall associate with the system $(f, \mathbf{V})$ the discrete dynamical system on the confluence set $\mathcal{R}^m$ defined by the set-valued map $\Phi : \mathcal{R}^m \rightsquigarrow \mathcal{R}^m$ associating with any qualitative state b the subset*

$$\Phi(b) \; := \; \{ c \in \mathcal{D}(f, \mathbf{V}) \mid \Gamma(b \wedge c) \; \subset \; c \vee 0 \}$$

We begin with necessary conditions for a qualitative state c to be a successor of b:

Proposition 10.2.4 *Let us assume that f is continuously differentiable and that the m functions V_j are twice continuously differentiable around the viability domain K. If $c \in \mathcal{D}(f, \mathbf{V})$ is a successor of b, then c belongs to $\Phi(b)$.*

Before proving this proposition, we need the following:

Lemma 10.2.5 *Let us assume that f is continuously differentiable and that the m functions V_j are twice continuously differentiable around the viability domain K. If v belongs to the contingent cone to the cell $\overline{K}_a$ at x, then the condition*

$$v \in T_K(x) \quad \text{and} \quad \forall\, i \in I_0(x), \quad \text{sign of } (g'(x)v)_i = a_i \quad \text{or} \quad 0$$

is satisfied.

The converse is true if we posit the transversality assumption:

$$\forall\, x \in \overline{K}_a, \quad g'(x)C_K(x) - a\mathbf{R}_+^{I_0(x)} = \mathbf{R}^m$$

Proof Since the large qualitative cell $\overline{K}_a$ is the intersection of K with the inverse image by g of the convex cone $a\mathbf{R}_+^m$, we know that the contingent cone to $\overline{K}_a$ at some $x \in \overline{K}_a$ is contained in

$$T_K(x) \cap g'(x)^{-1} T_{a\mathbf{R}_+^m}(g(x))$$

and is equal to this intersection provided that the "transversality assumption"

$$g'(x)C_K(x) - C_{a\mathbf{R}_+^m}(g(x)) = \mathbf{R}^m$$

is satisfied. On the other hand, we know that because $a\mathbf{R}_+^m$ is convex,

$$C_{a\mathbf{R}_+^m}(y) = T_{a\mathbf{R}_+^m}(y) = aT_{\mathbf{R}_+^m}(ay) \supset a\mathbf{R}_+^m$$

Recall that $v \in T_{\mathbf{R}_+^m}(z)$ if and only if

$$\text{whenever } z_j = 0, \text{ then } v_j \geq 0$$

Consequently, $v \in T_{a\mathbf{R}_+^m}(g(x))$ if and only if

$$\text{whenever } g(x)_j = 0, \text{ then sign of } v_j = a_j \text{ or } 0$$

that is, $T_{a\mathbf{R}_+^m}(g(x)) = a\mathbf{R}_+^{I_0(x)}$. Hence v belongs to the contingent cone to $\overline{K}_a$ at x if and only if v belongs to $T_K(x)$ and $g'(x)v$ belongs to $T_{a\mathbf{R}_+^m}(g(x))$, that is, the sign of $(g'(x)v)_j$ is equal to a_j or 0 whenever j belongs to $I_0(x)$. □

Proof of Proposition 10.2.4 Let c be a successor of b. Take any initial state x_0 in $\overline{K}_b \cap \overline{K}_c$ and set $x(t) := s(t)x_0$. We observe that the intersection of two qualitative cells $\overline{K}_b$ and $\overline{K}_c$ is equal to

$$\overline{K}_b \cap \overline{K}_c := \overline{K}_{b \wedge c}$$

Since the solution $x(t)$ to the differential equation crosses the intersection $\overline{K}_{b\wedge c}$ toward $\overline{K_c}$, $f(x_0)$ belongs to the contingent cone $T_{\overline{K}_c}(x_0)$ because

$$\liminf_{h\to 0+} d_{K_c}(x_0 + hf(x_0))/h \leq \liminf_{h\to 0+} \left\| x'(0) - \frac{x(h)-x_0}{h} \right\| = 0$$

By Lemma 10.2.5, this implies that

$$\forall\, x_0 \in \overline{K}_{b\wedge c},\ \ \forall\, i \in I_0(x_0),\ \ \text{sign of } h_i(x_0) \ = \ c_i \ \text{or} \ 0$$

or, equivalently, that

$$\Gamma(b \wedge c) \ \subset \ c \vee 0$$

Hence c belongs to $\Phi(b)$, as it was announced. □

10.3 Qualitative Equilibrium and Repeller

We can characterize the qualitative equilibria of differential equation (10.2).

Theorem 10.3.1 *Let us assume that f is continuously differentiable and that the m functions V_j are twice continuously differentiable around the viability domain K. We posit the transversality assumption*

$$\forall\, x \in \overline{K}_a,\ \ g'(x)C_K(x) - a\mathbf{R}_+^{I_0(x)} \ = \ \mathbf{R}^m$$

Then a is a qualitative equilibrium if and only if a belongs to $\Phi(a)$.

Proof We already know that if a is a qualitative equilibrium, then a belongs to $\Phi(a)$. We shall prove the converse statement, and, for that purpose, observe that saying that a is a qualitative equilibrium amounts to saying that the large qualitative cell $\overline{K}_a$ enjoys the viability property (or is invariant by f). By the Nagumo theorem, this is equivalent to saying that $\overline{K}_a$ is a viability domain, namely, that

$$\forall\, x \in \overline{K}_a,\ \ f(x) \ \in \ T_{\overline{K}_a}(x)$$

By Lemma 10.2.5, knowing that $f(x)$ belongs to the contingent cone $T_K(x)$ by assumption, this amounts to saying that

$$\forall\, x \in \overline{K}_a,\ \ \forall\, i \in I_0(x),\ \ \text{sign of } \ (g'(x)f(x))_i = a_i \ \text{or} \ 0$$

namely, that $\Gamma(a \wedge a) = \Gamma(a) \subset a \vee 0$. Hence, a is a fixed point of Φ. □

When a large qualitative cell $\overline{K}_a$ is not a viability domain of f (i.e., if a is not a qualitative equilibrium), at least a solution will leave the qualitative cell in finite time and thus will reach the boundary of this cell in finite time.

We infer the following from the definition of the viability kernel.

Proposition 10.3.2 *Let us assume that f is continuously differentiable and that the m functions V_j are twice continuously differentiable around the viability domain K. We posit the transversality assumption*

$$\forall\, x \in \overline{K}_a,\ \ g'(x)C_K(x) - a\mathbf{R}_+^{I_0(x)} \ = \mathbf{R}^m$$

- *The qualitative state a is a qualitative repeller if and only if the viability kernel of $\overline{K}_a$ is empty.*
- *If for some $b \in a \vee 0$, the qualitative cell $\overline{K}_b$ is contained in the viability kernel* $\mathrm{Viab}(\overline{K}_a)$*, then a is the only successor of b.*

Proof First, to say that some $x_0 \in \overline{K}_a$ does not belong to the viability kernel of $\overline{K}_a$ means that for some $t > 0$, $s(t)x_0 \notin \overline{K}_a$. If this happens for all $x_0 \in \overline{K}_a$, then obviously a is a qualitative repeller. Second, if $\overline{K}_b \subset \mathrm{Viab}(\overline{K}_a)$, then, for all $x_0 \in \overline{K}_b$, $s(t)x_0 \in \overline{K}_a$ for all $t \geq 0$. Hence a is the only successor of b. □

10.4 The QSIM Algorithm

We shall now distinguish the 2^n "full qualitative states" $a\#0$ from the other qualitative states, the "transition states." When I is a nonempty subset of $N := \{\ 1, \dots, m\ \}$, we associate with a full state $a\#0$ the transition state a^I defined by

$$a_i^I \ := \ \begin{cases} 0 & \text{if } i \in I \\ a_i & \text{if } i \notin I \end{cases}$$

What are the successors, if any, of a given transition state a^I? This question does not always receive an answer, because starting from some initial state $x \in K_{a^I}$, there may exist two sequences $t_n > 0$ and $s_n > 0$ converging to $0+$ such that $x(t_n) \in \overline{K}_a$ and $x(s_n) \notin \overline{K}_a$. We can exclude this pathological phenomenon in two instances. One obviously happens when either a or the transition state a^I is an equilibrium, that is, when

$$\Gamma(a)_i = 0 \ \text{ for } \ i \in I \ \text{ and } \ \Gamma(a)_i \subset \{a_i, 0\} \ \text{ for } \ i \notin I$$

This also happens in the following situation:

Lemma 10.4.1 *Let $a \# 0$ be a full transition state. If $\Gamma(a) \# 0$ (and thus is reduced to a point), then for any transition state a^I there exists a unique successor $b := \Phi(a^I) \# 0$: that is, for all initial states x in the transition cell K_{a^I} there exists $t_2 > 0$ such that, for all $t \in]0, t_2[$, the solution $x(t)$ remains in the full qualitative cell K_b.*

Proof We consider an initial state $x \in K_{a^I}$. If $i \notin I$, then the sign of $g(x)_i$ is equal to $a_i \neq 0$, and thus there exists $\eta_i > 0$ such that the sign of $(g(x(t)))_i$ remains equal to $a_i^I = a_i$ when $t \in [0, \eta_i[$.

If $i \in I$, then $g(x)_i = 0$, and we know that the sign of the derivative $(d/dt) g_i(x(t))_{|t=0} = h_i(x)$ is equal to $\Gamma(a)_i$ and is different from 0. Hence there exists $\eta_i > 0$ such that the sign of $h(x(t))_i$ remains equal to b_i when $t \in]0, \eta_i[$, so that the sign of

$$g_i(x(t)) = \int_{t_1}^{t} h_i(x(\tau))\, d\tau$$

remains equal to $\Gamma(a)_i$ on the interval $]0, \eta_i[$.

Hence we have proved that there exists some $\eta > 0$ such that $x(t) \in K_b$ for $t \in]0, t_2[$, where

$$b_i := \begin{cases} \Gamma(a)_i & \text{when } i \in I \\ a_i & \text{when } i \notin I \end{cases}$$

and where $t_2 := \min_i \eta_i > 0$. □

Definition 10.4.2 *We shall say that the system $(f, \mathbf{V})$ is "strictly filterable" if and only if for all full states $a \in \mathcal{D}(f, \mathbf{V}) \# 0$, either $\Gamma(a) \# 0$ or a is a qualitative equilibrium or all the transition states a^I $(I \neq \emptyset)$ are qualitative equilibria.*

We deduce from Definition 10.4.2 and the preceding observations the following consequence:

Theorem 10.4.3 *Let us assume that f is continuously differentiable, that the m functions V_j are twice continuously differentiable around the viability domain K, and that the system $(f, \mathbf{V})$ is "strictly filterable." Let $a \in \mathcal{R}^m$ be an initial full qualitative state. Then for any initial state $x \in K_a$, the sign vector*

$$a_x(t) := s_m\left(\frac{d}{dt}(\mathbf{V}s(t)x)\right)$$

is a solution to the QSIM algorithm defined in the following way: There exist a sequence of qualitative states a_k satisfying

$$a_0 := a \quad \text{and} \quad a_{k+1} \in \Phi(a_k \vee 0) \tag{10.3}$$

and a sequence of landmarks $t_0 := 0 < t_1 < \ldots < t_n < \ldots$ such that

$$\begin{array}{rcl} \forall\, t \in]t_k, t_{k+1}[, \;\; a(t) & = & a_k \\ a(t_{k+1}) & = & a_k \wedge a_{k+1} \end{array} \tag{10.4}$$

In other words, we know that the vector signs of the variations of the observations of the solutions to differential equation (10.2) evolve according to the set-valued dynamical system (10.3) and stop when a_k is either a qualitative equilibrium or all its transition states a_k^I are qualitative equilibria.

Remark The solutions to the QSIM algorithm (10.3) do not necessarily represent the evolution of the vector signs of the variations of the observations of a solution to the differential equation. Further studies must bring answers allowing one to delete impossible transitions from one full qualitative cell $\overline{K}_a$ to some of its transition cells $K_{a'}$.

Therefore, the QSIM algorithm requires definition of the set-valued map $\Gamma : \mathcal{R}^m \rightsquigarrow \mathcal{R}^m$ by computing the signs of the m functions $h_i(\cdot)$ on the qualitative cells K_a^i for all $i \in N$ and $a \in \mathcal{D}(f, \mathbf{V})\#0$. If by doing so we observe that the system is strictly filterable, then we know that the set-valued dynamical system (10.3) contains the evolutions of the vector signs of the m observations of solutions to the differential equation (10.2).

10.5 Replicator Systems

We begin by studying the viability property of the probability simplex

$$S^n := \left\{ x \in \mathbf{R}_+^n \mid \sum_{i=1}^{n} x_i = 1 \right\}$$

This is the most important example, because in many problems it is too difficult to describe mathematically the state of the system. Then, assuming there is a finite number n of states, one can rather study the evolution of their frequencies, probabilities, concentrations, mixed strategies (in games), and so forth, instead of the evolution of the state itself. We shall provide examples later in this section.

We refer to the first chapter of *Viability Theory* (Aubin 1991) for more details about the replicator systems, which are studied in depth in *The Theory of Evolution and Dynamical Systems* (Hofbauer and Sigmund 1988).

The contingent cone $T_{S^n}(x)$ to S^n at $x \in S^n$ is the cone of elements $v \in \mathbf{R}^n$ satisfying

$$\sum_{i=1}^{n} v_i = 0 \quad \text{and} \quad v_i \geq 0 \ \text{ whenever } \ x_i = 0 \tag{10.5}$$

(see Appendix A, Section A.7).

We shall investigate now how to make viable the evolution of a system for which we know the growth rates $g_i(\cdot)$ of the evolution without constraints (also called "specific growth rates"):

$$\forall\, i = 1, \ldots, n, \ \ x_i'(t) = x_i(t) g_i(x(t))$$

There is no reason[2] for the solutions to this system of differential equations to be viable in the probability simplex. But we can correct it by subtracting from each initial growth rate the common "feedback control $\tilde{u}(\cdot)$" (also called "global flux" in many applications), defined as the weighted mean of the specific growth rates

$$\forall\, x \in S^n, \ \ \tilde{u}(x) := \sum_{j=1}^{n} x_j g_j(x)$$

Indeed, the probability simplex S^n is obviously a viability domain of the new dynamical system, called the "replicator system" (or system "under constant organization"):

$$\begin{aligned} \forall\, i = 1, \ldots, n, \quad x_i'(t) \ &= x_i(t)(g_i(x(t)) - \tilde{u}(x(t))) \\ &= x_i(t) \left(g_i(x(t)) - \sum_{j=1}^{n} x_j(t) g_j(x(t)) \right) \end{aligned} \tag{10.6}$$

An equilibrium α of the replicator system (10.6) is a solution to the system

$$\forall\, i = 1, \ldots, n, \ \ \alpha_i(g_i(\alpha) - \tilde{u}(\alpha)) = 0$$

[2]By Nagumo's theorem, the functions g_i should be continuous and should satisfy

$$\forall\, x \in S^n, \quad \sum_{i=1}^{n} x_i g_i(x) \ = \ 0$$

(such an equilibrium does exist, thanks to the equilibrium theorem). These equations imply that either $\alpha_i = 0$, or $g_i(\alpha) = \tilde{u}(\alpha)$, or both, and that $g_{i_0}(\alpha) = \tilde{u}(\alpha)$ holds true for at least one i_0. We shall say that an equilibrium α is *nondegenerate* if

$$\forall\, i = 1, \ldots, n, \quad g_i(\alpha) \;=\; \tilde{u}(\alpha) \tag{10.7}$$

Equilibria α that are strongly positive (this means that $\alpha_i > 0$ for all $i = 1, \ldots, n$) are naturally nondegenerate.

We associate[3] with any $\alpha \in S^n$ the function V_α defined on the simplex S^n by

$$V_\alpha(x) \;:=\; \prod_{i=1}^{n} x_i^{\alpha_i} \;:=\; \prod_{i \in I_\alpha} x_i^{\alpha_i}$$

where we set $0^0 := 1$ and $I_\alpha := \{i = 1, \ldots, n \mid \alpha_i > 0\}$. Let us denote by S^I the subsimplex of elements $x \in S^n$ such that $x_i > 0$ if and only if $i \in I$.

Theorem 10.5.1 *Let us consider n continuous growth rates g_i. For any initial state $x_0 \in S^n$, there exists a solution to replicator system (10.6) starting from x_0 that is viable in the subsimplex $S^{I_{x_0}}$. The viable solutions satisfy*

$$\forall\, t \geq 0, \; \sum_{i=1}^{n} g_i(x(t)) x_i'(t) \;\geq\; 0 \tag{10.8}$$

and, whenever $\alpha \in S^n$ is a nondegenerate equilibrium,

$$\frac{d}{dt} V_\alpha(x(t)) \;=\; -V_\alpha(x(t)) \sum_{i=1}^{n} (x_i(t) - \alpha_i)(g_i(x(t)) - g_i(\alpha)) \tag{10.9}$$

[3] The reason we introduce this function is that α is the unique maximizer of V_α on the simplex S^n. This follows from the convexity of the function $\varphi := -\log$: Setting $0 \log 0 = 0 \log \infty = 0$, we get

$$\sum_{i=1}^{n} \alpha_i \log \frac{x_i}{\alpha_i} = \sum_{\alpha_i > 0} \alpha_i \log \frac{x_i}{\alpha_i} \leq \log\Big(\sum_{\alpha_i > 0} x_i\Big) \leq \log 1 = 0$$

so that

$$\sum_{i=1}^{n} \alpha_i \log x_i \;\leq\; \sum_{i=1}^{n} \alpha_i \log \alpha_i$$

and thus, $V_\alpha(x) \leq V_\alpha(\alpha)$, with equality if and only if $x = \alpha$.

Proof We first observe that

$$\forall\, x \in S^{I_{x_0}}, \quad \sum_{i \in I_{x_0}} x_i(g_i(x) - \tilde{u}(x)) \; = \; 0$$

because, $x_i = 0$ whenever $i \notin I_{x_0}$, that is, whenever $x_{0_i} = 0$. Therefore, the subsimplex $S^{I_{x_0}}$ is a viability domain of the replicator system (10.6).

Inequality (10.8) follows from Cauchy-Schwarz inequality, because

$$\left(\sum_{i=1}^{n} x_i g_i(x)\right)^2 \leq \left(\sum_{i=1}^{n} x_i\right)\left(\sum_{j=1}^{n} x_i g_i(x)\right)^2 = \sum_{i=1}^{n} x_i g_i(x)^2$$

We deduce formula (10.9) from

$$\begin{aligned} \tfrac{d}{dt} V_\alpha(x(t)) \; &= \; \textstyle\sum_{i \in I_\alpha} \tfrac{\partial}{\partial x_i} V_\alpha(x(t)) x_i'(t) \\ &= \; V_\alpha(x(t)) \sum_{i \in I_\alpha} \alpha_i \frac{x_i'(t)}{x_i(t)} \; = \; V_\alpha \sum_{i=1}^{n} \alpha_i \frac{x_i'(t)}{x_i(t)} \end{aligned}$$

and from

$$\sum_{i=1}^{n} \alpha_i \frac{x_i'(t)}{x_i(t)} \; = \; \sum_{i=1}^{n} (\alpha_i - x_i(t)) g_i(x(t))$$

Then we take into account that α is a nondegenerate equilibrium, because inequality (10.7) implies that

$$\sum_{i=1}^{n} (\alpha_i - x_i(t)) g_i(\alpha) \; = \; 0$$

□

Remark When the specific growth rates are derived from a differentiable potential function U by

$$\forall\, i = 1, \ldots, n, \quad g_i(x) \; := \; \frac{\partial U}{\partial x_i}(x)$$

condition (10.8) implies that

$$\forall\, t \geq 0, \quad \frac{dU}{dt}(x(t)) \; \geq \; 0$$

because

$$\frac{dU}{dt}(x(t)) = \sum_{i=1}^{n} \frac{\partial U}{\partial x_i}(x(t)) x_i'(t) = \sum_{i=1}^{n} g_i(x(t)) x_i'(t) \geq 0$$

Therefore, the potential function U does not decrease along the viable solutions to the replicator system (10.6). Furthermore, when this potential function U is homogeneous with degree p, Euler's formula implies that

$$\tilde{u}(x) \;=\; pU(x)$$

[because $\sum_{i=1}^n x_i(\partial/\partial x_i)U(x) = pU(x)$], so that in this case the global flux $\tilde{u}(x(t))$ also does not decrease along the viable solutions to the replicator system (10.6).

On the other hand, if we assume that the growth rates g_i are "decreasing" in the sense that

$$\forall\, x, y \in S^n, \quad \sum_{i=1}^{n}(x_i - y_i)(g_i(x) - g_i(y)) \;\leq\; 0$$

then inequality (10.9) implies that for any nondegenerate equilibrium $\alpha \in S^n$,

$$\forall\, t \geq 0, \qquad \frac{dV_\alpha}{dt}(x(t)) \;\geq\; 0$$

When $g(x) := U'(x)$ is derived from a concave differentiable potential U, it is decreasing, so that for a concave potential, both $U(x(\cdot))$ and $V_\alpha(x(\cdot))$ are increasing. □

Example: Replicator Systems for Constant Growth Rates The simplest example is the one where the specific growth rates $g_i(\cdot) \equiv a_i$ are constant. Hence we correct the constant-growth systems $x_i' = a_i x_i$ whose solutions are equal to exponential $x_{0_i} e^{a_i t}$ by the zero-order replicator system

$$\forall\, i = 1, \ldots, n, \;\; x_i'(t) \;=\; x_i(t)(a_i - \sum_{j=1}^{n} a_j x_j(t))$$

whose solutions are given explicitly by

$$x_i(t) \;=\; \frac{x_{0_i} e^{a_i t}}{\sum_{j=1}^{n} x_{0_j} e^{a_j t}} \;\; \text{whenever } x_{0_i} > 0$$

[and $x_i(t) \equiv 0$ whenever $x_{0_i} = 0$]. □

Example: Replicator Systems for Linear Growth Rates The next class of examples is provided by linear growth rates

$$\forall\, i = 1, \ldots, n, \;\; g_i(x) \;:=\; \sum_{j=1}^{n} a_{ij} x_j$$

Let A denote the matrix whose entries are the foregoing a_{ij}'s. Hence the global flux can be written

$$\forall\, x \in S^n, \quad \tilde{u}(x) \;=\; \sum_{k,l=1}^{n} a_{kl} x_k x_l \;=\; < Ax, x >$$

Hence, first-order replicator systems can be written[4]

$$\forall\, i = 1, \ldots, n, \quad x_i'(t) \;=\; x_i(t) \left(\sum_{j=1}^{n} a_{ij} x_j(t) - \sum_{k,l=1}^{n} a_{kl} x_k(t) x_l(t) \right)$$

Such systems have been investigated independently in

- population genetics (allele frequencies in a gene pool)
- theories of prebiotic evolution of self-replicating polymers (concentrations of polynucleotides in a dialysis reactor)
- sociobiological studies of evolutionary stable traits in animal behavior (distributions of behavioral phenotypes in a given species)
- population ecology (densities of interacting species)

10.6 Qualitative Simulation of Replicator Systems

Qualitative analysis of replicator systems had been carried out by Olivier Dordan, who has designed software that provides the transition matrix, qualitative equilibria, and repellers for any first-order replicator system. In the three-dimensional case, the computer program draws the qualitative cells in the two-dimensional simplex S^3.

Let A denote the matrix whose entries are a_{ij}. First-order replicator systems can be written

$$\forall\, i = 1, \ldots, n, \quad x_i'(t) \;=\; x_i(t) \left(\sum_{j=1}^{n} a_{ij} x_j(t) - \sum_{k,l=1}^{n} a_{kl} x_k(t) x_l(t) \right) \tag{10.10}$$

We infer that the boundaries of the qualitative cells are *quadratic manifolds*, since they are given by the equations

$$\forall\, i = 1, \ldots, n, \quad \sum_{j=1}^{n} a_{ij} x_j - \sum_{k,l=1}^{n} a_{kl} x_k x_l \;=\; 0$$

When the matrix A is entered in the software, it computes the qualitative

[4]Observe that if for each i, all the a_{ij} are equal to b_i, we find zero-order replicator systems.

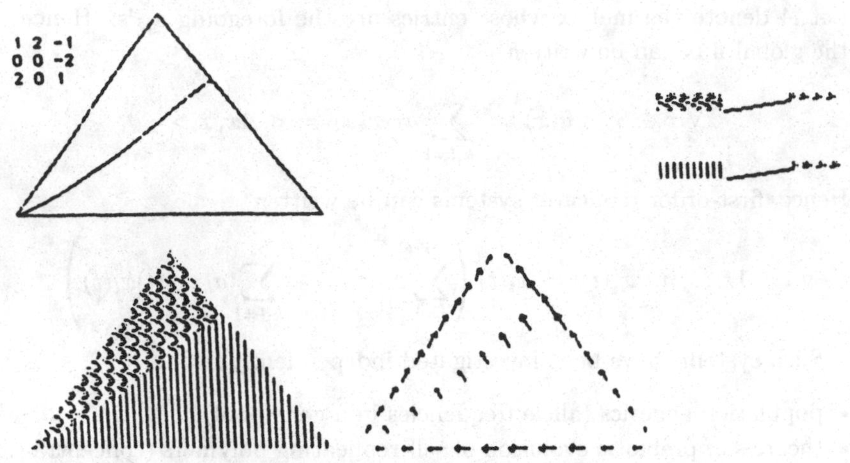

Fig. 10.1 Qualitative Simulation of Replicator Systems # 1. Example of qualitative cells of a replicator system for the matrix

$$A = \begin{pmatrix} 1.00 & 2.00 & -1.00 \\ .00 & .00 & -2.00 \\ 2.00 & .00 & 1.00 \end{pmatrix}$$

cells (and thus supplies all the landmarks), singles out the qualitative equilibria, and furnishes symbolically the qualitative transition map Φ. It also delivers $\LaTeX$ reports such as these:

Example 1 Let the matrix A involved in the replicator system (10.10) be (Figure 10.1.)

$$A = \begin{pmatrix} 1.00 & 2.00 & -1.00 \\ .00 & .00 & -2.00 \\ 2.00 & .00 & 1.00 \end{pmatrix}$$

Qualitative results: There are 2 nonempty "full" qualitative cells.

Computation of the qualitative system Φ:

$$\begin{bmatrix} \Phi(-,-,+) & = & (-,-,+) \\ \Phi(+,-,+) & = & (-,-,+) \end{bmatrix}$$

The following qualitative set is a qualitative equilibrium:

$$(-,-,+)$$

Computation of the set-valued map Γ:

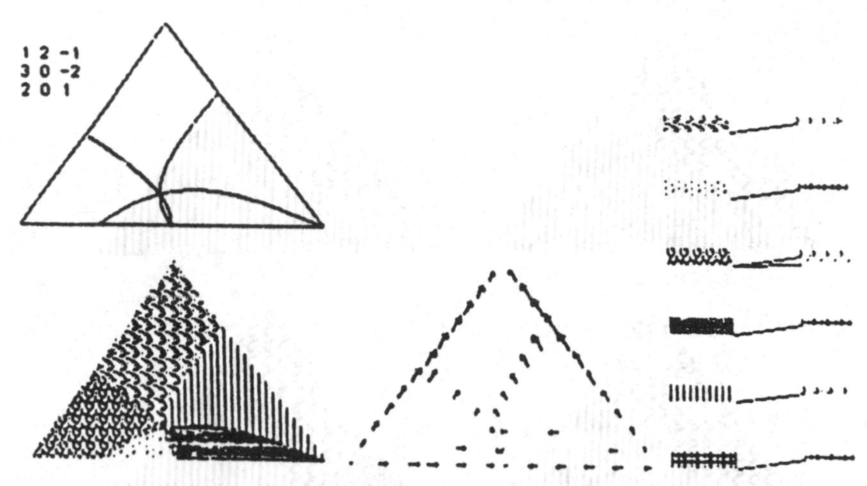

Fig. 10.2 Qualitative Simulation of Replicator Systems # 2 Example of qualitative cells of a replicator system for the matrix

$$A = \begin{pmatrix} 1.00 & 2.00 & -1.00 \\ 3.00 & .00 & -2.00 \\ 2.00 & .00 & 1.00 \end{pmatrix}$$

$$\begin{aligned} \Gamma(-,-,+) &= (\{-\},\{\emptyset\},\{\emptyset\}) \\ \Gamma(0,-,+) &= (\{-\},\{\emptyset\},\{\emptyset\}) \\ \Gamma(0,0,0) &= (\{0\},\{0\},\{\emptyset\}) \\ \Gamma(+,-,+) &= (\{-\},\{\emptyset\},\{\emptyset\}) \end{aligned}$$

Example 2 Let the matrix A involved in the replicator system (10.10) be (Figure 10.2)

$$A = \begin{pmatrix} 1.00 & 2.00 & -1.00 \\ 3.00 & .00 & -2.00 \\ 2.00 & .00 & 1.00 \end{pmatrix}$$

In Figure 10.3 we provide the qualitative cells for different values of an entry of the matrix A. Qualitative results: There are 6 nonempty "full" qualitative cells.

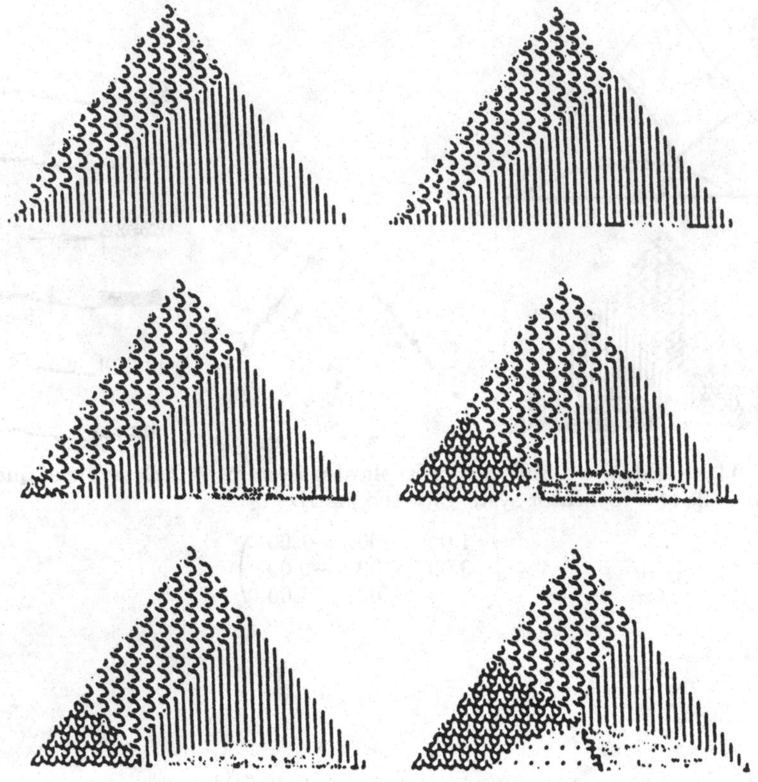

Fig. 10.3 Qualitative Simulation of Replicator Systems # 3. Examples of qualitative cells of replicator systems when the entry $a_{2,1}$ of the matrix

$$A = \begin{pmatrix} 1.00 & 2.00 & -1.00 \\ a_{2,1} & .00 & -2.00 \\ 2.00 & .00 & 1.00 \end{pmatrix}$$

varies from 0 to 3. One can observe cells appearing one after the other and that the transition properties from on cell to the others are preserved.

Computation of the qualitative system Φ:

$$\left[\begin{array}{lcl} \Phi(-,-,+) & = & (-,-,+) \\ \Phi(+,-,+) & = & (-,-,+) \\ \Phi(-,+,-) & = & (+,+,-) \\ \Phi(-,+,+) & = & \left\{\begin{array}{l}(-,-,+) \\ (-,+,-)\end{array}\right. \\ \Phi(+,-,-) & = & (+,+,-) \\ \Phi(+,+,-) & = & (+,+,-) \end{array}\right]$$

The following qualitative sets are qualitative equilibria:

$$(-,-,+)$$
$$(+,+,-)$$

Computation of the set-valued map Γ:

$$\begin{array}{rcl}
\Gamma(-,-,+) & = & (\{-,0\},\{-,0\},\{0\}) \\
\Gamma(-,0,+) & = & (\{0\},\{-,0\},\{0\}) \\
\Gamma(-,+,-) & = & (\{0,+\},\{0\},\{-,0\}) \\
\Gamma(-,+,0) & = & (\{0\},\{0\},\{-,0\}) \\
\Gamma(-,+,+) & = & (\{0\},\{-,0\},\{-,0\}) \\
\Gamma(0,-,+) & = & (\{-,0\},\{0\},\{0\}) \\
\Gamma(0,0,0) & = & (\{0\},\{0\},\{0\}) \\
\Gamma(0,+,-) & = & (\{0,+\},\{0\},\{0\}) \\
\Gamma(+,-,-) & = & (\{0\},\{0,+\},\{-,0,+\}) \\
\Gamma(+,-,0) & = & (\{0\},\{0\},\{-,0,+\}) \\
\Gamma(+,-,+) & = & (\{-,0\},\{0\},\{-,0,+\}) \\
\Gamma(+,0,-) & = & (\{0\},\{0,+\},\{0\}) \\
\Gamma(+,+,-) & = & (\{0,+\},\{0,+\},\{0\})
\end{array}$$

10.7 Nonemptiness and Singularity of Qualitative Cells

The question we answer now is whether or not these qualitative cells are nonempty.

Theorem 10.7.1 *Let us assume that f is continuously differentiable and that the m functions V_j are twice continuously differentiable around the viability domain K. Let $\bar{x}$ belong to the qualitative cell K_0. We posit the transversality condition:*

$$g'(\bar{x})C_K(\bar{x}) - a\mathbf{R}^m_+ = \mathbf{R}^m$$

Then the qualitative cell K_a is nonempty, and $\bar{x}$ belongs to its closure. In particular, if

$$g'(\bar{x})C_K(\bar{x}) = \mathbf{R}^m$$

then the 3^m qualitative cells K_a are nonempty. (We have a prechaotic situation, since every qualitative behavior can be implemented as an initial qualitative state.)

Proof We apply the constrained inverse-function theorem [see Theorem 4.3.1 in *Set-Valued Analysis* (Aubin and Frankowska 1990)] to the map $(x, y) \mapsto g(x) - y$ from $X \times Y$ to Y restricted to the closed subset $K \times a\mathbf{R}_+^m$ at the point $(\bar{x}, 0)$. Its Clarke tangent cone is equal to the product $C_K(\bar{x}) \times a\mathbf{R}_+^m$ since

$$C_{a\mathbf{R}_+^m}(0) \;=\; a\mathbf{R}_+^m$$

Therefore, we know that there exists $\varepsilon > 0$ such that, for all $z \in \varepsilon[-1, +1]^m$, there exist an element $x \in K$ and an element $y \in a\mathbf{R}_+^m$ satisfying $g(x) - y = z$ and $\|x - \bar{x}\| + \|y\| \leq l\|z\|$. Taking in particular $z_i = a_i\varepsilon$, we see that $g(x)_i = a_i\varepsilon + y_i$, and thus the sign of $g(x)_i$ is equal to a_i for all $i = 1, \ldots, m$. Hence x belongs to K_a, and $\|x - \bar{x}\| \leq l\varepsilon$. □

Let $\bar{x}$ belong to K_0. We shall say that the qualitative cell $\overline{K}_a$ is "singular" at $\bar{x}$ if $\bar{x}$ is locally the only point of the qualitative cell $\overline{K}_a$, that is, if there exists a neighborhood $N(\bar{x})$ of $\bar{x}$ such that

$$\forall\, x \in N(\bar{x}) \cap K, \;\; x \neq \bar{x}, \;\; g(x) \;\notin\; a\mathbf{R}_+^m$$

Theorem 10.7.2 *Let us assume that f is continuously differentiable and that the m functions V_j are twice continuously differentiable around the viability domain K. Let $\bar{x}$ belong to the qualitative cell K_0. We posit the following assumption:*

$$T_K(\bar{x}) \cap (g'(\bar{x})^{-1}(a\mathbf{R}_+^m)) \;=\; 0$$

Then the qualitative cell $\overline{K}_a$ is singular at $\bar{x}$.

Proof Assume the contrary: For all $n > 0$, there exists $x_n \in K \cap B(\bar{x}, 1/n)$, $x_n \neq \bar{x}$, such that $g(x_n)$ does belong to $a\mathbf{R}_+^m$. Let us set $h_n := \|x_n - \bar{x}\| > 0$, which converges to 0, and $v_n := \|(x_n - \bar{x})/h_n\|$. Since v_n belongs to the unit ball, which is compact, a subsequence, again denoted v_n, converges to some element v of the unit ball. This limit v belongs also to the contingent cone $T_K(\bar{x})$, because for all $n > 0$, $\bar{x} + h_n v_n = x_n$ belongs to K.

Finally, since $g(\bar{x} + h_n v_n) = g(x_n) \in a\mathbf{R}_+^m$ for all $n > 0$ and $g(\bar{x}) = 0$, we infer that the limit $g'(\bar{x})v$ of the difference quotients

$[g(\bar{x}+h_n v_n) - g(\bar{x})]/h_n \in a\mathbf{R}_+^m$ belongs to $a\mathbf{R}_+^m$. Hence we have proved the existence of a nonzero element

$$v \in T_K(\bar{x}) \cap (g'(\bar{x}))^{-1}(a\mathbf{R}_+^m)$$

a contradiction of the assumption. □

10.8 General Qualitative Cells

Let us consider the case in which K is covered by a finite family $\{K_a\}_{a\in\mathcal{A}}$ of arbitrary closed "qualitative cells" $K_a \subset K$ with nonempty interior:

$$K = \bigcup_{a\in\mathcal{A}} K_a$$

Let $f : K \mapsto X$ be a continuous function with linear growth enjoying the uniqueness property. We denote by $s_f(\cdot)x$ the solution to the differential equation $x' = f(x)$ starting at x when $t = 0$. It is possible to investigate the qualitative behavior of the system by introducing the following tools:

10.8.1 Characterization of Successors

We denote by

$$\widehat{K} := X\backslash\mathrm{Int}(K) = \overline{X\backslash K}$$

the complement of the interior of K and by

$$\partial K := \overline{K} \cap \widehat{K}$$

the boundary of K. We observe that K is the closure of its interior if and only if $X\backslash K$ is the interior of $\widehat{K}$. We introduce the Dubovitsky-Miliutin cone defined as follows:

Definition 10.8.1 *The Dubovitsky-Miliutin tangent cone $D_K(x)$ to K is defined by*

$$\begin{cases} v \in D_K(x) \text{ if and only if} \\ \exists\ \varepsilon > 0,\ \exists\ \alpha > 0 \text{ such that } x+]0,\alpha](v+\varepsilon B) \subset K \end{cases}$$

Lemma 10.8.2 *For any x in the boundary of K, the Dubovitsky-Miliutin cone $D_K(x)$ to K at x is the complement of the contingent cone $T_{X\backslash K}(x)$ to the complement $X\backslash K$ of K at $x \in \partial K$:*

$$\forall\, x \in \partial K,\ D_K(x) = X \setminus T_{X\backslash K}(x)$$

We need the following characterization of the contingent cone to the boundary:

Theorem 10.8.3 (Quincampoix) *Let K be a closed subset of a normed space and $\widehat{K}$ denote the closure of its complement. Then*

$$\forall\, x \in \partial K, \quad T_{\partial K}(x) \;=\; T_K(x) \cap T_{\widehat{K}}(x)$$

so that the whole space can be partitioned in the following way:

$$\forall\, x \in \partial K, \;\; D_{\mathrm{Int}(K)}(x) \,\cup\, D_{X\backslash K}(x) \,\cup\, T_{\partial K}(x) \;=\; X$$

Proof If the interior of K is empty, $\partial K = K$, so that the formula holds true. Assume that the interior of K is not empty, and take any $x \in \partial K$. Since inclusion $T_{\partial K}(x) \subset T_K(x) \cap T_{\widehat{K}}(x)$ is obviously true, we have to prove that any u in the intersection $T_K(x) \cap T_{\widehat{K}}(x)$ is contingent to the boundary ∂K at x. Indeed, there exist sequences $k_n > 0$ and $l_n > 0$ converging to $0+$ and sequences $v_n \in X$ and $w_n \in X$ converging to u such that

$$\forall\, n \geq 0, \quad x + k_n v_n \in K \quad \text{ and } \quad x + l_n w_n \in \widehat{K}$$

We shall prove that there exists $\lambda_n \in [0,1]$ such that, setting

$$h_n \;:=\; \lambda_n k_n + (1-\lambda_n) l_n \in [\min(k_n, l_n), \max(k_n, l_n)]$$

and

$$u_n \;:=\; \frac{\lambda_n k_n v_n + ((1-\lambda_n) l_n) w_n}{\lambda_n k_n + (1-\lambda_n) l_n}$$

we have

$$\forall\, n \geq 0, \quad x + h_n u_n \in \partial K$$

Indeed, we can take λ_n either 0 or 1 when either $x + k_n v_n$ or $x + l_n w_n$ belongs to the boundary. If not, $x + k_n v_n \in \mathrm{Int}(K)$ and $x + l_n w_n \in X\backslash K$. Since the interval $[0,1]$ is connected, it cannot be covered by the two nonempty disjoint open subsets

$$\Omega_+ \;:=\; \{\lambda \in [0,1] \mid x + \lambda k_n v_n + (1-\lambda) l_n w_n \;\in \mathrm{Int}(K)\}$$

and

$$\Omega_- \;:=\; \{\lambda \in [0,1] \mid x + \lambda k_n v_n + (1-\lambda) l_n w_n \;\in X\backslash K\}$$

Then there exists $\lambda_n \in [0,1]\backslash(\Omega_+ \cup \Omega_-)$ so that

$$x + \lambda_n k_n v_n + (1-\lambda_n) l_n w_n = x + h_n u_n \;\in\; \partial K$$

Because $h_n > 0$ converges to $0+$ and u_n converges to u, we infer that u belongs to the contingent cone to ∂K. □

This formula and Lemma 10.8.2 imply the decomposition formula.

We then can split the boundary of ∂K into three areas depending on f:

$$K_{\Leftarrow} := \{ x \in \partial K \mid f(x) \in D_{\mathrm{Int}(K)}(x) \} \quad \text{(the inward area)}$$

$$K_{\Rightarrow} := \{ x \in \partial K \mid f(x) \in D_{X \setminus K}(x) \} \quad \text{(the outward area)}$$

$$K_{\Leftrightarrow} := \{ x \in \partial K \mid f(x) \in T_{\partial K}(x) \neq \emptyset \}$$

Proposition 10.8.4

1. *Whenever $x \in K_{\Leftarrow}$, the solution starting at x must enter the interior of K on some open time interval $]0,T[$, and whenever $x \in K_{\Rightarrow}$, the solution starting at x must leave the subset K on some $]0,T[$.*
2. *If $x \in K_{\Leftrightarrow}$, if $\partial K \cap (x + rB) \subset K_{\Leftrightarrow}$ for some $r > 0$ and if f is Lipschitz around x, then the solution starting at x remains in the boundary ∂K on some $[0,T]$.*

Proof

1. Let $x \in K_{\Rightarrow}$, for instance. Then we shall prove that there exist $\rho_x > 0$ and $T_x > 0$ such that

$$\forall\, t \in [0, T_x], \ \ d(s_f(t)x, \partial K) \ \geq \ \rho_x t$$

Indeed, because $f(x) \in D_K(x)$, we associate

$$\rho_x \ := \ \liminf_{h \to 0+} \frac{d\left(x + hf(x), \widehat{K}\right)}{2h} > 0$$

This implies that there exists $\tau_x > 0$ such that

$$\forall\, h \in]0, \tau_x], \ \ d(x + hf(x), \widehat{K}) \ \geq \ 2\rho_x h$$

and thus that

$$\forall\, h \in]0, t_x], \ \ d(x + h(f(x) + \rho_x B), \widehat{K}) \ \geq \ \rho_x h$$

Let us consider now the solution $s_f(\cdot)x$. Because f is continuous, we know that $f(z) \subset f(x) + \rho_x B$ whenever $\|z - x\| \leq \eta_x$ for some η_x. Since f is bounded by a constant $c > 0$ on the ball $B(x, \eta_x)$, we infer that

$$\|x(t) - x\| \ \leq \ \int_0^t \|f(x(s))\|\, ds \ \leq \ ct \leq \eta_x$$

when $t \leq T_x := \min\{t_x, \eta_x/c\}$. In this case, we observe that $x(t) - x \in t(f(x) + \rho_x B)$, so that for any $t \in]0, T_x]$,

$$d(x(t), \widehat{K}) = d(x + x(t) - x, \widehat{K}) \geq d(x + t(f(x) + \rho_x B), \widehat{K}) \geq \rho_x t$$

In the same way, we deduce that when $x \in K_{\Leftarrow}$, the solution $s(\cdot)x$ belongs to $X \backslash K$ for $t \in]0, T_x]$.

2. Take now $x \in K_{\Leftrightarrow}$. We set $g(t) := d_{\partial K}(x(t))$. Because it is Lipschitz, it is differentiable almost everywhere. Let t be such that $g'(t)$ exists. There exists $\varepsilon(h)$ converging to zero with h such that

$$x(t+h) = x(t) + hf(x(t)) + h\varepsilon(h)$$

and

$$g'(t) = \lim_{h \to 0+} \frac{d_{\partial K}(x(t) + hx'(t) + h\varepsilon(h)) - d_{\partial K}(x(t))}{h}$$

Lemma A1.7.3 of Appendix A implies that

$$g'(t) \leq d(x'(t), T_{\partial K}(\Pi_{\partial K}(x(t))))$$

We denote by $\lambda > 0$ the Lipschitz constant of f, and we choose y in $\Pi_{\partial K}(x(t))$. We deduce that

$$\begin{aligned}
& d(x'(t), T_{\partial K}(\Pi_{\partial K}(x(t)))) \\
& \leq d(x'(t), T_{\partial K}(y)) \leq \|x'(t) - f(y)\| \qquad \text{(since } f(y) \in T_{\partial K}(y) \text{)} \\
& \leq \|x'(t) - f(x(t))\| + \lambda \|y - x(t)\| \qquad \text{(since } f \text{ is Lipschitz)} \\
& = 0 + \lambda d_{\partial K}(x(t)) = \lambda g(t)
\end{aligned}$$

Then g is a solution to

$$\text{for almost all } t \in [0, T], \ g'(t) \leq \lambda g(t) \quad \text{and} \quad g(0) = 0$$

We deduce that $g(t) = 0$ for all $t \in [0, T]$, and thus $x(t)$ is viable in ∂K on $[0, T]$. □

As a consequence, we obtain a criterion for a cell to be a successor of another one:

Proposition 10.8.5 *If $K_b \cap K_c \subset K_{c \Rightarrow}$, then the qualitative cell K_c is a successor of K_b, in the sense that for any $x \in K_b \cap K_c$, there exists τ*

such that the solution $s(t)x$ remains in K_c for $t \in [0, \tau]$. Conversely, if the qualitative cell K_c is a successor of K_b, then

$$K_b \cap K_c \subset K_{c\Rightarrow} \cup K_{c\Leftrightarrow}$$

10.8.2 Hitting and Exit Tubes

So far, we have defined the successors of the qualitative cells on the basis of the behaviors of the dynamical systems on the boundaries of the cells. We shall now investigate what happens to the solutions starting from the interiors of the qualitative cells. For that purpose we need to introduce the hitting and exit functionals on a continuous function $x(\cdot) \in \mathcal{C}(0, \infty; X)$.

Definition 10.8.6 *Let $M \subset X$ be a closed subset and $x(\cdot) \in \mathcal{C}(0, \infty; X)$ be a continuous function. We denote by*

$$\varpi_M : \mathcal{C}(0, \infty; X) \mapsto \mathbf{R}_+ \cup \{+\infty\}$$

the hitting functional associating with $x(\cdot)$ its hitting time $\varpi_M(x(\cdot))$, defined by

$$\varpi_M(x(\cdot)) \ := \ \inf\{t \in [0, +\infty[\mid x(t) \in M\}$$

The function $\varpi_M^\flat : K \mapsto \mathbf{R}_+ \cup \{+\infty\}$ defined by

$$\varpi_M^\flat(x) \ := \ \varpi_M(s_f(\cdot)x)$$

is called the hitting function. In the same way, when $K \subset X$ is a closed subset, the functional $\tau_K : \mathcal{C}(0, \infty; X) \mapsto \mathbf{R}_+ \cup \{+\infty\}$ associating with $x(\cdot)$ its exit time $\tau_K(x(\cdot))$, defined by

$$\tau_K(x(\cdot)) \ := \ \inf \{t \in [0, \infty[\mid x(t) \notin K\}$$

is called the exit functional. The function $\tau_K^\sharp : K \mapsto \mathbf{R}_+ \cup \{+\infty\}$ defined by

$$\tau_K^\sharp(x) \ := \ \tau_K(s_f(\cdot)x)$$

is called the exit function.

We then note that

$$\varpi_{\widehat{K}}(x(\cdot)) \ \le \ \tau_K(x(\cdot))$$

that

$$\forall\, t \in [0, \varpi_{\widehat{K}}(x(\cdot))[, \ \ x(t) \in \mathrm{Int}(K) \text{ and } \forall\, t \in [0, \tau_K(x(\cdot))[, \ \ x(t) \in K$$

and that, when $\varpi_{\widehat{K}}(x(\cdot))$ [respectively $\tau_K(x(\cdot))$] is finite,

$$x(\varpi_{\widehat{K}}(x(\cdot))) \in \partial K \quad \text{and} \quad x(\tau_K(x(\cdot))) \in \partial K$$

respectively. We remark also that $\varpi_{\widehat{K}}(x(\cdot)) \equiv 0$ when the interior of K is empty.

We continue to use the convention $\inf\{\emptyset\} := +\infty$, so that $\varpi_{\widehat{K}}(x(\cdot))$ being infinite means that $x(t) \in \text{Int}(K)$ for all $t \in [0, +\infty[$ and that $\tau_K(x(\cdot)) = +\infty$ means that $x(t) \in K$ for all $t \geq 0$.

Lemma 10.8.7 *Let $K \subset X$ be a closed subset. The functional τ_K and the exit function $\tau_K^\sharp$ are upper-semicontinuous when $\mathcal{C}(0,\infty;X)$ is supplied with the pointwise convergence topology. The functional ϖ_M and the hitting function $\varpi_M^\flat$ are lower-semicontinuous when $\mathcal{C}(0,\infty;X)$ is supplied with the compact convergence topology.*

Proof By the maximum theorem, the upper-semicontinuity of τ_K follows from the lower-semicontinuity of the set-valued map $x(\cdot) \rightsquigarrow \Xi(x(\cdot)) \subset \mathbf{R}_+$, where

$$\Xi(x(\cdot)) := \{t \in [0, \infty[\mid x(t) \notin K\}$$

since $\tau_K(x(\cdot)) = \inf\{\Xi(x(\cdot))\}$. Indeed, for any $t \in \Xi(x(\cdot))$ and any sequence $x_n(\cdot)$ converging pointwise to $x(\cdot)$, we see that $t \in \Xi(x_n(\cdot))$ for n large enough, because $x_n(t)$ belongs to the open set $X \backslash K$ [since $x(t) \in X \backslash K$].

Let us check now that the function ϖ_M is lower-semicontinuous for the compact convergence topology: Take any $T \geq 0$ and any sequence $x_n(\cdot)$ satisfying $\varpi_M(x_n(\cdot)) \leq T$ converging to $x(\cdot)$ uniformly over compact subsets and show that $\varpi_M(x(\cdot)) \leq T$. Let us introduce the subsets

$$\Theta_{T'}(x(\cdot)) := \{t \in [0, T'] \mid x(t) \notin \text{Int}(K)\}$$

By construction, for any $T' > T$, the subsets $\Theta_{T'}(x_n(\cdot))$ are not empty. We also observe that the graph of the set-valued map $x(\cdot) \rightsquigarrow \Theta_{T'}(x(\cdot))$ is closed in the Banach space $\mathcal{C}(0, T'; X) \times [0, T']$: Indeed, if $(x_n(\cdot), t_n) \in \text{Graph}(\Theta_{T'})$ converges to $(x(\cdot), t)$, then $x_n(t_n) \in M$ converges to $x(t)$, which belongs to the closed subset M, so that $(x(\cdot), t) \in \text{Graph}(\Theta_{T'})$. Taking its values in the compact interval $[0, T']$, the set-valued map $x(\cdot) \rightsquigarrow \Theta_{T'}(x(\cdot))$ is actually upper-semicontinuous. Therefore, for any given $\varepsilon > 0$, $\Theta_{T'}(x_n(\cdot)) \subset \Theta_{T'}(x(\cdot)) + [-\varepsilon, +\varepsilon]$. We thus infer that $\varpi_M(x(\cdot)) \leq \varpi_M(x_n(\cdot)) + \varepsilon \leq T + \varepsilon$ for every $\varepsilon > 0$. □

We are thus led to single out the following subsets:

Definition 10.8.8 *We associate with any $T \geq 0$ the subsets*

$$\begin{cases} \text{(i)} & \text{Hit}_f(M,T) := \{x \in X \mid \varpi_M^\flat(x) \leq T\} \\ \text{(ii)} & \text{Exit}_f(K,T) := \left\{x \in K \mid \tau_K^\sharp(x) \geq T\right\} \end{cases} \quad (10.11)$$

We shall say that the set-valued map $T \rightsquigarrow \text{Hit}_f(M,T)$ is the hitting tube and that the set-valued map $T \rightsquigarrow \text{Exit}_f(K,T)$ is the exit tube.

Lemma 10.8.7 implies that the graphs of the hitting and exit tubes are closed.

Proposition 10.8.9 *Let $K \subset X$ be a closed subset.*

Then $\text{Hit}_f(M,T)$ is the closed subset of initial states x such that the closed subset M is reached before T by the solution $s_f(\cdot)x$ to the differential equation starting at x.

The closed subset $\text{Exit}_f(K,T)$ is the subset of initial states $x \in K$ such that the solution $s_f(\cdot)x$ to the differential equation starting at x remains in K for all $t \in [0,T]$. Actually, such a solution satisfies

$$\forall\, t \in [0,T], \quad s_f(t)x \in \text{Exit}_f(K,T-t)$$

In particular, for $T = +\infty$,

$$\text{Viab}_f(K) = \text{Exit}_f(K,+\infty) = \bigcap_{T>0} \text{Exit}_f(K,T)$$

The subset

$$\text{Entr}_f(K) := \bigcup_{T>0} \text{Exit}_f(K,T) := \left\{x \in K \mid \tau_K^\sharp(x) > 0\right\}$$

is the subset of elements $x \in K$ from which the solution is viable in K on some nonempty interval $[0,T]$.

We observe that if $T_1 \leq T_2$,

$$\partial K = \text{Hit}_f(K,0) \subset \text{Hit}_f(K,T_1) \subset \text{Hit}_f(K,T_2) \subset \ldots$$

and

$$\text{Viab}_f(K) \subset \text{Exit}_f(K,T_2) \subset \text{Exit}_f(K,T_1) \subset \ldots \subset \text{Entr}_f(K) \subset K$$

The interior of K is contained in $\text{Entr}_f(K)$. If the interior of K is nonempty, then the Baire theorem implies that the interior of some $\text{Exit}_f(K,T)$ is also nonempty.

Proof Because the subset of initial states x such that the subset M is reached before T by the solution $x(\cdot)$ to the differential equation starting at x is obviously contained in $\mathrm{Hit}_f(M,T)$, consider an element $x \in \mathrm{Hit}_f(M,T)$, and prove that it satisfies the foregoing property.

By definition of the hitting functional, we can associate a time $t_\varepsilon \leq T + \varepsilon$ such that $x(t_\varepsilon) \in M$. A subsequence, again denoted by t_ε, converges to $t \in [0, T+\varepsilon]$, so that the limit $x(t)$ of $x(t_\varepsilon) \in M$ belongs to the closed subset M. This implies that $\varpi_M(s_f(\cdot)x) \leq T + \varepsilon$ for every $\varepsilon > 0$.

In the same way, let $T \geq 0$ be finite or infinite. We observe that the subset of initial states $x \in K$ such that a solution $x(\cdot)$ to the differential equation starting at x remains in K for all $t \in [0, T[$ is contained in $\mathrm{Exit}_f(K,T)$, so that it is enough to prove that for any $x \in \mathrm{Exit}_f(K,T)$, the solution $s_f(\cdot)x$ satisfies the foregoing property.

By definition of the exit function, we know that $x(t) \in K$ for any $t \leq \tau_K(s_f(\cdot)x)$ and thus for any $t \leq T$. □

We deduce from Proposition 10.8.9 a characterization of the successors of a qualitative cell:

Proposition 10.8.10 *A qualitative cell K_c is a successor of K_b if and only if*

$$K_b \cap K_c \subset \mathrm{Entr}_f(K_c)$$

Let us mention also the following observations:

Proposition 10.8.11 *Let $K_a \subset K$ be a closed qualitative cell. The complement $K_a \backslash \mathrm{Exit}_f(K_a, T)$ is equal to the set*

$$\{x \in K_a \mid \exists\, t \in [0,T] \quad \textit{such that} \ \ s_f(t)x \notin K_a\}$$

of initial states x from which the solution $s_f(\cdot)x$ leaves K_a at some $t \leq T$.

Consequently, if $M \subset K_a \backslash \mathrm{Viab}_f(K_a)$ is compact, there exists $T \geq 0$ such that, for every $x \in M$, there exists $t \in [0,T]$ such that $s_f(t)x \notin K_a$. In particular, if K_a is a compact repeller, there exists $T < +\infty$ such that for every $x \in K_a$, $s_f(t)x \notin K_a$ for some $t \in [0,T]$.

Proposition 10.8.9 implies also the following result:

Proposition 10.8.12 *Let us consider qualitative cells K_a and K_b. Then*

$$K_a \cap \mathrm{Hit}(K_b, T)$$

is the subset of elements of the qualitative cell K_a that reach the qualitative cell K_b before time T, and

$$K_a \cap \bigcup_{T \geq 0} \mathrm{Hit}(K_b, T)$$

is the subset of elements of the qualitative cell K_a that reach the qualitative cell K_b in finite time.

Lemma 10.8.13 *Let us assume that the interior of each qualitative cell is not empty, that the qualitative cells are equal to the closure of their interiors, and that*

$$\forall\, a \in \mathcal{A},\ \widehat{K_a} \;=\; \bigcup_{b \in \mathcal{A}} K_b$$

Then

$$\forall\, x \in K_a,\ \varpi^{b}_{\widehat{K_a}}(x) \;=\; \min_{b \in \mathcal{A}} \varpi^{b}_{K_b}(x)$$

Therefore, we can cover the qualitative cell K_a by its viability kernel and the closed subcells

$$K_a^b \;:=\; \{x \in K_a \mid \tau^{\sharp}_{K_a}(x) \;\geq\; \varpi^{b}_{K_b}(x)\}$$

of elements of K_a from which the solution reaches K_b before leaving K_a.

Indeed, either $\tau^{\sharp}_{K_a}(x)$ is infinite, and x belongs to the viability kernel of the qualitative cell, or it is finite, and thus there exists at least one qualitative cell K_b such that $\tau^{\sharp}_{K_a}(x) \;\geq\; \varpi^{b}_{K_b}(x)$, that is, such that $x \in K_a^b$.

10.9 Sufficient Conditions for Chaos

Let $f : K \mapsto X$ be a continuous function with linear growth enjoying the uniqueness property. We denote by $s(\cdot)x$ the solution to the differential equation $x' = f(x)$ starting at x when $t = 0$ and by $L(s(\cdot)x)$ its limit set.

Theorem 10.9.1 *Let us assume that a closed viability domain K of f is covered by a family of compact subsets K_a ($a \in \mathcal{A}$) such that the following "controllability assumption"*

$$\forall\, a \in \mathcal{A}, \forall\, y \in K,\ \ \exists\, x \in K_a,\ t \in [0, \infty[\ \text{ such that }\ s(t)x = y$$

holds true.

Then, for any sequence $a_0, a_1, \ldots, a_n, \ldots$, there exist at least an initial state $x \in K_{a_0}$ and a nondecreasing sequence of elements $t^j \in [0, \infty]$ such that

$$\begin{cases} \text{(i)} & s(t^j)x \in K_{a_j} \text{ if } t^j < \infty \\ \text{(ii)} & L(s(\cdot)x) \cap K_{a_j} \neq \emptyset \text{ if } t^j = +\infty \end{cases}$$

The t^j's are finite when we strengthen the controllability assumption by assuming that there exists $T \in]0, \infty[$ such that

$$\forall\, a \in \mathcal{A}, \forall\, y \in K, \; \exists\, x \in K_a, \; t \in [0, T] \text{ such that } s(t)x = y$$

Proof Let $M \subset K$ be any closed subset. We associate with any $x \in K$ the number $\varpi_M := \inf_{s(t)x \in M} t$, which is nonnegative and finite thanks to the "controllability assumption." We associate with the sequence $a_0, a_1, \ldots$ the subsets $M_{a_0 a_1 \cdots a_n}$ defined by induction by $M_{a_n} := K_n$,

$$M_{a_{n-1} a_n} \; := \; \{x \in K_{a_{n-1}} \mid s(t_{M_{a_n}})x \in K_{a_n}\}$$

and, for $j = n-2, \ldots, 0$, by

$$M_{a_j a_{j+1} \cdots a_n} \; := \; \{x \in K_{a_j} \mid s(t_{M_{a_{j+1} \cdots a_n}})x \in M_{a_{j+1} \cdots a_n}\}$$

The subset $M_{a_0 a_1 \cdots a_n}$ is then the subset of teh initial states $x \in K_{a_0}$ such that the solution $s(t)X$ bisits the cells $K_{a_1}, \cdots, K_{a_n}$ in this order. They are nonempty closed subsets and form a nonincreasing family. Because K_{a_0} is compact, the intersection $K_\infty := \bigcap_{n=0}^{\infty} M_{a_0 a_1 \cdots a_n}$ is therefore nonempty.

Let us take an initial state x in K_∞ and fix n. We set $t_n^j := \sum_{k=1}^{j} \varpi_{M_{a_k \cdots a_n}}$ for any $j = 1, \ldots, n$. We see at once that $s(t_n^j)x \in M_{a_j \cdots a_n} \subset K_{a_j}$.

On the other hand, we observe that $\varpi_{M_2} \le \varpi_{M_1}$ whenever $M_1 \subset M_2$. Because $M_{a_1 \cdots a_{n+1}} \subset M_{a_1 \cdots a_n}$, we deduce that $t_n^j \le t_{n+1}^j$ for any $j = 1, \ldots, n$.

Therefore, j being fixed, the nondecreasing sequence t_n^j (for $n \ge j$) converges to some $t^j \in [0, \infty]$. Furthermore, the sequence t^j is not decreasing, and if for some index J, $t^{J-1} < \infty$ and $t^J = \infty$, all the t^j's are equal to $+\infty$ for $j \ge J$.

Because $s(t_n^j)$ belongs to K_{a_j} for all $n \ge j$, we infer that $s(t^j)x$ belongs to K_{a_j} if $j < J$ and that, for $j \ge J$, the intersection $L(s(\cdot)x) \cap K_{a_j}$ is not empty.

If we assume that the stronger assumption holds true, we know that the t_n^j remain in the interval $[0, jT]$, so that the limits t^j are finite. □

Appendix 1

Convex and Nonsmooth Analysis

We devote this appendix to some results in convex and nonsmooth analysis that are relevant for solving the minimization problems involved in neural networks.

A1.1 Fenchel's Theorem

Suppose we have two finite-dimensional vector spaces X and Y, together with

$$\begin{cases} \text{(i)} & \text{a continuous, linear operator } A \in \mathcal{L}(X,Y) & (1.1) \\ \text{(ii)} & \text{two nontrivial, convex, lower-semicontinuous functions} \\ & f : X \to \mathbf{R} \cup \{+\infty\} \text{ and } g : X \to \mathbf{R} \cup \{+\infty\}. & (1.2) \end{cases}$$

We shall study the minimization problem

$$v := \inf_{x \in X} [f(x) + g(Ax)]. \tag{1.3}$$

Note that the function $f + g \circ A$ that we propose to minimize is nontrivial only if $A\,\text{Dom} f \cap \text{Dom}\, g \neq \emptyset$, that is to say, if

$$0 \in A\,(\text{Dom}\, f) - \text{Dom}\, g \tag{1.4}$$

In this case, we have $v < +\infty$.

Now we introduce the dual minimization problem

$$v^\star := \inf_{q \in Y^\star} [f^\star(-A^\star q) + g^\star(q)] \tag{1.5}$$

where $A^\star \in \mathcal{L}(Y^\star, X^\star)$ is the transpose of A, $f^\star : X^\star \to \mathbf{R} \cup \{+\infty\}$ is the conjugate of f, and $g^\star : Y^\star \to \mathbf{R} \cup \{+\infty\}$ is the conjugate of g. This makes sense only if we assume that

$$0 \in A^\star \operatorname{Dom} g^\star + \operatorname{Dom} f^\star \tag{1.6}$$

and in this case, $v^\star < +\infty$.

Note that we still have the inequality

$$v + v^\star \geq 0 \tag{1.7}$$

since, by virtue of Fenchel's inequality,

$$f(x) + g(Ax) + f^\star(-Aq) + g^\star(q) \geq \langle -A^\star q, x\rangle + \langle q, Ax\rangle = 0$$

Consequently, conditions (1.4) and (1.6) imply that v and $v^\star$ *are finite.*

Theorem A1.1.1 *Suppose that X and Y are finite-dimensional vector spaces, that $A \in \mathcal{L}(X, Y)$ is a continuous, linear operator from X to Y, and that $f : X \to \mathbf{R} \cup \{+\infty\}$ and $g : Y \to \mathbf{R} \cup \{+\infty\}$ are nontrivial, convex, lower-semicontinuous functions. We consider the case in which $0 \in A(\operatorname{Dom} f) - \operatorname{Dom} g$ and $0 \in A^\star(\operatorname{Dom} g^\star) + \operatorname{Dom} f^\star$ (which is equivalent to the assumption that v and $v^\star$ are finite).*

If we suppose that

$$0 \in \operatorname{Int}(A \operatorname{Dom} f - \operatorname{Dom} g) \tag{1.8}$$

then

$$\begin{cases} \text{(i)} & v + v^\star = 0 \\ \text{(ii)} & \exists \bar{q} \in Y^\star \text{ such that } f^\star(-A^\star\bar{q}) + g^\star(\bar{q}) = v^\star \end{cases} \tag{1.9}$$

If we suppose that

$$0 \in \operatorname{Int}(A^\star \operatorname{Dom} g^\star + \operatorname{Dom} f^\star) \tag{1.10}$$

then

$$\begin{cases} \text{(i)} & v + v^\star = 0 \\ \text{(ii)} & \exists \bar{x} \in X \text{ such that } f(\bar{x}) + g(A\bar{x}) = v \end{cases} \tag{1.11}$$

Proof We introduce the mapping ϕ from $\operatorname{Dom} f \times \operatorname{Dom} g$ to $Y \times \mathbf{R}$ defined by

$$\phi(x, y) = \{Ax - y, f(x) + g(y)\} \tag{1.12}$$

together with

$$\left\{\begin{array}{ll}\text{(i)} & \text{the vector } (0,v) \in Y \times \mathbf{R} \\ \text{(ii)} & \text{the cone } Q = \{0\} \times]0, \infty[\subset Y \times \mathbf{R}\end{array}\right. \tag{1.13}$$

It is easy to show that the linearity of A and the convexity of the functions f and g imply that

$$\phi(\operatorname{Dom} f \times \operatorname{Dom} g) + Q \text{ is a convex subset of } Y \times \mathbf{R} \tag{1.14}$$

Furthermore, if we suppose that $(0, v)$ belongs to $\phi(\operatorname{Dom} f \times \operatorname{Dom} g) + Q$, we can deduce the existence of $x \in \operatorname{Dom} f$ and $y \in \operatorname{Dom} g$ such that $Ax - y = 0$ and $v > f(x) + g(y) = f(x) + g(Ax)$, which would contradict the definition of v. Thus,

$$(0, v) \notin \phi(\operatorname{Dom} f \times \operatorname{Dom} g) + Q \tag{1.15}$$

Because Y is a finite-dimensional space, we can use the separation theorem to show that there exists a linear form $(p, a) \in Y^\star \times \mathbf{R}$ such that

$$\left\{\begin{array}{lll}\text{(i)} & (p,a) \neq 0 & (1.16) \\ \text{(ii)} & av = \langle (p,a), (0,v) \rangle \leq & \\ & \inf_{\substack{x \in \operatorname{Dom} f \\ y \in \operatorname{Dom} g}} [a(f(x) + g(y)) + \langle p, Ax - y \rangle] + \inf_{\theta > 0} a\theta & (1.17)\end{array}\right.$$

Because the number $\inf_{\theta>0} a\theta$ is bounded below, we deduce that it is zero and that a is positive or zero. We cannot have $a = 0$, since in that case the inequality (1.17) would imply that

$$0 \leq \inf_{\substack{x \in \operatorname{Dom} f \\ y \in \operatorname{Dom} g}} \langle p, Ax - y \rangle = \inf_{z \in A \operatorname{Dom} f - \operatorname{Dom} g} \langle p, z \rangle \tag{1.18}$$

Because the set $A \operatorname{Dom} f - \operatorname{Dom} g$ contains a ball of radius η and center 0, by virtue of (1.8) we deduce that $0 \leq -\eta \|p\|$ and thus that $p = 0$. This contradicts (1.16) Consequently, a is nontrivially positive. Dividing the inequality (1.17) by a and taking $\bar{p} = p/a$, we obtain

$$\begin{array}{rcl} v & \leq & \inf_{\substack{x \in \operatorname{Dom} f \\ y \in \operatorname{Dom} g}} [\langle A^\star \bar{p}, x \rangle - \langle \bar{p}, y \rangle + f(x) + g(y)] \\ & = & - \sup_{\substack{x \in X \\ y \in Y}} [\langle -A^\star \bar{p}, x \rangle + \langle \bar{p}, y \rangle - f(x) - g(y)] = -f^\star(-A^\star \bar{p}) - g^\star(\bar{p}) \end{array}$$

whence $f^\star(-A^\star \bar{p}) + g^\star(\bar{p}) = -v \leq v^\star$, which proves that $\bar{p}$ is a solution of the dual problem and that $v^\star = -v$.

The second assertion is proved by replacing f by $g^\star$, g by $f^\star$, and A by $-A^\star$. □

Corollary A1.1.2 *Let $L \subset X$ and $M \subset Y$ be closed convex subsets and $A \in \mathcal{L}(X,Y)$ a linear operator linked by the condition*

$$0 \in \text{Int}(AL - M)$$

Then the normal cone to $L \cap A^{-1}(M)$ is given by

$$N_{L\cap A^{-1}(M)}(x) = N_L(x) + A^\star N_M(Ax)$$

and the tangent cone by

$$T_{L\cap A^{-1}(M)}(x) = T_L(x) \cap A^{-1}T_M(Ax)$$

Proof Because $\psi_{L\cap A^{-1}(M)}(x) = \psi_L(x) + \psi_M(Ax)$ and because

$$N_{L\cap A^{-1}(M)}(x) = \partial\psi_{L\cap A^{-1}(M)}(x)$$

we deduce from Fenchel's theorem the formula for normal cones. The formula for tangent cones is obtained by polarity and transposition. □

A1.2 Properties of Conjugate Functions

First, we note the following elementary propositions:

Proposition A1.2.1 (a) *If $f \leq g$, then $g^\star \leq f^\star$.* (b) *If $A \in \mathcal{L}(X,X)$ is an isomorphism, then*

$$(f \circ A)^\star = f^\star \circ A^{\star-1}$$

(c) *If $g(x) := f(x - x_0) + \langle p_0, x\rangle + a$, then*

$$g^\star(p) = f^\star(p - p_0) + \langle p, x_0\rangle - (a + \langle p_0, x_0\rangle)$$

(d) *If $g(x) := f(\lambda x)$, then $g^\star(p) = f^\star(p/\lambda)$, and if $h(x) := \lambda f(x)$, then $h^\star(p) = \lambda f^\star(p/\lambda)$.*

Proof The first assertion is evident. The second assertion can be proved by showing that

$$\sup_{x\in X}[\langle p, x\rangle - f(Ax)] = \sup_{y\in X}[\langle A^{\star-1}p, y\rangle - f(y)] = f^\star(A^{\star-1}p)$$

For the third assertion, we observe that

$$\begin{aligned}
\sup_{x\in X}[\langle p,x\rangle - g(x)] &= \sup_{x\in X}[\langle p-p_0,x\rangle - f(x-x_0)] - a \\
&= \sup_{x\in X}[\langle p-p_0,y\rangle - f(y)] - a + \langle p-p_0,x_0\rangle \\
&= f^\star(p-p_0) + \langle p,x_0\rangle - a - \langle p_0,x_0\rangle
\end{aligned}$$

□

Proposition A1.2.2 *Suppose that X and Y are two finite-dimensional vector spaces and that f is a nontrivial, convex function from $X \times Y$ to $\mathbf{R}\cup\{+\infty\}$. Set $g(y) := \inf_{x\in X} f(x,y)$. Then*

$$g^\star(q) = f^\star(0,q) \tag{1.19}$$

Proof

$$\begin{aligned}
g^\star(q) &= \sup_{y\in Y}[\langle q,y\rangle - \inf_{x\in X} f(x,y)] \\
&= \sup_{y\in Y}\sup_{x\in X}[\langle 0,x\rangle + \langle q,y\rangle - f(x,y)] = f^\star(0,q)
\end{aligned}$$

□

Proposition A1.2.3 *Suppose that X and Y are two finite-dimensional spaces, that $B \in \mathcal{L}(Y,X)$ is a continuous linear operator from Y to X, and that $f : X \to \mathbf{R}\cup\{+\infty\}$ and $g : Y \to \mathbf{R}\cup\{+\infty\}$ are two nontrivial functions. Set $h(x) := \inf_{y\in Y}(f(x-By)+g(y))$. Then*

$$h^\star(p) = f^\star(p) + g^\star(B^\star p) \tag{1.20}$$

Proof

$$\begin{aligned}
&\sup_{x\in X}[\langle p,x\rangle - \inf_{y\in Y} f((x-By)+g(y))] \\
&\qquad= \sup_{\substack{x\in X\\ y\in Y}}[\langle p,x\rangle - f(x-By) - g(y)] \\
&\qquad= \sup_{\substack{x\in X\\ y\in Y}}[\langle p,x+By\rangle - f(x) - g(y)] \\
&\qquad= f^\star(p) + g^\star(B^\star p)
\end{aligned}$$

□

Next we shall calculate the function conjugate to $f^\star + g^\star \circ B^\star$. We shall not recover the function h, because we do not know if the latter is lower-semicontinuous. For this, we need a slightly stronger assumption, namely,

$$0 \in \mathrm{Int}\,(B^\star \,\mathrm{Dom}\, f^\star - \mathrm{Dom}\, g^\star) \tag{1.21}$$

In fact, this is a consequence of the following proposition:

Proposition A.2.4. *Suppose that X and Y are two finite-dimensional spaces, that $A \in \mathcal{L}(X,Y)$ is a continuous linear operator, and that $f : X \to \mathbf{R} \cup \{+\infty\}$ and $g : Y \to \mathbf{R} \cup \{+\infty\}$ are two nontrivial, lower-semicontinuous functions. Suppose further that*

$$0 \in \mathrm{Int}\,(A\,\mathrm{Dom}\, f - \mathrm{Dom}\, g). \tag{1.22}$$

Then, for all $p \in A^\star \,\mathrm{Dom}\, g^\star + \mathrm{Dom}\, f^\star$, there exists $\bar{q} \in Y^\star$ such that

$$\begin{aligned} (f + g \circ A)^\star(p) &= f^\star(p - A^\star \bar{q}) + g^\star(\bar{q}) \\ &= \inf_{q \in Y^\star} (f^\star(p - A^\star q) + g^\star(q)) \end{aligned} \tag{1.23}$$

Proof We may write

$$\sup_{x \in X} [\langle p, x\rangle - f(x) - g(Ax)] = -\inf[f(x) - \langle p, x\rangle + g(Ax)]$$

We apply Fenchel's theorem with f replaced by $f(\cdot) - \langle p, \cdot\rangle$, the domain of which coincides with that of f, and the conjugate function of which is equal to $q \to f^\star(q + p)$. Thus, there exists $\bar{q} \in Y^\star$ such that

$$\begin{aligned} \sup_{x \in X} [\langle p, x\rangle - f(x) - g(Ax)] &= f^\star(p - A^\star \bar{q}) + g^\star(\bar{q}) \\ &= \inf_{q \in Y} [f^\star(p - A^\star q) + g^\star(q)] \end{aligned}$$

□

It is useful to state the following consequence explicitly:

Proposition A.2.5. *Suppose that X and Y are two finite-dimensional vector spaces, that $A \in \mathcal{L}(X,Y)$ is a continuous linear operator from X to Y, and that $g : Y \to \mathbf{R} \cup \{+\infty\}$ is a nontrivial, convex, lower-semicontinuous function. We suppose further that*

$$0 \in \mathrm{Int}\,(\mathrm{Im}\, A - \mathrm{Dom}\, g) \tag{1.24}$$

Then, for all $p \in A^\star \operatorname{Dom} g^\star$, there exists $\bar{q} \in \operatorname{Dom} g^\star$ satisfying

$$A^\star \bar{q} = p \quad \text{and} \quad (g \circ A)^\star(p) = g^\star(\bar{q}) = \min_{A^\star q = p} g^\star(q)$$

Proof We apply the previous proposition with $f = 0$, where the domain is the whole space X. Its conjugate function $f^\star$ is defined by $f^\star(p) = \{0\}$ if $p = 0$ and $f^\star(p) = +\infty$ otherwise. Consequently, $f^\star(p - A^\star q)$ is finite (and equal to zero) if and only if $p = A^\star q$. □

A1.3 Subdifferential Calculus

We can deduce easily from the calculus of conjugate functions a subdifferential calculus.

Theorem A1.3.1 *We consider two finite-dimensional spaces X and Y, a continuous linear operator $A \in \mathcal{L}(X, Y)$, and two nontrivial, convex, lower-semicontinuous functions $f : X \to \mathbf{R} \cup \{+\infty\}$ and $g : Y \to \mathbf{R} \cup \{+\infty\}$. We assume further that*

$$0 \in \operatorname{Int}(A \operatorname{Dom} f - \operatorname{Dom} g) \tag{1.25}$$

Then

$$\partial(f + g \circ A)(x) = \partial f(x) + A^\star \partial g(Ax) \tag{1.26}$$

Proof It is easy to check that $\partial f(x) + A^\star \partial g(Ax)$ is always contained in $\partial(f + g \circ A)(x)$. The inverse inclusion follows from Proposition A.2.4. We take $p \in \partial(f + g \circ A)(x)$. There exists $\bar{q} \in Y^\star$ such that $(f + g \circ A)^\star(p) = f^\star(p - A^\star \bar{q}) + g^\star(\bar{q})$. Thus, from equation (1.23),

$$\begin{aligned} \langle p, x \rangle &= f(x) + g(Ax) + (f + g \circ A)^\star(p) \\ &= (f(x) + f^\star(p - A^\star \bar{q})) + (g(Ax) + g^\star(\bar{q})) \end{aligned}$$

Consequently,

$$0 = (\langle p - A^\star \bar{q}, x \rangle - f(x) - f^\star(p - A^\star \bar{q})) + (\langle \bar{q}, Ax \rangle - g(Ax) - g^\star(\bar{q}))$$

Because each of these two expressions is negative or zero, it follows that they are both zero, whence if follows that $\bar{q} \in \partial g(Ax)$ and $p - A^\star \bar{q} \in \partial f(x)$. Thus, we have shown that $p = p - A^\star \bar{q} + A^\star \bar{q} \in \partial f(x) + A^\star \partial g(Ax)$. □

Corollary A1.3.2 *If f and g are two nontrivial, convex, lower-semicon- linebreak tinuous functions from X to $\mathbf{R} \cup \{+\infty\}$, and if*

$$0 \in \operatorname{Int}(\operatorname{Dom} f - \operatorname{Dom} g) \tag{1.27}$$

then

$$\partial(f+g)(x) = \partial f(x) + \partial g(x) \tag{1.28}$$

If g is a nontrivial, convex, lower-semicontinuous function from Y to $\mathbf{R} \cup \{+\infty\}$, and if $A \in \mathcal{L}(X,Y)$ satisfies

$$0 \in \operatorname{Int}(\operatorname{Im} A - \operatorname{Dom} g) \tag{1.29}$$

then

$$\partial(g \circ A)(x) = A^\star \partial g(Ax) \tag{1.30}$$

Proposition A1.3.3 *Let g be a nontrivial convex function from $X \times Y$ to $\mathbf{R} \cup \{+\infty\}$. Consider the function $h : Y \to \mathbf{R} \cup \{+\infty\}$ defined by*

$$h(y) := \inf_{x \in X} g(x,y) \tag{1.31}$$

If $\bar{x} \in X$ satisfies $h(y) = g(\bar{x},y)$, then the following conditions are equivalent:

$$\begin{cases} \text{(i)} & q \in \partial h(y) \\ \text{(ii)} & (0,q) \in \partial g(\bar{x},y) \end{cases} \tag{1.32}$$

Proof Because $h^\star(q) = g^\star(0,q)$, following Proposition A1.2.2 we deduce that q belongs to $\partial h(y)$ if and only if $\langle q,y\rangle = h(y) + h^\star(q) = g(\bar{x},y) + g^\star(0,q)$, that is, if and only if $(0,q) \in \partial g(\bar{x},y)$. □

Proposition A1.3.4 *We consider a family of convex functions $x \to f(x,p)$ indexed by a parameter p running over a set P. We assume that*

$$\begin{cases} \text{(i)} & P \text{ is compact} \\ \text{(ii)} & \text{There exists a neighborhood } U \text{ of } x \text{ such that,} \\ & \text{for all } y \text{ in } U, p \to f(y,p) \text{ is upper-semicontinuous} \\ \text{(iii)} & \forall p \in P,\ y \to f(y,p) \text{ is continuous at } x \end{cases} \tag{1.33}$$

Consider the upper envelope k of the functions $f(\cdot,p)$, defined by $k(y) := \sup_{p \in P} f(y,p)$. Set

$$P(x) := \{p \in P \mid k(x) = f(x,p)\} \tag{1.34}$$

Then

$$Dk(x)(v) = \sup_{p \in P(x)} Df(x,p)(v) \tag{1.35}$$

and

$$\partial k(x) = \overline{\text{co}} \bigcup_{p \in P(x)} \partial f(x,p) \tag{1.36}$$

Proof Because when p belongs to $P(x)$ we can write

$$\frac{f(x+hv,p)-f(x,p)}{h} \leq \frac{k(x+hv)-k(x)}{h}$$

letting h tend to zero we obtain

$$\sup_{p \in P(x)} Df(x,p)(v) \leq Dk(x)(v) \tag{1.37}$$

We must establish the inverse inequality. Fix $\varepsilon > 0$; we shall show that there exists $p \in P(x)$ such that $Dk(x)(v) - \varepsilon \leq Df(x,p)(v)$. Because the function k is convex, we know that

$$Dk(x)(v) = \inf_{h>0} \frac{k(x+hv)-k(x)}{h}$$

Then, for all $h > 0$, the set

$$B_h := \left\{ p \in P \,\middle|\, \frac{f(x+hv,p)-k(x)}{h} \geq Dk(x)(v) - \varepsilon \right\} \tag{1.38}$$

is nonempty. Consider the neighborhood U mentioned in assumption (A1.3.4ii). There exists $h_0 > 0$ such that $x + hv$ belongs to U for all $h \leq h_0$. Because $p \to f(x+hv,p)$ is upper-semicontinuous, the set B_h is closed. On the other hand, if $h_1 \leq h_2$, then $B_{h_1} \subset B_{h_2}$; if p belongs to B_{h_1}, the convexity of f with respect to x implies that

$$\begin{aligned} & Dk(x)(v) - \varepsilon \\ & \quad \leq \frac{1}{h_1}\left[\left(1-\frac{h_1}{h_2}\right)(f(x,p)-k(x)) + \frac{h_1}{h_2}(f(x+h_2v,p)-k(x))\right] \\ & \quad \leq \frac{1}{h_2}(f(x+h_2v,p)-k(x)) \end{aligned}$$

because

$$x + h_1 v = \left(1-\frac{h_1}{h_2}\right)x + \frac{h_1}{h_2}(x+h_2v)$$

and because $f(x,p) - h(x) \leq 0$ for all p. Consequently, since P is compact, the intersection $\bigcap_{0<h\leq h_0} B_h$ is nonempty, and all elements $\bar{p}$ of this intersection satisfy

$$h(Dk(x)(v) - \varepsilon) \leq f(x + hv, \bar{p}) - k(x) \tag{1.39}$$

Letting h tend to zero, we deduce that $f(x,\bar{p}) - k(x) \geq 0$, whence $\bar{p}$ belongs to $P(x)$. Dividing by $h > 0$, we obtain the inequality

$$Dk(x)(v) - \varepsilon \leq Df(x,\bar{p})(v) \leq \sup_{p\in P(x)} Df(x,p)(v)$$

Thus, it is sufficient to let ε tend to zero.

Because $y \to f(y,p)$ is continuous at x, we know that $Df(x,p)(\cdot)$ is continuous for each p, whence we know that $Dk(x)(\cdot)$ is lower-semicontin uous. Equation (1.35) may be written as

$$\sigma(\partial k(x,v)) = \sup_{p\in P(x)} \sigma(\partial f(x,p),v)$$

which implies equation (1.36). □

Corollary A1.3.5 *Consider n convex functions f_i continuous at a point x. Then*

$$\partial\left(\sup_{i=1,\dots,n} f_i\right)(x) = \overline{\text{co}}\left(\bigcup_{i\in I(x)} \partial f_i(x)\right) \tag{1.40}$$

where $I(x) = \{i = 1,\dots,n | f_i(x) = \sup_{j=1,\dots,n} f_j(x)\}$.

A1.4 Duality Theory

We use the foregoing subdifferential calculus for implementing the Fermat rules and duality theory for the following general class of convex minimization problems. We consider

1. two finite dimensional spaces X and Y
2. two nontrivial, convex, lower-semicontinuous functions

$$\begin{cases} \text{(i)} & f : X \to \mathbf{R} \cup \{+\infty\} \\ \text{(ii)} & g : Y \to \mathbf{R} \cup \{+\infty\} \\ \text{iii)} & \text{a continuous linear operator } A \in \mathcal{L}(X,Y) \end{cases} \tag{1.41}$$

We shall choose elements $y \in Y$ and $p \in X^\star$ as parameters of the optimization problems

$$v := \inf_{x\in X} (f(x) - \langle p, x\rangle + g(Ax + y)) \tag{1.42}$$

and

$$v_\star := \inf_{q \in Y^\star} (f^\star(p - A^\star q) + g^\star(q) - \langle q, y \rangle) \tag{1.43}$$

which we shall solve at the same time. We shall say the minimization problems v and $v_\star$ are *dual.*

Theorem A1.4.1 (a) *We suppose that the conditions (1.41) are satisfied. If*

$$p \in \mathrm{Int}\,(\mathrm{Dom}\, f^\star + A^\star\, \mathrm{Dom}\, g^\star) \tag{1.44}$$

then there exists a solution $\overline{x}$ of the problem v and

$$v + v_\star = 0 \tag{1.45}$$

(b) *If we suppose further that*

$$y \in \mathrm{Int}\,(\mathrm{Dom}\, g - A\,\mathrm{Dom}\, f) \tag{1.46}$$

then the following conditions are equivalent:

$$\begin{cases} \text{(i)} & \overline{x} \text{ is a solution of the (primal) problem } v \qquad (1.47) \\ \text{(ii)} & \overline{x} \text{ is a solution of the inclusion } p \in \partial f(\overline{x}) + A^\star \partial g(A\overline{x} + y) \end{cases}$$

(c) *Similarly, assumption (1.46) implies that there exists a solution $\overline{q}$ of the dual problem $v_\star^\star$, and the two assumptions imply that the following conditions are equivalent:*

$$\begin{cases} \text{(iii)} & \overline{q} \text{ is a solution of the problem } v_\star \qquad (1.47) \\ \text{(iv)} & \overline{q} \text{ is a solution of the inclusion } y \in \partial g^\star(\overline{q}) - A\partial f^\star(p - A^\star \overline{q}) \end{cases}$$

(d) *The two assumptions imply that the solutions $\overline{x}$ and $\overline{q}$ of the problems v and $v_\star$ are solutions of the system of inclusions*

$$\begin{cases} \text{(i)} & p \in \partial f(\overline{x}) + A^\star(\overline{q}) \\ \text{(ii)} & y \in -A\overline{x} + \partial g^\star(\overline{q}) \end{cases} \tag{1.48}$$

Remark An optimal solution of the dual minimization problem $v_\star$ is usually called a *Lagrange* (or *Kuhn–Tucker*) *multiplier*, the inclusion (1.47ii) is usually called the *Euler–Lagrange inclusion*, and the inclusion (1.47iv) is the *Euler–Lagrange dual inclusion*. The system of inclusions (1.48) is usually called the *Hamiltonian system*. □

The correspondence $(x, q) \to (\partial f(x) + A^\star q) \times (-Ax + \partial g^\star(q))$ from $X \times Y^\star$ to its dual $X^\star \times Y$ can then be written symbolically in the suggestive matrix form:

$$\begin{pmatrix} \partial f & A^\star \\ -A & \partial g^\star \end{pmatrix} \tag{1.49}$$

The set of solutions $(\overline{x}, \overline{q})$ of the minimization problems v and $v_\star$ can be written in the suggestive form:

$$\begin{pmatrix} \partial f & A^\star \\ -A & \partial g^\star \end{pmatrix}^{-1} \begin{pmatrix} p \\ y \end{pmatrix}$$

This notation highlights the variation of the set of solutions as a function of the parameters $p \in X^\star$ and $y \in Y$.

Remark When assumptions (1.44) and (1.46) of Theorem A1.4.1 are satisfied, solution of the problem v is equivalent to solution of the inclusion (set-valued equation)

$$p \in \partial f(\overline{x}) + A^\star \partial g(A\overline{x} + y) \tag{1.50}$$

Theorem A1.4.1 indicates another way of solving this problem. This involves first solving the inclusion

$$y \in \partial g^\star(\overline{q}) - A\partial f^\star(p - A^\star \overline{q}) \tag{1.51}$$

and then choosing $\overline{x}$ in the set

$$\partial f^\star(p - A^\star \overline{q}) \cap A^{-1}(\partial g^\star(\overline{q}) - y) \tag{1.52}$$

This procedure is sensible only if the second inclusion is easier to solve than the first. This clearly depends on the functions f and g. If g is differentiable, it may be better to solve the inclusion (1.47ii). If, indeed, $f^\star$ is differentiable, it may be easier to solve the inclusion (1.47iv), which in this case can be written as

$$A\nabla f^\star(p - A^\star \overline{q}) + y \in \partial g^\star(\overline{q}) \tag{1.53}$$

or as

$$\forall q \in Y, \qquad \langle -A\nabla f^\star(p - A^\star \overline{q}) - y, \overline{q} - q\rangle + g^\star(\overline{q}) - g^\star(q) \leq 0 \quad (1.54)$$

Remark: Gradient Methods For solving the minimization problem v (or $v_\star$), we can use the gradient algorithm to either problem v or $v_\star$. They yield

$$\begin{cases} x_{n+1} - x_n \in -\delta_n \frac{p_n}{\|p_n\|} \quad \text{where} \\ p_n \in \partial f(x_n) + A^\star \partial g(Ax_n + y) \end{cases}$$

and

$$\begin{cases} q_{n+1} - q_n \in -\delta_n \frac{y_n}{\|y_n\|} \quad \text{where} \\ y_n \in \partial g^\star(q_n) - A\partial f^\star(p - A^\star q_n) \end{cases} \square$$

A1.5 Minimization Problems with Constraints

Let us consider

1. two finite-dimensional spaces X and Y
2. a continuous linear operator $A \in L(X, Y)$
3. a convex, closed subset $M \subset Y$
4. a nontrivial, convex, lower-semicontinuous function $f : X \to \mathbf{R} \cup \{+\infty\}$ and two elements $y \in Y$ and $p \in X^\star$

We consider the minimization problem

$$v := \inf_{Ax \in M - y} (f(x) - \langle p, x\rangle) \quad (1.55)$$

with its associated dual problem

$$v_\star := \inf_{q \in Y^\star} (f^\star(p - A^\star q) + \sigma_M(q) - \langle q, y\rangle) \quad (1.56)$$

Corollary A1.5.1 *If we suppose that*

$$p \in \text{Int}\,(\text{Dom}\, f^\star + A^\star \text{Dom}(\psi_M)) \quad (1.57)$$

then there exists a solution $\overline{x}$ (satisfying $A\overline{x} \in M - y$) of the problem v. If we suppose further that

$$y \in \text{Int}\,(M - A\,\text{Dom}\, f) \quad (1.58)$$

then the solutions $\overline{x}$ of the problem v are the solutions of the inclusion

$$p \in \partial f(\overline{x}) + A^\star N_M(A\overline{x} + y) \quad (1.59)$$

The following conditions are then equivalent:

1. *$\overline{q}$ is a solution of the inclusion $y \in \partial\sigma_M(\overline{q}) - A\partial f^\star(p - A^\star\overline{q})$.*
2. *The optimal solutions $\overline{x}$ and $\overline{q}$ of the problems v and $v_\star$ are related by*

$$p \in \partial f(\overline{x}) + A^\star\overline{q} \quad \text{and} \quad \overline{q} \in N_M(A\overline{x} + y) \tag{1.60}$$

The minimization problem

$$v := \inf_{Ax+y=0} (f(x) - \langle p, x\rangle) \tag{1.61}$$

which is a minimization problem with "constraints of equality," is obtained as the particular case in which $M = \{0\}$. Its dual problem is

$$v_\star := \inf_{q\in Y^\star} (f^\star(p - A^\star q) - \langle q, y\rangle) \tag{1.62}$$

Corollary A1.5.2 *If we suppose that*

$$p \in \operatorname{Int}(\operatorname{Dom} f^\star + \operatorname{Im} A^\star) \tag{1.63}$$

then there exists a solution $\overline{x}$ of the problem v.

If we suppose further that

$$-y \in \operatorname{Int}(A \operatorname{Dom} f) \tag{1.64}$$

then the solutions $\overline{x}$ of the problem v are the solutions of the inclusion

$$p \in \partial f(\overline{x}) + \operatorname{Im} A^\star, \qquad A\overline{x} + y = 0 \tag{1.65}$$

The following conditions are equivalent:

1. *$\overline{q}$ is a solution of the problem $v_\star$*
2. *$\overline{q}$ is a solution of the inclusion $y \in -A\partial f^\star(p - A^\star\overline{q})$*

The optimal solutions $\overline{x}$ and $\overline{q}$ of the problems v and $v_\star$ are related by

$$p \in \partial f(\overline{x}) + A^\star\overline{q} \tag{1.66}$$

Suppose that $P \subset Y$ is a convex, closed cone, and denote its negative polar cone by P^-. The cone P defines an order relation $\geq$ by

$$y_1 \geq y_2 \text{ if and only if } y_1 - y_2 \in P \tag{1.67}$$

and the cone P^- defines the order relation

$$q_1 \leq q_2 \text{ if and only if } q_1 - q_2 \in P^- \tag{1.68}$$

The minimization problem

$$v := \inf_{Ax+y\geq 0} (f(x) - \langle p, x\rangle) \tag{1.69}$$

which is a minimization problem with "inequality constraints," is obtained in the special case in which $M = P$. Its dual problem is

$$v_\star := \inf_{q \in P^-} (f^\star(p - Aq) - \langle q, y \rangle) \tag{1.70}$$

Corollary A1.5.3 *If we suppose that*

$$p \in \operatorname{Int}(\operatorname{Dom} f^\star + A^\star P^-) \tag{1.71}$$

then there exists a solution $\overline{x}$ of the problem v. If we suppose further that

$$y \in \operatorname{Int}(P - A \operatorname{Dom} f) \tag{1.72}$$

then the solutions $\overline{x}$ of the problem v are the solutions of the inclusion

$$p \in \partial f(\overline{x}) + A^\star N_P(A\overline{x} + y) \tag{1.73}$$

The following conditions are equivalent:

1. *$\overline{q}$ is a solution of $v_\star$*
2. *$\overline{q}$ is a solution of the inclusion $y \in N_{P^-}(\overline{q}) - A\partial f^\star(p - A^\star\overline{q})$.*

The solutions $\overline{x}$ and $\overline{q}$ of the problems v and $v_\star$ are related by

$$\begin{cases} \text{(i)} & p \in \partial f(\overline{x}) - A^\star\overline{q} \\ \text{(ii)} & A\overline{x} + y \geq 0, \quad \overline{q} \leq 0, \quad \text{and} \quad \langle \overline{q}, A\overline{x} + y \rangle = 0 \end{cases} \tag{1.74}$$

A1.6 Differentiability Properties of Convex Functions

Definition A1.6.1 *Consider a nontrivial function $f : X \mapsto \mathbf{R} \cup \{+\infty\}$. We shall say that f is Gâteaux differentiable at x if f is right-differentiable and $v \to Df(x)(v) = \langle \nabla f(x), v \rangle$ is linear and continuous. We shall call $f'(x) := \nabla f(x) \in X^\star$ the gradient of f at x. We shall say that f is continuously differentiable if for all $v \in U$, the function $y \to \langle \nabla f(y), v \rangle$ is continuous at x. We shall say that f is Fréchet-differentiable at x if*

$$\lim_{v \to 0} \left| \frac{f(x+v) - f(x) - \langle \nabla f(x), v \rangle}{\|v\|} \right| = 0 \tag{1.75}$$

and that f is strictly Fréchet-differentiable at x if

$$\lim_{\substack{y \to x \\ v \to 0}} \left| \frac{f(y+v) - f(y) - \langle \nabla f(x), v \rangle}{\|v\|} \right| = 0$$

We say that

$$D_\uparrow f(x)(u) \;=\; \liminf_{h\to 0+, u'\to u} \frac{f(x+hu')-f(x)}{h} \;\in\; \mathbf{R}\cup\{\pm\infty\}$$

is the contingent epiderivative of f *at* x *in the direction* u*. The function* f *is contingently epidifferentiable at* x *if and only if* $D_\uparrow f(x)(0) = 0$*.*

The subdifferential $\partial f(x)$ *of* f *at* x *is the following subset of subgradients* $p \in X^\star$*:*

$$\partial f(x) \;:=\; \{p \in X^\star \mid \forall\, v \in X,\;\; \langle p, v\rangle \le D_\uparrow f(x)(v)\}$$

Naturally, when f is differentiable at x, we have

$$D_\uparrow f(x)(u) \;=\; \langle f'(x), u\rangle \quad \text{and} \quad \partial f(x) \;=\; \{f'(x)\}$$

where $f'(x) := \nabla f(x)$ denotes the gradient of f at x.

The Fermat rule becomes as follows: If x minimizes f over X, then

$$\forall\, u \in X,\;\; 0 \le D_\uparrow f(\overline{x})(u)$$

Proposition A1.6.2 *When the function* $f : X \mapsto \mathbf{R}\cup\{+\infty\}$ *is convex, the contingent epiderivative is equal to*

$$D_\uparrow V(x)(u) \;=\; \liminf_{u'\to u} \left(\inf_{h>0} \frac{V(x+hu')-V(x)}{h} \right)$$

Furthermore, when x *belongs to the interior of the domain of* f*,*

$$\forall\, u \in X,\;\; D_\uparrow f(x)(u) \;=\; \inf_{h>0} \frac{f(x+hu)-f(x)}{h}$$

is finite. When f *is convex, the two concepts of subdifferential coincide.*

A1.7 Tangent Cones to a Subset

We use the concept of tangent cones to any subset of a finite-dimensional vector space in several chapters of this book. The contingent cone plays a crucial role in viability theory (Chapters 7, 8, and 10) and the Clarke tangent cone is involved in the transversality conditions for the inverse function theorem used in qualitative analysis (Chapter 9). They are also used in the definitions of *epiderivatives* (in Appendix A1, Section A1.8) and of *derivatives of set-valued maps* (in Section A1.10). We begin with the definitions of the contingent and Clarke tangent cones:

Definition A1.7.1 (Contingent Cones) *Let $K \subset X$ be a subset of a normed vector space X and $x \in \overline{K}$ belong to the closure of K. The contingent cone $T_K(x)$ is defined by*

$$T_K(x) := \{v \mid \liminf_{h \to 0+} d_K(x + hv)/h = 0\}$$

The Clarke[1] *tangent cone or circatangent cone $C_K(x)$ is defined by*

$$C_K(x) := \{v \mid \lim_{h \to 0+, x' \to_K x} d_K(x' + hv)/h = 0\}$$

where $\to_K$ denotes the convergence in K.

It is very convenient to have the following characterization of these cones in terms of sequences:

$$\begin{cases} v \in T_K(x) \text{ if and only if } \exists\, h_n \to 0+ \text{ and } \exists\, v_n \to v \\ \text{such that } \forall\, n,\ x + h_n v_n \in K \end{cases}$$

and

$$\begin{cases} v \in C_K(x) \text{ if and only if } \forall\ h_n \to 0+,\ \forall\, x_n \to_K x, \\ \exists\, v_n \to v \text{ such that } \forall\, n,\ x_n + h_n v_n \in K \end{cases}$$

We see at once that they are closed cones, that

$$C_K(x) \subset T_K(x)$$

that these tangent cones to K and the closure $\overline{K}$ of K do coincide, and that

$$\text{if } x \in \operatorname{Int}(K), \text{ then } C_K(x) = X$$

(One can prove that the converse is true when the dimension of X is finite.)

These tangent cones may all be different: In the example of the set shown in Figure A1.1, the contingent cone at zero coincides with the whole space, and the cone $C_K(0)$ is equal to the horizontal axis $\mathbf{R} \times \{0\}$. Elementary properties of contingent cones are summarized in Table A.1.

Let us mention an astonishing fact: The tangent cone $C_K(x)$ is always a closed convex cone.

Proposition A1.7.2 *The tangent cone $C_K(x)$ is a closed convex cone.*

[1] We shall use the adjective *circatangent* to describe properties derived from this tangent cone, e.g., circatangent derivatives and epiderivatives.

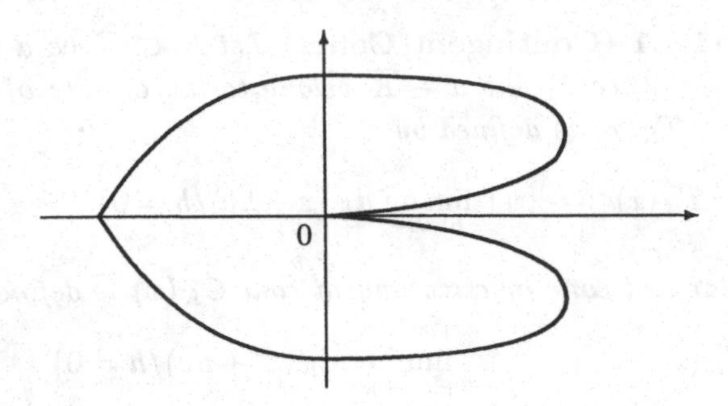

Fig. A1.1 Contingent Cone at a Boundary Point may be the Whole Space. Subset K such that $T_K(0) = X$

Proof Let v_1 and v_2 belong to $C_K(x)$. To prove that $v_1 + v_2$ belongs to this cone, let us choose any sequence $h_n > 0$ converging to zero and any sequence of elements $x_n \in K$ converging to x. There exists a sequence of elements v_{1n} converging to v_1 such that the elements $x_{1n} := x_n + h_n v_{1n}$ belong to K for all n. But because the sequence x_{1n} also converges to x in K, there exists a sequence of elements v_{2n} converging to v_2 such that

$$\forall\, n, \;\; x_{1n} + h_n v_{2n} \;=\; x_n + h_n(v_{1n} + v_{2n}) \;\in\; K$$

This implies that $v_1 + v_2$ belongs to $C_K(x)$, because the sequence of elements $v_{1n} + v_{2n}$ converges to $v_1 + v_2$. □

Unfortunately, the price to be paid for enjoying this convexity property of the Clarke tangent cones is that they often may be reduced to the trivial cone $\{0\}$.

We point out an easy but important relation between the external contingent cone at a point and the contingent cone at its projection:

Lemma A1.7.3 *Let K be a closed subset of a finite-dimensional vector space and $\Pi_K(y)$ be the set of projections of y onto K, that is, the subset of $z \in K$ such that $\|y - z\| = d_K(y)$. Then the inequalities*

$$D_\uparrow d_K(y)(v) \;\leq\; d(v, T_K(\Pi_K(y)))$$

hold true. Therefore,

$$T_K(\Pi_K(y)) \subset T_K(y)$$

Proof Choose $z \in \Pi_K(y)$ and $w \in T_K(z)$. Then

Table A1.1. *Properties of contingent cones*

(1)	▷	If $K \subset L$ and $x \in \overline{K}$, then $T_K(x) \subset T_L(x)$
(2)	▷	If $K_i \subset X$ $(i = 1, \ldots, n)$ and $x \in \overline{\bigcup_i K_i}$, then $T_{\bigcup_{i=1}^n K_i}(x) = \bigcup_{i \in I(x)} T_{K_i}(x)$, where $I(x) := \{i \mid x \in \overline{K_i}\}$
(3)	▷	If $K_i \subset X_i$ $(i = 1, \ldots, n)$ and $x_i \in \overline{K_i}$, then $T_{\prod_{i=1}^n K_i}(x_1, \ldots, x_n) \subset \prod_{i=1}^n T_{K_i}(x_i)$
(4)	▷	If $g \in \mathcal{C}^1(X, Y)$, if $K \subset X$, $x \in \overline{K}$, and $M \subset Y$, then $g'(x)(T_K(x)) \subset T_{g(K)}(g(x))$ $T_{g^{-1}(M)}(x) \subset g'(x)^{-1} T_M(g(x))$
(5)	▷	If $K_i \subset X$ $(i = 1, \ldots, n)$ and $x \in \overline{\bigcap_i K_i}$, then $T_{\bigcap_{i=1}^n K_i}(x) \subset \bigcap_{i=1}^n T_{K_i}(x)$

$$\frac{d_K(y+hv) - d_K(y)}{h} \leq \frac{\|y-z\| + d_K(z+hv) - d_K(y)}{h}$$

$$= \frac{d_K(z+hv)}{h} \leq \frac{d_K(z+hw)}{h} + \|v-w\|$$

Because z belongs to K and $w \in T_K(z)$, the foregoing inequality implies that

$$D_\uparrow d_K(y)(v) \leq d(v, T_K(z))$$

□

We recall the definition of sleek subsets:

Definition A1.7.4 *We shall say that a closed subset K is sleek at $x_0 \in K$ if the cone-valued map*

$$K \ni x \rightsquigarrow T_K(x)$$

is lower-semicontinuous at x_0 and that it is sleek *if it is sleek at every point of K.*

Convex subsets and smooth manifolds are sleek.

Theorem A1.7.5 (Tangent Cones to Sleek Subsets) *Let K be a closed subset of a Banach space. If K is sleek at $x \in K$, then the*

Table A1.2 *Properties of contingent cones to sleek sets in finite-dimensional spaces*

(5)	▷	If $L \subset X$ and $M \subset Y$ are *closed sleek* subsets, $f \in \mathcal{C}^1(X,Y)$ is a continuously differentiable map, and $x \in L \cap f^{-1}(M)$ satisfies the *transversality condition* $f'(x)C_L(x) - C_M(f(x)) = Y$, then $T_{L\cap f^{-1}(M)}(x) = T_L(x) \cap f'(x)^{-1}T_M(f(x))$
(5a)	▷	If $M \subset Y$ is a *closed sleek* subset, $f \in \mathcal{C}^1(X,Y)$ is a continuously differentiable map, and $x \in f^{-1}(M)$ satisfies $\mathrm{Im}(f'(x)) - C_M(f(x)) = Y$, then $T_{f^{-1}(M)}(x) = f'(x)^{-1}T_M(f(x))$
(5b)	▷	If K_1 and K_2 are *closed sleek* subsets contained in X, and if $x \in K_1 \cap K_2$ satisfies $C_{K_1}(x) - C_{K_2}(x) = X$, then $T_{K_1\cap K_2}(x) = T_{K_1}(x) \cap T_{K_2}(x)$
(5c)	▷	If $K_i \subset X$ $(i = 1,\ldots,n)$ are *closed sleek* subsets and $x \in \bigcap_i K_i$ satisfies $\forall v_i = 1,\ldots,n, \quad \bigcap_{i=1}^n (C_{K_i}(x) - v_i) \neq \emptyset$, then $T_{\bigcap_{i=1}^n K_i}(x) = \bigcap_{i=1}^n T_{K_i}(x)$

contingent and Clarke tangent cones to K at x coincide and consequently are convex.

The properties of tangent cones to sleek subsets are presented in Table A1.2.

Example: Tangent Cones to Subsets Defined by Equality and Inequality Constraints Consider a closed subset L of a Banach space X and two continuously differentiable maps

$$g := (g_1,\ldots,g_p) : X \mapsto \mathbf{R}^p \quad \text{and} \quad h := (h_1,\ldots,h_q) : X \mapsto \mathbf{R}^q$$

defined on an open neighborhood of L. Let K be the subset of L defined by the constraints

$$K := \{x \in L \mid g_i(x) \geq 0, \quad i = 1,\ldots,p \text{ and } h_j(x) = 0, j = 1,\ldots,q\}$$

We denote by $I(x) := \{i = 1,\ldots,p \mid g_i(x) = 0\}$ the subset of active constraints.

Table A1.3. *Properties of tangent cones to convex sets*

---

(1) ▷ If $x \in K \subset L \subset X$, then
$T_K(x) \subset T_L(x)$ and $N_L(x) \subset N_K(x)$

(3) ▷ If $x_i \in K_i \subset X_i$ $(i = 1, \ldots, n)$, then
$T_{\prod_{i=1}^n K_i}(x_1, \ldots, x_n) = \prod_{i=1}^n T_{K_i}(x_i)$
$N_{\prod_{i=1}^n K_i}(x_1, \ldots, x_n) = \prod_{i=1}^n N_{K_i}(x_i)$

(4a) ▷ If $A \in \mathcal{L}(X, Y)$ and $x \in K \subset X$, then
$T_{A(K)}(Ax) = \overline{A(T_K(x))}$
$N_{A(K)}(Ax) = A^{\star^{-1}} N_K(x)$

(4b) ▷ If K_1, $K_2 \subset X$, $x_i \in K_i$, $i = 1, 2$, then
$T_{K_1+K_2}(x_1 + x_2) = \overline{T_{K_1}(x_1) + T_{K_2}(x_2)}$
$N_{K_1+K_2}(x_1 + x_2) = N_{K_1}(x_1) \cap N_{K_2}(x_2)$
In particular, if $x_1 \in K$ and x_2 belongs to a subspace P of X, then
$T_{K+P}(x_1 + x_2) = \overline{T_{K_1}(x_1) + P}$
$N_{K+P}(x_1 + x_2) = N_K(x_1) \cap P^\perp$

(5) ▷ If $L \subset X$ and $M \subset Y$ are *closed convex* subsets and $A \in \mathcal{L}(X, Y)$ satisfies the *constraint qualification assumption* $0 \in \text{Int}(M - A(L))$, then, for every $x \in L \cap A^{-1}(M)$,
$T_{L \cap A^{-1}(M)} = T_L(x) \cap A^{-1} T_M(Ax)$
$N_{L \cap A^{-1}(M)} = N_L(x) + A^\star N_M(Ax)$

(5a) ▷ If $M \subset Y$ is *closed convex* and if $A \in \mathcal{L}(X, Y)$ satisfies $0 \in \text{Int}(\text{Im}(A) - M)$, then for $x \in A^{-1}(M)$
$T_{A^{-1}(M)}(x) = A^{-1} T_M(Ax)$
$N_{A^{-1}(M)}(x) = A^\star N_M(Ax)$

(5b) ▷ If K_1, $K_2 \subset X$ are *closed convex* and satisfy $0 \in \text{Int}(K_1 - K_2)$, then for any $x \in K_1 \cap K_2$
$T_{K_1 \cap K_2}(x) = T_{K_1}(x) \cap T_{K_2}(x)$
$N_{K_1 \cap K_2}(x) = N_{K_1}(x) + N_{K_2}(x)$

(5c) ▷ If $K_i \subset X$, $(i = 1, \ldots, n)$, are *closed* and *convex*, $x \in \bigcap_{i=1}^n K_i$ and if there exists $\gamma > 0$ satisfying $\forall x_i$ such that $\|x_i\| \leq \gamma$, $\bigcap_{i=1}^n (K_i - x_i) \neq \emptyset$, then
$T_{\bigcap_{i=1}^n K_i}(x) = \bigcap_{i=1}^n T_{K_i}(x)$
$N_{\bigcap_{i=1}^n K_i}(x) = \sum_{i=1}^n N_{K_i}(x)$

Proposition A1.7.6 *Let us posit the following transversality condition at a given $x \in K$:*

$$\begin{cases} \text{(i)} & h'(x)C_L(x) = \mathbf{R}^q \\ \text{(ii)} & \exists\, v_0 \in C_L(x) \text{ such that } h'(x)v_0 = 0 \text{ and} \\ & \forall\, i \in I(x),\ \langle g_i'(x), v_0\rangle > 0 \end{cases}$$

Then u belongs to the contingent cone to K at x if and only if u belongs to the contingent cone to L at x and satisfies the constraints

$$\forall\, i \in I(x),\ \langle g_i'(x), u\rangle \geq 0 \quad \text{and} \quad \forall\, j = 1, \ldots, q,\ h_j'(x)u = 0$$

Furthermore, if $u \in C_L(x)$ and satisfies the foregoing constraints, then u belongs to $C_K(x)$.

For convex subsets K, the situation is dramatically simplified by the fact that the Clarke tangent cones and the contingent cones coincide with the closed cone spanned by $K - x$ which are sleek.

We provide examples of tangent cones to some specific convex subsets:

Examples

1. We observe first that an element v belongs to $T_{\mathbf{R}^n_+}(x)$ if and only if $v_i \geq 0$ whenever $x_i = 0$.
2. We denote by

$$S^n := \left\{ x \in \mathbf{R}^n_+ \mid \sum_{i=1}^{n} x_i = 1 \right\}$$

the probability simplex.

Lemma A1.7.7 *The contingent cone $T_{S^n}(x)$ to S^n at $x \in S^n$ is the cone of elements $v \in \mathbf{R}^n$ satisfying*

$$\sum_{i=1}^{n} v_i = 0 \quad \text{and} \quad v_i \geq 0 \text{ whenever } x_i = 0 \tag{1.76}$$

Proof Let us take $v \in T_{S^n}(x)$. There exist sequences $h_p > 0$ converging to zero and v_p converging to v such that $y_p := x + h_p v_p$ belongs to S^n for any $p \geq 0$. Then

$$\sum_{i=1}^{n} v_{p_i} = \frac{1}{h_p}\left(\sum_{i=1}^{n} y_{p_i} - \sum_{i=1}^{n} x_{p_i}\right) = 0$$

so that $\sum_{i=1}^{n} v_i = 0$. On the other hand, if $x_i = 0$, then

$$v_{p_i} = y_{p_i}/h_p \geq 0$$

so that $v_i \geq 0$.

Conversely, let us take v satisfying (1.76) and deduce that

$$y := x + hv$$

belongs to the simplex for h small enough. First, the sum of the y_i is obviously equal to 1. Second, $y_i \geq 0$ when $x_i = 0$, because in this case v_i is nonnegative, and when $x_i > 0$, it is sufficient to take $h < x_i/|v_i|$ for having $y_i \geq 0$. Hence y does belong to the simplex. □

A1.8 Generalized Gradient

Before defining the contingent epiderivatives of a function by taking the contingent cones to its epigraph, we need to prove the following statement:

Proposition A1.8.1 *Let $f : X \mapsto \mathbf{R} \cup \{\pm\infty\}$ be a nontrivial extended function and x belong to its domain. Then the contingent cone to the epigraph of f at $(x, f(x))$ is the epigraph of an extended function denoted $D_\uparrow f(x)$:*

$$\mathcal{E}p(D_\uparrow f(x)) = T_{\mathcal{E}p(f)}(x, f(x))$$

which is equal to

$$\forall\, u \in X, \;\; D_\uparrow f(x)(u) = \liminf_{h \to 0+, u' \to u} (f(x + hu') - f(x))/h$$

Proof Indeed, to say that

$$(u, v) \in T_{\mathcal{E}p(f)}(x, f(x))$$

amounts to saying that there exist sequences $h_n > 0$ converging to 0+ and (u_n, v_n) converging to (u, v) satisfying

$$\forall\, n \geq 0, \;\; \frac{f(x + h_n u_n) - f(x)}{h_n} \leq v_n$$

This is equivalent to saying that

$$\forall\, u \in X, \;\; \liminf_{h \to 0+, u' \to u} (f(x + hu') - f(x))/h \leq v$$

□

Definition A1.8.2 *Let $f : X \mapsto \mathbf{R}\cup\{\pm\infty\}$ be an extended function and $x \in \mathrm{Dom}(f)$. We shall say that the function $D_\uparrow f(x)$ is the contingent epiderivative of f at x and that the function f is contingently epidifferentiable at x if for any $u \in X$, $D_\uparrow f(x)(u) > -\infty$ [or, equivalently, if $D_\uparrow f(x)(0) = 0$]. A function f is episleek (at x) if its epigraph is episleek [at $(x, f(x))$].*

Consequently, the epigraph of the contingent epiderivative at x is a closed cone. It is then lower-semicontinuous and positively homogeneous whenever f is contingently epidifferentiable at x.

Remark There are other intimate connections between contingent cones and contingent epiderivatives. Let ψ_K be the *indicator* of a subset K. Then it is easy to check that

$$D_\uparrow(\psi_K)(x) \;=\; \psi_{T_K(x)} \quad \text{and} \quad \partial\psi_K(x) \;=\; N_K(x)$$

Therefore, we can either derive properties of the epiderivatives from properties of the tangent cones to epigraphs or take the opposite approach by using the foregoing formula. □

There is also an obvious link between the contingent cone and the contingent epiderivative of the distance function to K, because we can write, for every $x \in K$,

$$T_K(x) \;=\; \{v \in X \mid D_\uparrow d_K(x)(v) \;=\; 0\}$$

By using the Clarke tangent cone, we define in the same way the concept of *Clarke epiderivative*:

Definition A1.8.3 *Let f be a function from X to $\mathbf{R}\cup\{+\infty\}$ with a nonempty domain. The Clarke tangent cone to the epigraph of f at $(x, f(x))$ is the epigraph of an extended function denoted $C_\uparrow f(x)$:*

$$\mathcal{E}p(C_\uparrow f(x)) = C_{\mathcal{E}p(f)}(x, f(x))$$

called the Clarke epiderivative of f at x. The subset $\partial f(x)$ of X defined by

$$\partial f(x) = \{p \in X^* | \langle p, v\rangle \leq C_\uparrow f(x)(v),\ \forall v \in X\} \tag{1.77}$$

is called the generalized gradient of f at x.

The introduction of the Clarke epiderivatives and generalized gradients allows us to implement the Fermat rule for locally Lipschitz functions:

Proposition A1.8.4 *Let $f : X \to \mathbf{R} \cup \{+\infty\}$ be a nontrivial extended function. Suppose that $x \in \text{Int Dom} f$ is a local minimum of f. Then, for all $v \in X$, $C_\uparrow f(x)(v) \geq 0$, and thus*

$$0 \in \partial f(x) \qquad \text{(Fermat's rule)} \tag{1.78}$$

One can define Clarke epiderivatives as adequate limits of differential quotients. But this is reasonable only for the class of locally Lipschitz functions. This is an interesting class of functions, because both continuously differentiable functions and convex continuous functions are locally Lipschitz. Furthermore, the upper envelope of Lipschitz functions is Lipschitz.

Proposition A1.8.5 *Let f be a locally Lipschitz function from X to $\mathbf{R} \cup \{+\infty\}$ with a nonempty domain. Then the Clarke epiderivative of f at x in the direction v is equal to*

$$C_\uparrow f(x)(v) := \limsup_{\substack{h \to 0+ \\ y \to x}} \frac{f(y+hv) - f(y)}{h} \tag{1.79}$$

For all $x \in \text{Int Dom} f$,

$$v \to C_\uparrow f(x)(v) \text{ is positively homogeneous, convex, and continuous} \tag{1.80}$$

Moreover,

$$\{x, v\} \in \text{Int Dom} f \times U \to D_x f(x)(v) \text{ is upper-semicontinuous} \tag{1.81}$$

We note that the function $v \to C_\uparrow f(x)(v)$ is *positively homogeneous* and that we have

$$D_\uparrow f(x)(v) \leq C_\uparrow f(x)(v) \tag{1.82}$$

Suppose that f is continuously differentiable at x. Then f is Clarke-differentiable, and

$$\langle f'(x), v\rangle = C_\uparrow f(x)(v) \tag{1.83}$$

Proposition A1.8.6 *Suppose the function $f : X \to \mathbf{R} \cup \{+\infty\}$ is locally Lipschitz. Then it has a nonempty generalized gradient $\partial f(x)$ at any point x in the interior of* Dom f, *which is convex, closed, and bounded and has a support function $\sigma(\partial f(x), v) := \sup\{\langle p, v\rangle | p \in \partial f(x)\}$ that satisfies*

$$\sigma(\partial f(x), v) = C_\uparrow f(x)(v) \tag{1.84}$$

Next we shall establish some elementary properties of Clarke derivatives of locally Lipschitz functions.

Proposition A1.8.7 *Suppose that f and g are two locally Lipschitz functions from X to $\mathbf{R} \cup \{+\infty\}$ and that $x \in \operatorname{Int} \operatorname{Dom} f \cap \operatorname{Int} \operatorname{Dom} g$. Then*

$$C_{\uparrow}(\alpha f + \beta g)(x)(v) \leq \alpha C_{\uparrow} f(x)(v) + \beta C_{\uparrow} g(x)(v) \tag{1.85}$$

if $\alpha, \beta > 0$. If $x \in \operatorname{Int} \operatorname{Dom} f$, then

$$C_{\uparrow}(-f)(x)(v) = C_{\uparrow} f(x)(-v) \tag{1.86}$$

Proof To prove the second formula, we write

$$\frac{-f(y + \lambda v) - (-f(y))}{\lambda} = \frac{f(z + \lambda(-v)) - f(z)}{\lambda} \tag{1.87}$$

where $z = y + \lambda v$ converges to x when y converges to x and $\lambda > 0$ tends to zero. Taking the upper limits as y and z converge to x and λ converges to zero, the term on the left converges to $C_{\uparrow}(-f)(x)(v)$, and that on the right to $C_{\uparrow} f(x)(-v)$. □

Next, we shall study the differentiability of the composition $g = f \circ G$, where G maps an open subset Ω of a Hilbert space Y into $\operatorname{Dom} f$, the domain of f.

Proposition A1.8.8 *Suppose that f is locally Lipschitz. If G is strictly Fréchet-differentiable at x, then*

$$C_{\uparrow} g(x)(v) \leq (C_{\uparrow} f)(G(x))(G'(x) \cdot v) \tag{1.88}$$

Proposition A1.8.9 *Suppose that f and g are two locally Lipschitz functions from X to $\mathbf{R} \cup \{+\infty\}$ and that $x \in \operatorname{Int} \operatorname{Dom} f \cap \operatorname{Int} \operatorname{Dom} g$. Then*

$$C_{\uparrow}(\alpha f + \beta g)(x)(v) \leq \alpha C_{\uparrow} f(x)(v) + \beta C_{\uparrow} g(x)(v) \tag{1.89}$$

if $\alpha, \beta > 0$, and thus

$$\partial(\alpha f + \beta g)(x) \subset \alpha \partial f(x) + \beta \partial g(x) \tag{1.90}$$

If $x \in \operatorname{Int} \operatorname{Dom} f$, then

$$C_{\uparrow}(-f)(x)(v) = C_{\uparrow} f(x)(-v) \tag{1.91}$$

and thus

$$\partial_c(-f)(x) = -\partial_c f(x) \tag{1.92}$$

We consider m functions $f_i : X \to \mathbf{R}\cup\{+\infty\}$ and their upper envelope g defined by $g(x) = \max_{i\in I} f_i(x)$, where $I = \{1,\dots,n\}$. We denote $I(x) = \{i \in I \mid g(x) = f_i(x)\}$. We note that if the functions f_i are locally Lipschitz, the same is true of their upper envelope. [If $|f_i(y) - f_i(z)| \le L_i\|y-x\|$ and $y, z \in x+\eta_i B$, then $|g(y)-g(z)| \le L\|y-z\|$ if $y, z \in x+\eta B$, where $\eta = \min_{i\in I} \eta_i > 0$ and $L = \max_{i\in I} L_i > 0$.] Whence the functions f_i and g are Clarke-differentiable.

Proposition A1.8.10 *Suppose that the m functions f_i are locally Lipschitz and that $x \in \cap_{i\in I}\mathrm{Int\,Dom}\, f_i$. Then*

$$C_\uparrow g(x)(v) \le \max_{i\in I(x)} C_\uparrow f_i(x)(v) \tag{1.93}$$

and thus

$$\partial g(x) \subset \overline{\mathrm{co}} \cup_{i\in I(x)} \partial_c f_i(x) \tag{1.94}$$

Proof (a) We first note that there exists $\alpha_1 > 0$ such that if $\|x - y\| \le \alpha_1$ then $I(y) \subset I(x)$. [Suppose $a = g(x) - \max_{j\notin I(x)} f_j(x)$, $\varepsilon = a/3$, and $\alpha_1 > 0$ are such that for all $i \in I$, $|f_i(y) - f_i(x)| \le \varepsilon$ whenever $\|y - x\| \le \alpha_1$. Then if $j \in I(y)$,

$$\begin{aligned} f_j(x) &\ge f_j(y) - \varepsilon = g(y) - \varepsilon \ge g(x) - 2\varepsilon \\ &= a - 2\varepsilon + \max_{i\notin I(x)} f_i(x) > \max_{i\notin I(x)} f_i(x) \end{aligned}$$

Thus $j \in I(x)$.]

(b) If $\alpha \le \alpha_1/2$ and $\beta \le \alpha_1/2\|v\|$, we obtain the inequality

$$\frac{g(y+\theta v)-g(y)}{\theta} \le \max_{i\in I(y+\theta v)} \frac{f_i(y+\theta v)-f_i(y)}{\theta}$$

$$\le \max_{i\in I(x)} \frac{f_i(y+\theta v)-f_i(v)}{\theta}$$

whence

$$C_\uparrow g(x)(v) \le \inf_{\alpha,\beta>0} \max_{i\in I(x)} \sup_{\substack{\|y-x\|\le\alpha \\ 0\le\beta}} \frac{f_i(y+\theta v)-f_i(y)}{\theta}$$

Moreover, for all $\varepsilon > 0$, and for all $i \in I$, there exist $\alpha_i > 0$ and $\beta_i > 0$ such that

$$\sup_{\substack{\|y-x\|\le\alpha_i \\ \theta\le\beta_i}} \frac{f_i(y+\theta v)-f_i(y)}{\theta} \le \max_{i\in I(x)} C_\uparrow f_i(x)(v) + \varepsilon$$

Taking $\alpha = \min_{i\in I(x)} \alpha_i > 0$ and $\beta = \min_{i\in I(x)} \beta_i > 0$, it then follows that

$$C_{\uparrow}g(x)(v) \leq \max_{i\in I(x)} \sup_{\substack{\|y-x\|\leq\alpha \\ \theta\leq\beta}} \frac{f_i(y+\theta v) - f_i(y)}{\theta}$$

$$\leq \max_{i\in I(x)} \sup_{\substack{\|y-x\|\leq\alpha \\ \theta\leq\beta}} \frac{f_i(y+\theta v) - f_i(y)}{\theta} \leq \max_{i\in I(x)} C_{\uparrow}f_i(x)(v) + \varepsilon$$

Letting ε tend to zero completes the proof of the proposition. □

A1.9 Set-valued Maps

We recall in this section the basic definitions dealing with set-valued maps, also called *multifunctions*, *multivalued functions*, *point-to-set maps*, or *correspondences*.[2]

Definition A1.9.1 *Let X and Y be metric spaces. A set-valued map F from X to Y is characterized by its* graph, *the subset* Graph(F) *of the product space* $X \times Y$ *defined by*

$$\text{Graph}(F) := \{(x,y) \in X \times Y \mid y \in F(x)\}$$

We shall say that $F(x)$ is the image or the value of F at x. A set-valued map is said to be nontrivial if its graph is not empty, that is, if there exists at least an element $x \in X$ such that $F(x)$ is not empty. We say that F is strict if all images $F(x)$ are not empty. The domain of F is the subset of elements $x \in X$ such that $F(x)$ is not empty:

$$\text{Dom}(F) := \{x \in X \mid F(x) \neq \emptyset\}$$

The image of F is the union of the images (or values) $F(x)$, when x ranges over X:

$$\text{Im}(F) := \bigcup_{x\in X} F(x)$$

The inverse F^{-1} of F is the set-valued map from Y to X defined by

$$x \in F^{-1}(y) \iff y \in F(x) \iff (x,y) \in \text{Graph}(F)$$

The domain of F is thus the image of F^{-1} and coincides with the projection of the graph onto the space X, and in a symmetric way, the

[2]For more details, refer to the book *Set-Valued Analysis* (Aubin and Frankowska 1990).

image of F is equal to the domain of F^{-1} and to the projection of the graph of F onto the space Y.

If K is a subset of X, we denote by $F|_K$ the *restriction* of F to K, defined by

$$F|_K(x) := \begin{cases} F(x) & \text{if } x \in K \\ \emptyset & \text{if } x \notin K \end{cases}$$

Let $\mathcal{P}$ be a property of a subset (closed, convex, etc.) We shall say as a general rule that a set-valued map satisfies property $\mathcal{P}$ if and only if its graph satisfies it. For instance, a set-valued map is said to be closed (respectively convex, closed convex) if and only if its graph is closed (respectively convex, closed convex). If the images of a set-valued map F are closed, convex, bounded, compact, and so on, we say that F is closed-valued, convex-valued, bounded-valued, compact-valued, and so on.

When $\star$ denotes an operation on subsets, we use the same notation for the operation on set-valued maps, which is defined by

$$F_1 \star F_2 : x \rightsquigarrow F_1(x) \star F_2(x)$$

We define in that way $F_1 \cap F_2$, $F_1 \cup F_2$, $F_1 + F_2$ (in vector spaces), and so forth. Similarly, if α is a map from the subsets of Y to the subsets of Y, we define

$$\alpha(F) : x \rightsquigarrow \alpha(F(x))$$

For instance, we shall use $\overline{F}$, $\text{co}(F)$, and so forth, to denote the set-valued maps $x \rightsquigarrow \overline{F}(x)$, $x \rightsquigarrow \text{co}(F(x))$, and so forth. We shall write

$$F \subset G \iff \text{Graph}(F) \subset \text{Graph}(G)$$

and say that G is an *extension of* F.

Let us mention the following elementary properties:

$$\begin{cases} \text{(i)} & F(K_1 \cup K_2) & = & F(K_1) \cup F(K_2) \\ \text{(ii)} & F(K_1 \cap K_2) & \subset & F(K_1) \cap F(K_2) \\ \text{(iii)} & F(X \backslash K) & \supset & \text{Im}(F) \backslash F(K) \\ \text{(iv)} & K_1 \subset K_2 & \Longrightarrow & F(K_1) \subset F(K_2) \end{cases}$$

There are two ways to define the inverse image by a set-valued map F of a subset M:

$$\begin{cases} \text{(i)} & F^{-1}(M) := \{x \mid F(x) \cap M \neq \emptyset\} \\ \text{(ii)} & F^{+1}(M) := \{x \in \text{Dom}(F) \mid F(x) \subset M\} \end{cases}$$

The subset $F^{-1}(M)$ is called the *inverse image* of M by F, and $F^{+1}(M)$ is called the *core* of M by F. They naturally coincide when F is single-valued. We observe at once that

$$F^{+1}(Y \setminus M) = X \setminus F^{-1}(M) \quad \text{and} \quad F^{-1}(Y \setminus M) = X \setminus F^{+1}(M)$$

One can conceive, as well, two dual ways for defining composition products of set-valued maps (which coincide when the maps are single-valued):

Definition A1.9.2 *Let X, Y, Z be metric spaces and*

$$G : X \rightsquigarrow Y \quad \textit{and} \quad H : Y \rightsquigarrow Z$$

be set-valued maps.

1. *The usual composition product (called simply the product) $H \circ G : X \rightsquigarrow Z$ of H and G at x is defined by*

$$(H \circ G)(x) := \bigcup_{y \in G(x)} H(y)$$

2. *The square product $H \square G : X \rightsquigarrow Z$ of H and G at x is defined by*

$$(H \square G)(x) := \bigcap_{y \in G(x)} H(y)$$

Let 1 denote the identity map from one set to itself. We deduce the following formulas:

$$\begin{array}{rcl} \text{Graph}(H \circ G) & = & (G \times \mathbf{1})^{-1}(\text{Graph}(H)) \\ & = & (\mathbf{1} \times H)(\text{Graph}(G)) \\ & & \\ \text{Graph}(H \square G) & = & (G \times \mathbf{1})^{+1}(\text{Graph}(H)) \end{array} \tag{1.95}$$

Indeed, to say that a pair (x, z) belongs to the graph of $H \circ G$ amounts to saying that there exists $y \in G(x) \cap H^{-1}(z)$. This means that $(G \times \mathbf{1})(x, z) \cap \text{Graph}(H) \neq \emptyset$ and that

$$(x, z) \in (\mathbf{1} \times H)(x, y) \in (\mathbf{1} \times H)(\text{Graph}(G))$$

The following formulas state that the inverse of a product is the product of the inverses (in reverse order):

$$\left\{ \begin{array}{lrcl} \text{(i)} & (H \circ G)^{-1}(z) & = & G^{-1}(H^{-1}(z)) = \left(G^{-1} \circ H^{-1}\right)(z) \\ \text{(ii)} & (H \square G)^{-1}(z) & = & G^{+1}(H^{-1}(z)) \end{array} \right.$$

The first formula is obvious. The second one follows from the fact that $z \in (H \square G)(x)$ if and only if z belongs to $H(y)$ for all $y \in G(x)$, that is, if and only if $G(x) \subset H^{-1}(z)$. Therefore, we also observe that

$$\begin{cases} \text{(i)} & x \in (H \square G)^{-1}(z) \iff G(x) \subset H^{-1}(z) \\ \text{(ii)} & x \in (G^{-1} \square H^{-1})(z) \iff H^{-1}(z) \subset G(x) \end{cases}$$

and thus

$$G(x) = H^{-1}(z) \iff x \in (G^{-1} \square H^{-1})(z) \cap (H \square G)^{-1}(z)$$

Let us also point out the following relations: When M is a subset of Z, then

$$\begin{cases} \text{(i)} & (H \circ G)^{-1}(M) = G^{-1}(H^{-1}(M)) \\ \text{(ii)} & (H \circ G)^{+1}(M) = G^{+1}(H^{+1}(M)) \end{cases}$$

A1.10 Derivatives of Set-valued Maps

By coming back to the original point of view proposed by Fermat, we are able to define, geometrically, derivatives of set-valued maps from the choice of tangent cones to the graphs, even though they yield very strange limits of differential quotients.

Definition A1.10.1 *Let $F : X \rightsquigarrow Y$ be a set-valued map from a normed space X to another normed space Y, and let $y \in F(x)$. The contingent derivative $DF(x,y)$ of F at $(x,y) \in$ Graph(F) is the set-valued map from X to Y defined by*

$$\text{Graph}(DF(x,y)) := T_{\text{Graph}(F)}(x,y)$$

and the circatangent derivative $CF(x,y)$ is defined by

$$\text{Graph}(CF(x,y)) := C_{\text{Graph}(F)}(x,y)$$

When $F := f$ is single-valued, we set $Df(x) := Df(x, f(x))$ and $Cf(x) := Cf(x, f(x))$. We shall say that F is sleek at $(x,y) \in$ Graph(F) if and only if the map

$$(x',y') \in \text{Graph}(F) \rightsquigarrow \text{Graph}(DF(x',y'))$$

is lower-semicontinuous at (x,y) [i.e., if the graph of F is sleek at (x,y)]. The set-valued map F is sleek if it is sleek at every point of its graph.

Naturally, the circatangent derivative $CF(x,y)$ is a closed convex process that coincides with the contingent derivative $DF(x,y)$ whenever F is sleek at (x,y).

We can easily compute the derivative of the inverse of a set-valued map F (or even of a noninjective single-valued map): The contingent derivative of the inverse of a set-valued map F is the inverse of the contingent derivative:

$$D(F^{-1})(y,x) \;=\; DF(x,y)^{-1}$$

If K is a subset of X and f is a single-valued map that is Fréchet-differentiable around a point $x \in K$, then the contingent derivative of the restriction of f to K is the restriction of the derivative to the contingent cone:

$$D(f|_K)(x) = D(f|_K)(x, f(x)) = f'(x)|_{T_K(x)}$$

These contingent derivatives can be characterized by adequate limits of differential quotients:

Proposition A1.10.2 *Let $(x,y) \in \mathrm{Graph}(F)$ belong to the graph of a set-valued map $F : X \rightsquigarrow Y$ from a normed space X to a normed space Y. Then*

$$\begin{cases} v \;\in\; DF(x,y)(u) \text{ if and only if} \\ \liminf_{h\to 0+, u'\to u} d\left(v, \frac{F(x+hu')-y}{h}\right) = 0 \end{cases}$$

If $x \in \mathrm{Int}(\mathrm{Dom}(F))$ and F is Lipschitz around x, then

$$v \;\in\; DF(x,y)(u) \text{ if and only if } \liminf_{h\to 0+} d\left(v, \frac{F(x+hu)-y}{h}\right) = 0$$

If, moreover, the dimension of Y is finite, then

$$\mathrm{Dom}(DF(x,y)) \;=\; X, \quad \text{and } DF(x,y) \text{ is Lipschitz}$$

Proof The first two statements being obvious, let us check the last one. Let u belong to X and l denote the Lipschitz constant of F on a neighborhood of x. Then, for all $h > 0$ small enough and $y \in F(x)$,

$$y \;\in\; F(x) \;\subset\; F(x+hu) \;+\; lh\|u\|B$$

Hence there exists $y_h \in F(x+hu)$ such that $v_h := (y_h - y)/h$ belongs to $l\|u\|B$, which is compact. Therefore the sequence v_h has a cluster point v, which belongs to $DF(x,y)(u)$. □

We shall use this example of the derivative of a set-valued map:

Proposition A1.10.3 *Let X and Y be normed spaces, f a single-valued map from an open subset $\Omega \subset X$ to Y, $M : X \rightsquigarrow Y$ a set-valued map and $L \subset X$. Consider the set-valued map $F : X \rightsquigarrow Y$ defined by*

$$F(x) := \begin{cases} f(x) - M(x) & \text{when} \quad x \in L \\ \emptyset & \text{when} \quad x \notin L \end{cases}$$

If f is continuously differentiable at $x \in \Omega \cap \mathrm{Dom}(F)$ and M is Lipschitz at x, then for every $y \in F(x)$ the circatangent derivative of F at (x, y) is equal to

$$CF(x,y)(u) = \begin{cases} f'(x)u - CM(x, f(x) - y)(u) & \text{when} \quad u \in C_L(x) \\ \emptyset & \text{when} \quad u \notin C_L(x) \end{cases}$$

[See Proposition 5.2.3 in *Set-Valued Analysis* (Aubin and Frankowska 1990).]

Appendix 2
Control of an AUV

Introduction

We present in this appendix the tests of the external and internal algorithms conducted by Nicolas Seube at Thomson-SINTRA to control the tracking of an exosystem by an autonomous underwater vehicle (AUV). This system has three degrees of freedom (planar motion), six state variables (positions, heading, and their derivatives), and three controls (thruster forces). The dynamics of an AUV are highly *nonlinear, coupled, and sometimes fully interacting*, thus making it difficult to control by the usual methods. Moreover, the dynamics are poorly known, because only approximate hydrodynamic models are available for real-world vehicles. Finally, we need to involve the marine currents that can significantly perturb the dynamics of the AUV.

In addition, the problem of controlling an AUV cannot be linearized about a single velocity axis[1] because all vehicle velocities usually have the same range; conventional linear control techniques clearly are unable to provide adequate performance by the control systems.

We shall present three different learning rules that address the problems of uniform minimization and adaptive learning by a set-valued feedback control map. The three classes of algorithms presented here have been tested in the case of the Japanese Dolphin AUV.

In particular, it is shown that the gradient step size is critical for the external rule, but is not critical for the uniform external algorithm. The latter could also be applied to pattern-classification problems, and may provide a plausible alternative method to stochastic gradient algorithms.

Both the external algorithm and the uniform external algorithm enable the neural network to adapt its behavior in order to resist dynamical

[1] as it can for aircraft or large submarines.

perturbations caused by marine currents, even if it learned its feedback control without being disturbed by these currents. These robustness properties remain to be proved mathematically.

The internal algorithm has been presented for the case in which we suppose perfect knowledge of the system dynamics. The internal algorithm is an adaptive one. It allows the neural network to modify its internal state in order to adapt to modifications of state constraints: The differential equation governing the evolution of the neural-network synaptic matrix allows the neural-network controller to perform adaptive tracking control. But this knowledge is not stored, and as it now stands, this algorithm is not robust to resist perturbations. It is planned to test the robustness of the method in the case in which the dynamics are modeled as a differential inclusion, taking account of all contingent uncertainties (currents, estimation errors on some coefficients).

A2.1 Dynamics of the AUV

The dynamics of the Japanese *Dolphin* vehicle are of the following form:

$$M\frac{dX}{dt} = V(X) + n(X) + U \tag{2.1}$$

where $X = (u, v, r, x, y, \psi)$ denotes the state of the AUV (motion in the horizontal plane), M is the inertia matrix, $V(X)$ denotes the idealized dynamics, $n(X)$ accounts for dynamic perturbation due to currents, and $U = (F_u, F_v, F_r)$ the set of commands. In this appendix we denote by

F_u the surging thrust
F_v the swaying thrust
F_r the thrust torque
u the surging velocity
v the swaying velocity
r the heading angular velocity

The pair (x, y) denotes the absolute location of the vehicle, and ψ is the heading of the vehicle with respect to a horizontal axis. The pair (x_c, y_c) designates the position of the center of buoyancy, and (u_a, v_a) denotes the current velocity component in the components u and v.

The expressions of M, V, and n are given by the following formulas:

$$M = \begin{pmatrix} (m + X_u) & 0 & -my_c \\ 0 & (m + Y_v) & mx_c \\ -my_c & mx_c & (I_z + N_r) \end{pmatrix} \tag{2.2}$$

$$V(u,v,r) = \begin{pmatrix} mx_c r^2 + (m + X_{vr})vr \\ my_c r^2 + (m + Y_{ur})ur \\ -mr(x_c u + y_c v) - N_{rr} \mid r \mid r \end{pmatrix} \quad (2.3)$$

$$n(u,v,r) = \begin{pmatrix} -X_{uu} \mid u - u_a \mid (u - u_a) - (m_f + X_{vr})v_a r \\ -Y_{vv} \mid v - v_a \mid (v - v_a) - (m_f + Y_{ur})u_a r \\ m_f r(x_b u_a + y_b v_a) \end{pmatrix} \quad (2.4)$$

Therefore, the dynamics of the system are described by the system of differential equations

$$\begin{pmatrix} u' \\ v' \\ r' \\ x' \\ y' \\ \psi' \end{pmatrix} = \begin{pmatrix} \left(M^{-1}(V(u,v,r) + n(u,v,r) + U)\ c \right) \\ u\cos\psi - v\sin\psi \\ u\sin\psi + v\cos\psi \\ r \end{pmatrix} \quad (2.5)$$

summarized by the differential equation

$$X'(t) = f(X(t), U(t))$$

We consider now the problem of tracking a reference trajectory, the output of an exosystem whose evolution law is given by

$$Y'(t) = g(Y(t))$$

Given an observation map H (possibly set-valued), the tracking property can be expressed as a viability constraint:

$$\forall t \geq 0, \quad (Y(t), X(t)) \in \text{Graph}(H) \quad (2.6)$$

Under the assumptions of the viability theorem, we can show that controls governing the evolution of solutions to the tracking problem are given by the regulation law

$$U(t) \in RKH(X(t), Y(t)) = \{U | f(X(t), U) \in DH(X(t), Y(t))g(Y(t))\}$$

where $DH(\cdot,\cdot)$ denotes the contingent derivative of the set-valued map H.

A2.2 The External Algorithm

The external algorithm tends to find controls that coerce the state to remain in the viability domain K. Let us consider the discrete dynamics of the vehicle $X_{n+1} = f(X_n, U_n)$. For the AUV tracking-control problem, Nicolas Seube used a three-layer neural network, composed of six input cells, six hidden cells, and three output cells. The network is fully connected. The input of the neural net is the tracking error.

When K is closed and convex, and Π_K denotes the projector of best approximation, we set

$$E_{n+1} := (I - \Pi_{H(X_{n+1})})X_{n+1}$$

and the external algorithm is given by

$$\begin{aligned} w2_{ij}^{n+1} &= w2_{ij}^{n} - \varepsilon(x^1)_i(\phi_2'(x^2)^\star \tfrac{\partial f}{\partial u}(X_n, x^2)^\star E_{n+1})_j \\ w1_{ij}^{n+1} &= w1_{ij}^{n} - \varepsilon(x^0)_i(\phi_1'(x^1)^\star W_2^\star \phi_2'(x^2)^\star \tfrac{\partial f}{\partial u}(X_n, x^2)^\star E_{n+1})_j \end{aligned} \tag{2.7}$$

where W_i denotes the network synaptic matrices, ϕ_i the threshold functions, and x^1 and x^2 the internal state of the network (nonthresholded values of layers 2 and 3).

The learning phase was composed of 337 trials for reaching a global tracking relative error, with respect to the length of the trajectory, lower than 0.005.

It is important to note that the gradient step ε needs to be changed during learning in order to ensure, or to speed up, the convergence rate. In Figure A2.1, where the actual and reference trajectories are represented, one can check that tracking is perfect.

The learning rule enables the network controller to be robust with regard to dynamic perturbations due to current. (Figure A2.2. After a spurious oscillatory phase (of very low amplitude, which decreases very quickly), the network can perform tracking control with good accuracy (the tracking error was close to 0.006, instead of 0.005).

A2.3 The Uniform External Algorithm

Learning the tracking-control feedback law by the uniform external algorithm occurs off-line. The main result obtained experimentally is that *the gradient step size can be kept constant until the learning algorithm converges.* As a result of the uniform minimization procedure, the *global*

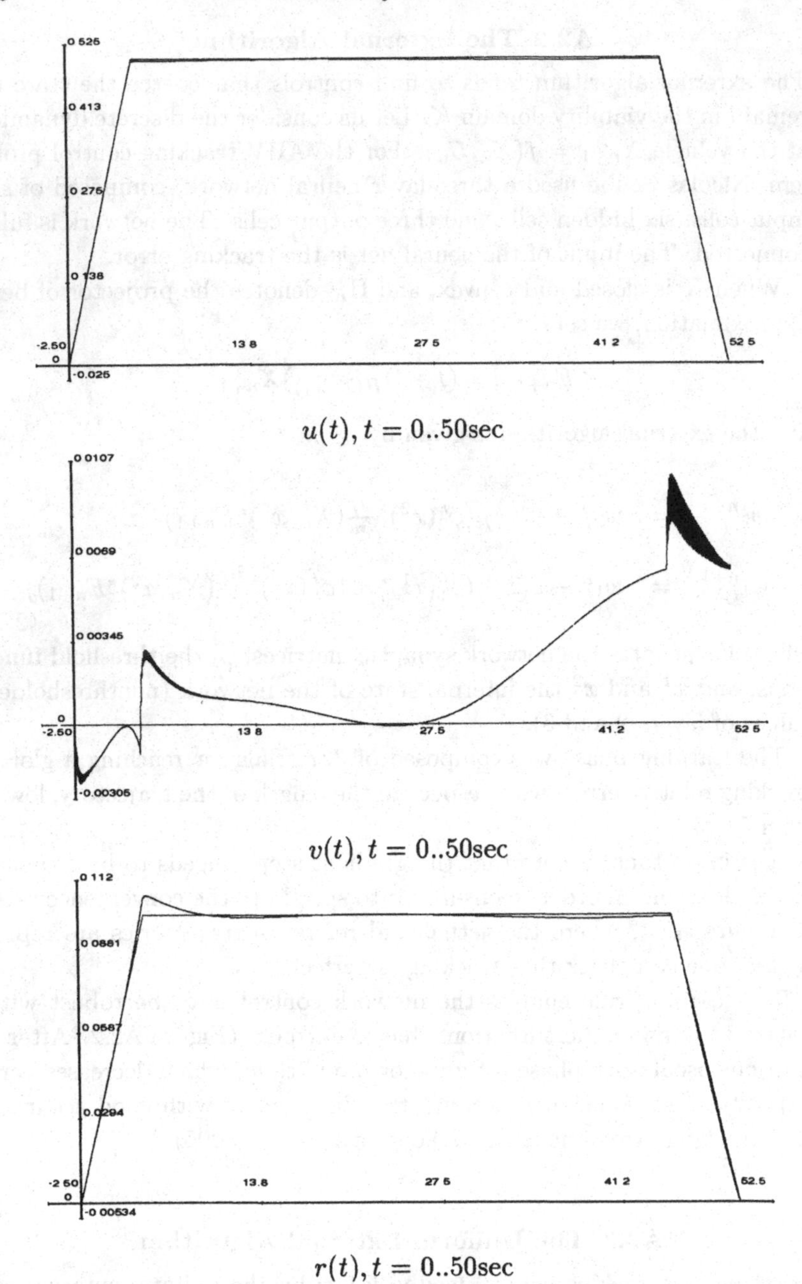

Fig. A2.1 External Rule. Case without current. Velocities of the vehicle, after learning. $u(t)$ is the surging velocity, $v(t)$, the swaying velocity, and $r(t)$, the heading velocity.

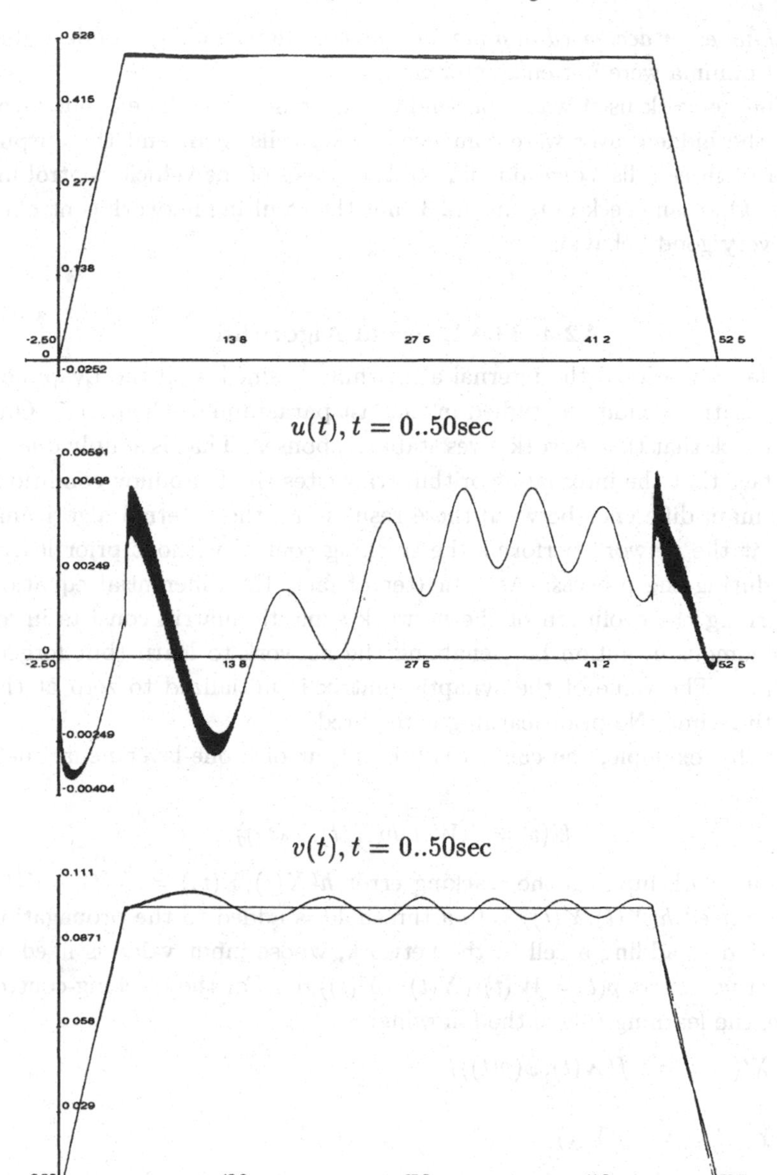

Fig. A2.2 External Rule. Case with North-South current. Velocities of the vehicle, after learning. The North-South current disturbs the system without disturbing the AUV swaying velocity. After a short oscillatory phase of the surging velocity $u(t)$ (dark part, during the first 25 secondes), the network adapts its control. Dark parts are due to high frequency oscillations of low amplitude of the swaying velocity $v(t)$. But they tend to disappear during the adaptation phase. The heading $r(t)$ velocity oscillates as the other velocities. But this oscillatory behavior tends to damp out.

tracking error decreased in a monotonic way. In particular, nondesirable local minima were not encountered.

The network used was a one-hidden layer network: The input layer and the hidden layer were composed of six cells each, and the output layer of three cells, corresponding to the values of the vehicle control inputs. One can check in figure A2.3 that the resulting network controller has very good behavior.

A2.4 The Internal Algorithm

Nicolas Seube used the internal algorithm obtained with the Lyapunov stabilization condition studied in the last paragraph of Chapter 7. One can check that the network gives stable responses. That is simply due to the fact that the internal algorithm translates the Lyapunov condition. The main difference between these results and the external algorithms is that the network performs the tracking control without prior learning during the process: As a matter of fact, the differential equation governing the evolution of the network synaptic matrix consists in an autonomous evolution law, enabling the network to learn (but forget) on-line. The value of the synaptic matrix is initialized to zero at the starting time. No prior learning is required.

In this example, the control is the output of a one-layer neural network.

$$U(t) = \phi(W(t)h(X(t), Y(t)))$$

The network input is the tracking error $h(X(t), Y(t)) = X(t) - Y(t)$. As we need $h(X(t), Y(t)) \neq 0$, a threshold is added to the propagation function in adding a cell to the network, whose input value is fixed at 1. Let us denote $p(t) = W(t)((X(t) - Y(t)), 1)$. For the tracking-control case, the learning rule is the following:

$$\begin{aligned} X'(t) &= f(X(t), \phi(p(t))) \\ Y'(t) &= g(Y(t)) \\ w'_{ij}(t) &= \left(\tfrac{h(X(t),Y(t))}{\|h(X(t),Y(t))\|}\right)_i(\phi'(p(t))^+(U'(X(t))(f(X(t), \phi(p(t))) \\ &\quad -g(Y(t))) - \phi'(p(t))W(t)h'(X(t))(f(X(t), \phi(p(t))) \\ &\quad -g(Y(t)))))_j) \end{aligned} \tag{2.8}$$

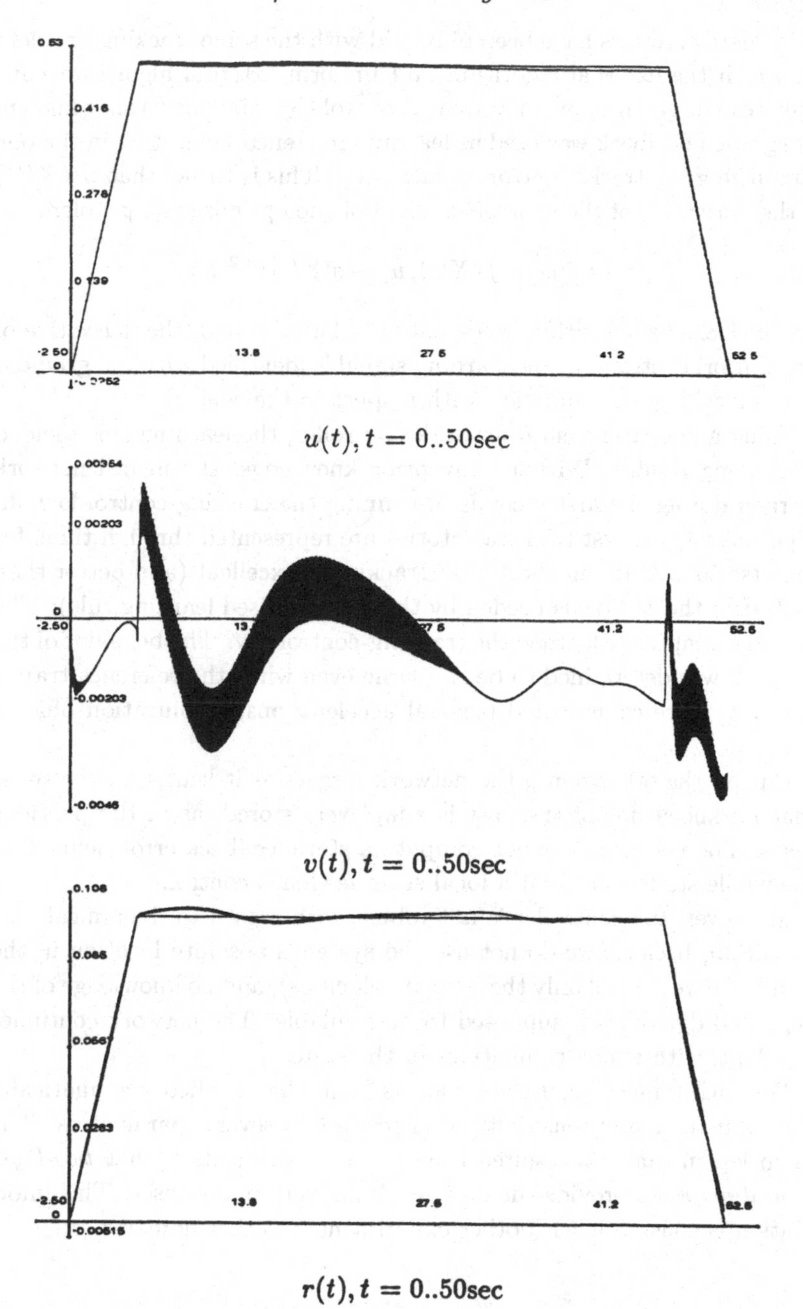

Fig. A2.3 The Uniform External Algorithm. Velocities of the vehicle. $u(t)$ is the surging velocity, $v(t)$, the swaying velocity, and $r(t)$, the heading velocity.

Numerical results have been obtained with the same tracking problem on which the external algorithm and uniform external algorithm have been tested. As the a priori feedback control $U(\cdot)$ obviously is not known, a degraded feedback was used as learning reference, consisting in the one minimizing the tracking error at each step. This is to say that $U'(X(t))$ is the derivative of the control solution of the optimization problem

$$\min_{u \in Z} \frac{1}{2} \parallel f(X(t), u) - g(Y(t)) \parallel^2$$

As our learning algorithm needs only the knowledge of the derivative of the a priori control law, the learning signal is identified with the gradient of the tracking-error function with respect to the control.

Using a one-layer neural-network controller, the learning rule yielded interesting results: Without any prior knowledge, the neural network learned during the first seconds of running the tracking-control law. In Figure A2.4, the first trial trajectories are represented through time, for the first 50 s. One can check that tracking is excellent (and better than that after the 337 trials needed by the unsupervised learning rule): The network adaptively learned the tracking-control law. The behavior of the network was determined to be the same even when the reference trajectory was more complicated (several accelerations, stabilization phases, etc.).

But on the other hand, the network forgets as it learns, in the sense that no knowledge of the past is adaptively stored, as in the previous cases: The network does not compute a state feedback error defined on the whole state space, but a local state feedback control.

Moreover, the controller is not robust with regard to dynamical perturbation, because we do not use the system's absolute location in the feedback (we control only the system velocities), and no knowledge of the degraded dynamics is supposed to be available. The network continues to adapt, with wrong estimations of the state.

For each type of algorithm, Nicolas Seube has studied systematically the robustness and sensitivity with respect to several parameters. The main lesson that has resulted from those experiments is that *two-layer neural networks* provide the best results in very many cases. This information is consistent with other experiments in other domains.

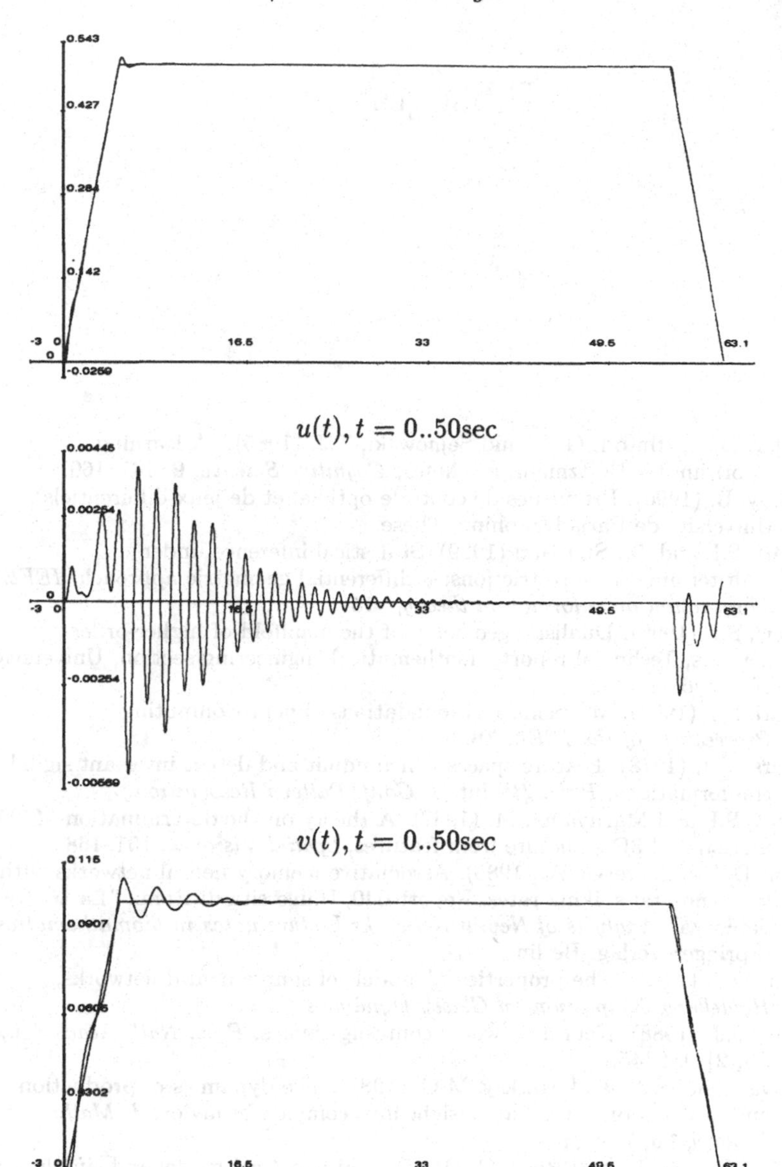

$r(t), t = 0..50\text{sec}$

Fig. A2.4 The Internal Algorithm. Without prior learning, the surging velovity $u(t)$ is tracked with a very good accuracy. The swaying velocity $v(t)$ is kept close to zero with better accuracy than the external algorithms and the heading velocity $r(t)$ is tracked with the same accuracy than the external algorithms.

Bibliography

Ackley, D.H., Hinton, G.E., and Sejnowski, T.J. (1985). A learning algorithm for Boltzmann machines, *Cognitive Science*, **9**, 147–169.

Alziary, B. (1990). Problèmes de contrôle optimal et de jeux différentiels., Université de Paris-Dauphine, Thèse.

Amari, S.I. and Te, Sun Han (1989). Statistical inference under multiterminal rate restrictions: a differential geometric approach, *IEEE Transaction on information theory,* **35**.

Amari, S.I. (1990). Dualistic geometry of the manifold of higher-order neurons, Technical reports, mathematical engineering section, University of Tokyo.

Amari, S.I. (1990). Mathematical foundations of neurocomputing, *Proceedings of the IEEE*, **78**.

Amarsi, S.I. (1978). Feature spaces which admit and detect invariant signal tranformations, *Proc. 4th Int. J. Conf. Pattern Recognition.*

Amari, S.I. and Maruyama, M. (1987). A theory on the determination of 3D motion and 3D structure from features, *Spatial Vision* **2**, 151–168.

Amit, D.J. and Treves, A. (1988). Associative memory neural networks with low temporal spiking rates, Report 640, Universita di Roma "La Sapienza" *Analysis of Neural Networks Lecture notes in biomathematics* (Springer-Verlag, Berlin).

Amit, D.J. (1986). The properties of models of simple neural networks, *Heidelberg Symposium on Glassy Dynamics.*

Amit, D.J. (1988). Neural networks counting chimes, *Proc. Natl. Acad. Sci.*, **85**, 2141–2145.

An Der Heiden, U. and Mackey, M.C. (1982). The dynamics of production and destruction: analytic hinsight into complex behavior, *J. Math. Biology*, **16**, 75–101.

An Der Heiden, U. and Roth, G. (1987). Mathematical model and simulation of retina and tectum opticum of lower vertebrates, *Acta Biotheoretica* **36**, 179–212.

An Der Heiden, U. (1980). *Analysis of neural networks, Lecture notes in biomathematics* (Springer-Verlag, Berlin, Heidelberg, New York).

Anderson, J.A. and Rosenfeld, E. (1988). *Neurocomputing* (MIT Press, Cambridge, MA).

Anderson, J.A. (1983). Cognitive and psychological computation with neural models, *AEEE Trans. on Syst. Man and Cybern.* **13**, 799–815.

Anderson, J.A. (1983). *The architecture of cognition* (Harvard University Press, Cambridge, MA).
Anderson, J.A. (1986). Cognitive capabilities of a parallel system, in *Disordered Systems and Biological Organization, Systems and Computer Science* (Springer-Verlag, Berlin).
Andler, D. (1992). *Introduction aux sciences cognitives* (Folio Essais, Paris).
Arbib, M.A. and Hanson, A.R. (1985). *Vision, brain and cooperative computation* (MIT Press, Cambridge, MA).
Asada and Slotine J.-J. (1986). *Robot Analysis and Control* (Wiley, New York).
Atkins, G.L. (1974). *Multicomportmental models in biological systems* (Chapman and Hall, London).
Atlan, L. and Meyer, J-A (1990). Application des techniques de programmation à l'optimisation de réseaux de neurones, *Report de stage de D.E.A. de Biomathématiques de l'université Paris* **VII**.
Aubin, J.-P. and Cellina, A. (1984). Differential inclusions, Grundlehren der math. Wiss. **264** (Springer-Velag, Berlin).
Aubin, J.-P. and Ekeland, I. (1984). *Applied nonlinear analysis* (Wiley-Interscience, New York).
Aubin, J.-P. and Frankowska, H. (1984). Trajectoires lourdes de systèmes contrôlés, *C.R. Acad. Sci. Paris*, **298**, 521–524 .
Aubin, J.-P. and Frankowska, H. (1990). *Set-valued analysis* (Birkhäuser, Boston).
Aubin, J.-P. and Najman, L. (1994). L'algorithme des montagnes russes pour l'optimisation globale, *C. R. Acad. Sci.*
Aubin, J.-P. and Seube, N. (1991). Apprentissage adaptatif de lois de rétroaction de systèmes contrôlés par réseaux de neurones, *C. R. Acad. Sci.,* **314**, 957–963.
Aubin J.-P. and Seube, N. (1993). Apprentissage des feedbacks contrôland des sysèmes sous sontraintes d'état par des réseaux de neurones, *Contract* 911411 *passé par la Direction Recherche Etudes et Techniques.*
Aubin, J.-P. (1982). An alternative mathematical description of a player in game theory, *IIASA WP*, **82**, 122.
Aubin, J.-P. (1988). Equations qualitatives aux confluences, *C. R. Acad. Sci.*, **307**, 679–682.
Aubin, J.-P. (1988). Problèmes mathématiques posés par l'algorithme qsim (qualitative simulation), *C. R. Acad. Sci.*, **307**, 731–734.
Aubin, J.-P. (1989). Qualitative simulation of differential equations, *J. Differential and Integral Equations*, **2**, 183–192.
Aubin, J.-P. (1989). Règles d'apprentissage de systémes cognitifs, *C. R. Acad. Sci.*, **308**, 147–150.
Aubin, J.-P. (1990). A survey of viability theory, *SIAM J. on Control and Optimization*, **28**, 749–788.
Aubin, J.-P. (1990). Fuzzy differential inclusions, *Problems On Control and Information Theory*, **19**, 55–67.
Aubin, J.-P. (1991). *Viability theory* (Birkhäuser, Boston).
Aubin, J.-P. (1993). Beyond Neural Networks: Cognitive Systems, in *Mathemaitcs Applied to Biology and Medicine*, ed. Demongeot, J. and Capasso, (Wuers, Winnipeg).
Aubin, J.-P. (in preparation). Mutational and morphological analysis: tools for shape regulation and optimization.

Azencot, R. (1990). Synchronous Boltzmann machines : learning rules, in *Neural Networks* (Springer-Verlag, Berlin).

Azencot, R. (1990). General Bolzmann machines with multiple interactions, *IEEE Trans. Pattern Analysis Machine Intelligence.*

Baars, B. L. (1989). A cognitive theory of consciousness (Cambridge University Press, London).

Babloyantz, A. and Sepulchre, J. A. (1991). Target and spiral aves in oscillatory media in the presence of obstacles, *Physica D*, **49**, 52–60.

Babloyantz, A. (1986). Molecules, dynamics and life.

Barhen, J., Dress, W.B. and Jorgensen, C.C (1987). Applications of concurent neuromorphic algorithms for autonomous robots., *Report Oak Ridge National Laboratory.*

Barr, A., Cohen, P.R. and Feigenbaum, E.A. (1982). *The Handbook of Artificial Intelligence* (Pitman, London).

Barron, R.L. (1975). Learning network improve computer aided prediction and control., *Computer Design*, 65–70.

Barrow, H. G. (1984). Verify: a program for proving correctness of digital hardware design, *Artificial Intelligence*, **24**, 434–491.

Barto, A. G., Bradtke, S.J., and Satinder, Pal Singh (1991). Real-time learning and control using asynchronous dynamic programming, *Computer and Information Science.*

Barto, A. G. and Satinger Pal Singh (1990). On the computational economics of reinforcement learning, Presented at *1990 Connectionist Models Summer School.*

Barto, A. G., Sutton, R. S., and Watkins, J.C.H. (1990). Learning and sequential decision making.

Barto, A.G. (1986). Game theoretic cooperativity in networks of self interested units, Presented at *AIP Conference Neural Networks for Computing*, Snowbird, 41–46.

Basseville, M., Benveniste, A., and Moustakides, G.V. (1987). Detection and diagnosis of changes in the eigenstructure of nonstationary multivariable systems, *Automatica*, **23**, 479–489.

Basseville, M., Benveniste, A., and Moustakides, G.V. (1987). Optimal sensor location for detecting changes in dynamical behavior, *IEEE Transactions on Automatic control*, **32**, 1067–1075.

Bersini, H., Saerens, M., and Soquet, A. (1989). Connexionism for low level and high level process control., *Report Universite Libre de Bruxelles.*

Bersini, H., Gonzales Sotelino, L., and Decossaux, E. (1991). Hopfield net generation and encoding of trajectories in contrained environment, *IRIDIA Université Libre de Bruxelles.*

Bersini, H. (1989). Connexionism vs Gofai for modelling and supporting process plant operators, Presented at the *2d European Meeting on Cognitive Science Approaches to Process Control.*

Bersini, H. (1991). Immune network and adaptive control, *IRIDIA Université Libre de Bruxelles.*

Bersini, H. (1991). Q learning and adaptive distributed control, *IRIDIA Université Libre de Bruxelles.*

Bertsekas, D. and Tsitsiklis, J.N. (1989). *Parallel and Distributed Computation* (Prentic Hall, Englewood Cliffs, New Jersey).

Bienenstock, E. and Doursat, R. (1988). Elastic matching and pattern recognition in neural networks, in *Neural Networks: From Models to*

Applications ed. L. Personnaz and G. Dreyfus.
Bienenstock, E., Fogelman, F., and Weisbuch, G. (1986). Disordered systems and biological organization, *NATO ASI Series in System and Computer Sciences* (Springer-Verlag, Berlin).
Bienenstock, E. and Von Der Malsburg, C. (1987). A neural network for invariant pattern recognition, *Europhysics Letters*, **4**, 121–126.
Bienenstock, E. (1988). Relational models in natural and artificial vision, *NATO ASI Series*, F 41, Neural Computers, 61–70.
Blum, E. K. (1991). Approximation theory and feedforward networks, *Neural Networks*, **4**, 511–516.
Bobylev, V.N. (1990). A possibilistic argument for irreversibility, *Fuzzy sets and systems*, **34**, 73–80.
Bruck J. and Sanz J. (1988). A study on neural networks, *Int'l. J. Intelligent Systems*, **3**, 59–75.
Burnod, G. (1988). *An Adaptative Neural Network the Cerebral Network: the Cerebral Cortex* (Masson, Paris).
Bylander, T. (1987). Primitives for reasoning about the behavior of devices, *Workshop on Qualitative Reasoning*, Urbana-Champaign IL.
Bylander, T. (1987). Using consolidation for reasoning about devices, *Technical Report*, Lab. for AI Research, the Ohio State University at Colombus.
Carpenter, G.A. and Grossberg, S. (1987). A massively parallel architecture for a self-organizing neural pattern recognition machine, *Computer Vision, Graphics and Image Processing*, **37**, 54–115.
Cerdeira, H.A. and Huberman, B.A. (1987). Apparent randomness in quantum dynamics, *Physical Review A*, **36**, 3–27.
Changues, J-P, Courrege, P. and Danchin, A. (1973). A theory of the epigenesis of neuronal networks by selective stabilization of synapses, *Proc. Nat. Acad. Sci.*, **70**, 2974–2978.
Changeus, J-P, Courrege, P., and Danchin, A. (1981). Un mécanisme biochimique pour l'épigénèse de la jonction neuromusculaire, *C.R. Acad. Sci.*, **292**, 449–453.
Changeus, J-P and Danchin, A. (1976). Selective stabilization of developing synapses as a mechanism for the specification of neuronal networks, *Nature*, **264**, 705–712.
Changuex, J-P (1979). Molecular interactions in adult and developing neuromuscular junctions, in *Neuroscience: Fourth Study Program*, (MIT Press, Cambridge, MA).
Changues, J-P (1983). *L'homme neuronal*, (Fayard, Paris).
Cochet, Y. and Paget, G. (1988). ZZENN (Zig Zag Epigenetic Neural Networks) a new approach to connectionist machine learning, Presented at the *International Computer Science Conference 88*, Hong-Kong.
Cochet, Y. (1988). Réseaux de neurones, *IRISA*, **389**.
Coffman, Jr, E.G., Gelenbe, E., and Gilbert, E.N. (1988). Analysis of a conveyor queue in a flexible manufacturing system, *Eur. J. of Operational Research*, **35**, 382–392.
Cohen, M.A. and Grossberg, S. (1983). Absolute stability of global pattern formation and parallel memory storage by competitive neural networks, *IEEE Trans. on Syst. Man and Cyb.*, **13**, 815–826.
Coiffet, P. (1986). *La robotique* (Hermès, Paris).
Commenges, D., Pinatel, F. and Seal, J. (1986). A program for analysing

single neuron activity by methods based on estimation of a change-point, *Computer Methods and Programs in Biomedicine*, **23**, 123–132.

Commenges, D., Seal, J. and Pinatel, F. (1986). Inference about a change point in experimental neurophysiology, *Mathematical Biosciences*, **80**, 81–108.

Commenges, D. and Seal, J. (1985). The analysis of neuronal discharge sequences: change-point estimation and comparison of variances, *Statistics in Medicine*, **4**, 91–104.

Commenges, D. and Seal, J. (1986). The formulae-relating slopes, correlation coefficients and variance ratios used to determine stimulus - or movement-related neuronal activity, *Brain Research*, **383**, 350–352.

Cotterill, M. and Fort, J.C (1986). A stochastic model of retinotopy : a self organizing process, *Biol. Cyb.*, **53**, 405–411.

Cottrell, M. and Fort, J.C (1987). Etude d'un processus d'auto-organisation, *Ann. Institut Henri Poincaré*, **23**, 1–20.

Cottrell, M. (1991). Bases mathématiques pour les réseaux de neurones artificiels, Support de cours NEURAL-COMET, Université Panthéon-Sorbonne.

Cottrell, M. (1992). Mathematical analysis of a neural network with inhibitory coupling, *Stochastic Processes and their Applications*, **40**, 103–126.

Cottrell, M. (1988). Stability and attractivity in associative memory networks, *Biol. Cybern.*, **58**, 123–139.

Cowan, W.M (1973). Neuronal death as regulative mechanism in the control of cell number in the nervous system, in *Development and Aging in the Nervous System* (Academic Press, New York).

Cybenko, G. (1988). Approximation by superpositions of sigmoidal functions, *Math. Contr. Signal Systems*, **2**, 303–324.

Cybenko, G. (1988). Continuous valued neural networks: approximation theoretic results, *Computing Sciences and Statistics*, 174–183.

Danchin, A. (1987). Biological foundations of language: a comment on noam chomsky's approach of syntactic structures, in *Noam Chomsky: Consensus and Controversy*, ed. S. Modgil and C. Modgil (Falmer Press).

Davies, P. (1989). *The new physics*, (Cambridge University Press, London).

Davis, R. (1983). Reasoning from first principles in electronic troubleshooting, *Int. J. Man Machine Studies*, **19**, 403–423.

Davis, R. (1984). Diagnostic reasoning based on structure and behavior, *Artificial Intelligence*, **24**, 347–410.

Davis, R. (1986). Knowledge-based systems, *Science*, **231**, 957–963.

De Kleer, J. and Bobrow, D.J (1984). Qualitative reasoning with higher order derivatives, *Artificial Intelligence*, **24**.

De Kleer, J. and Brown, J.S (1986). Theories of causal ordering, *Artificial Intelligence*, **29**.

De Kleer, J. and Brown, J.S. (1984). A qualitative physics based on confluences, *Artificial Intelligence*, **24**.

De Kleer, J. and Williams, B.C. (1987). Diagnosing multiple faults, *Artificial Intelligence*, **32**, 97–130.

De Kleer, J. (1984). How circuits work, *Artificial Intelligence*, **24**, 205–280.

Deangelis, D. L., Post, W., and Travis, C. C. (1986). *Positive Feedback in Natural Systems* (Springer-Verlag, Berlin).

Dehaene, S., Changeux, J-P., and Nadal, J-P. (1987). Neural networks that learn temporal sequences by selection, *Proc. Natl. Acad. Sci.*, **84**, 2727–2731.

Deledicq, A. and Diener, M. (1989). *Leçones de Calcul Infinitésimal* (Armand Colin, Paris).

Delgado, M. and Moral, S. (1987). A definition of inclusion for evidences, *Fuzzy Mathematics*, **6**, 81–87.

Demongeot, J., Goles, E., and Tchuente, M. (1985). *Dynamical systems and cellular automata* (Academic Press, New York).

Dempster, A.P. (1967). Upper and lower probabilities induced by a multivalued mapping, *Ann. Math. Stat.*, **38**, 325–339.

Derin, H. and Elliott, H. (1987). Modeling and segmentation of noisy and textured images using Gibbs random fields, *IEEE Trans. Pattern Anal. Machine Intell.*, **9**, 39–55.

Derrida, B. and Bessis, D. (1988). Statistical properties of valleys in the annealed random map model, *J. Phys. A: Math. Gen.* **21**, 509–515.

Derrida, B. and Flyvbjerg, H. (1987). Distribution of local magnetisations in random networks of automata, *J. Phys. A: Math. Gen.* **20**, 1107–1112.

Derrida, B. and Nadal, J-P. (1987). Learning and forgetting on asymmetric, diluted neural networks, *J. Statistical Physics*, **49**, 993–1009.

Derrida, B. (1988). Dynamics of automata, spin glasses and neural network models, *Nonlinear evolution and chaotic phenomenia*, 213–244.

Diener, M. (1983). *Une initiation aux outils non standard fondamentaux* (Office des Publications Universitaires, Paris).

Diener, F. and Reeb, G. (1989). *Analyse Nonstandard* (Hermann, Paris).

Dordan, O. (1988). *Differential qualitative simulation: a numerical approach*, (Cahiers de Mathématiques de la Décision, Université de Paris-Dauphine).

Dordan, O. (1990). Algorithme de simulation qualitative d'une équation différentielle sur le simplexe, *C. R. Acad. Sci.*, **310**, 479–482.

Dordan, O. (1990). Analyse qualitative, Université de Paris-Dauphine Thèse.

Dordan, O. (1992). Mathematival problems arising in qualitative simulation of a differential equation, *Artificial Intelligence*, **55**, 61–86.

Dordan, O. (in press) *Analyse qualitative* (Masson, Paris).

Dormoy, J.L and Raiman, O. (1986). Physique qualitative: la représentation des connaissances en physique pour l'intelligence artificielle, Presented at *Actes des journées nationales du PRC-IA*, Aix-les-Bains.

Dormoy, J.L (1984). Représentation des connaissances en physique: tour d'horizon et recherches actuelles, Presented at *Colloque intelligence artificielle du groupe C-F Picard*, Aix-en-Provence.

Dormoy, J.L (1984). Représentation des connaissances en physique et systèmes experts, Presented at *Journées internationales sur les systèmes experts et leurs applications*, Avignon.

Doyle, J. and Sacks, E.P. (1989). Stochastic analysis of qualitative dynamics, *Report.*

Dreyfus, G., Personnaz, L., and Bienenstock, E. (1989). What's wrong with neural networks, *Neuro-Nimes*, **89**.

Dreyfus, S.E. (1965). *Dynamic programming and the calculus of variations* (Academic Press, New York).

Dubois, D. and Prade, H. (1980). *Fuzzy sets and systems: theory and applications* (Academic Press, New York).

Dubois, D. and Prade, H. (1982). On several representations of an uncertain body of evidence, *Fuzzy Information and Decision Processes*, 167–181.

Dubois, D. and Prade, H. (1986). A set-theoretic view of belief functions: logical operations and approximations by fuzzy sets, *Int. J. General systems*, **12**, 193–226.

Dubois, D. and Prade, H. (1986). On the unicity of Dempster rule of combination, *Inter. J. Intelligent Systems*, **1**, 133–142.

Dubois, D. and Prade, H. (1987). Analysis of fuzzy information, *Mathematics and Logic*, **1**, 3–39.

Dubois, D. and Prade, H. (1987). Defense et illustration des approches non-probabilistes de l'imprecis et de l'incertain, *Report 269, Langages et systèmes informatiques*, Université de Toulouse.

Dubois, D. and Prade, H. (1987). On several definitions of the differential of a fuzzy mapping, *Fuzzy Sets and Systems Journal*, **24**, 117–120.

Dubois, D. and Prade, H. (1987). The mean value of a fuzzy number, *Fuzzy Sets and Systems Journal*, **24**, 279–300.

Dubois, D. and Prade, H. (1987). Upper and lower images of a fuzzy set induced by a fuzzy relation, *Int. Tech. Report LSI*, **265**.

Dubois, D. and Prade, H. (1988). Conditionnement et induction avec des probabilités non-additives, *Report 301, Langages et systèmes informatiques* , Université de Toulouse.

Dubois, D. and Prade, H. (1988). Ensembles flous 87, algébre, integrations, calcul d'intervalles et approximations, *Report 302, Langages et systèmes informatiques*, Université de Toulouse.

Dubois, D. and Prade, H. (1988). *Théorie des possibilités* (Masson, Paris).

Duda, R.O. and Hart, P.E. (1973). *Pattern Classification and Scene Analysis* (Wiley, New York).

Edelman, G.M. and Reeke, G.N. (1982). Selective networks capable of representative transformations, limited generalizations and associative memory, *Proc. Nat'l. Acad. Sci.*, **79**, 2091–2095.

Edelstein-Keshel, L. (1987). *Mathematical models in biology* (Random House, New York).

Ellis, R. (1987). Topological dynamics and ergotic theory, *Ergod. Th. and Dynam. Sys.*, **7**, 25–47.

Elman, J.L. and Zipser, D. (1987). Learning the hidden structure of speech, *Report 8701, Institute for Cognitive Science, Universityo of California at San Diego.*

Ermoliev, Y. (1966). On numerical methods for solving nonlinear optimization problems, *Kibernetika*, **4**, 1–17.

Ermoliev, Y. and Nurminski, E. (1972). Limiting extremal problems, *Kibernetika*, **10**.

Fahlan, S.E., Hinton, G.E. (1987). Connectionist architectures for artificial intelligence, *Computer*, IEEE, 100–109.

Falkenhainer, B., Forbus, K.D., and Gentner, D. (1986). The structure-mapping engine, *Report*, UIUCDCS-R-86-1275.

Falkenhainer, B., Forbus, K.D., and Gentner, D. (1987). The structure-mapping engine: algorithm and examples, *Report*, UIUCDCS-R-87-1361.

Faltings, B. (1987). Qualitative place vocabularies for mechanisms in configuration space, *Report*, UIUCDCS-R-87-1360.

Farmer, J. D., Toffoli, T., and Wolfram, S. (1984). Cellular automata, *Physica D*, **10**.

Farmer, J. D. and Sidorowich, J. J. (1988). Exploiting chaos to predict the future and reduce noise, in *Evolution, Learning and Cognition*, ed. Lee Y. C. (Wold Scientific Press).

Farmer, J.D., Kauffmann, S.A., and Packard, N.H. (1986). Autocatalyptic replication of polymers, *Physica D*, **22**, 50–67.

Farmer, J.D. (1990). A rosetta stone for connectionism, *Physica D*, **42**, 153–187.

Feldman, J.A. and Ballard, D.H. (1982). Connectionist models and their properties, *Cognitive Science*, **6**, 205–254.

Feldman, J.A. (1982). Dynamic connections in neural networks, *Biol. Cybern.*, **46**, 27–39.

Fogelman Soulie, F., Gallinari, P., and Le Cun, Y. (1986). Learning in automata networks, Actes du 2ème Colloque International d'Intelligence Artificielle CIIAM 86, (Hermés, Paris).

Fogelman Soulie, F., Gallinari, P., and Le Cun, Y. (1987). Evaluation of networks architectures on test learning tasks, Presented at *IEEE, First Int. Conf. on Neural Netwoks.*

Fogelman Soulie, F., Gallinari, P., and Le Cun, Y. (to appear). Automata networks and artificial intelligence, *Automata Networks in Computer Science, Theory and Applications.*

Fogelman Soulie, F., Gallinari, P., and Le Cun, Y. (to appear). Network learning, *Machine Learning* .

Fogelman Soulie, F., Gallinari, P., and Thiria, S. (1987). *Learning and associative memory* (Springer-Verlag, Berlin).

Fogelman Soulie, F. and Weisbuch, G. (1987). Random iterations of threshold networks and associative memory, *SIAM J. on Computing.*

Fogelman Soulie, F. (1984). Frustration and stability in random boolean networks, *Discrete Applied Mathematics*, **9**, 139–156.

Fogelman Soulie, F. (1985). Contributionà une théorie du calcul sur réseaux, Thèse d'Etat, Grenoble.

Forbus, D.K., Nielsen, P., and Faltings, B. (1987). Qualitative kinematics: a framework, *Report*, UIUCDCS-R-87-1352.

Forbus, D.K. and Gentner, D. (1986). Learning physical domains: toward a theoretical framework, *Report*, UIUCDCS-R-86-1247.

Forbus, D.K. and Stevens, A. (1981). Using qualitative simulation to generate explanations, *Report*, 4490 (Bolt Beranek and Newman Inc, Cambridge, MA).

Forbus, D.K. (1985). The problem of existence, *Report*, UIUCDCS-R-85-1239.

Forbus, D.K. (1986). Interpreting observations of physical systems, *Report*, UIUCDCS-R-86-1248.

Forbus, D.K. (1986). The logic occurrence, *Report*, UIUCDCS-R-86-1300.

Forbus, D.K. (1988). Commonsense physics: a review, *Ann. Rev. Comput. Sci.*, **3**, 197–232.

Forbus, D.K. (1988). QPE: using assumption-based truth maintenance for qualitative simulation, *Int. Journal of AI in Engineering.*

Forbus, D.K. (1988). Qualitative physics: past, present, and future, in *Exploring Artificial Intelligence* (Morgan Kauffman Publisher).

Forbus, K.D. (1984). Qualitative process theory, *Artificial Intelligence*, **24**, 85–168.
Forbus, K.D. (1988). Introducing actions into qualitative simulation, *Report*, UIUCDCS-R-88-1452.
Fort, J.C. (1991). Stabilité de l'algorithme de séparation de sources de Jutten et Hérault., *Traitement du signal*, **8**, 35–42.
Frankowska, H. (1990). Some inverse mapping theorems, *Ann. Inst. Henri Poincaré* **8**, 35– 42.
Frankowska, H. (1995). *Set-Valued Analysis and Control Theory* (Birkhäuser, Boston, Basel, Berlin).
Frankowska, H., and Da Prato, G. (To Appear). A sotchastic Filippov theorem.
Frankowska, H., and Quincampoix, M. (1991). Viability kernels of differential inclusions with constraints, *Mathematics of Systems, Estimation and Control.*
Frankowska, H., and Quincampoix, M. (1991). L'algorithme de viabilité, *C.R. Acad. Sci.* **312**, 31–36.
Fukushima, K. and Miyake, S. (1978). A self-organizing neural network with a function of associative memory: feedback-type cognition, *Biol. Cyber.*, **28**, 201–208.
Gaivoronski, A. (1978). *Stochatic programming problems* (Kibernetika).
Gallez, D. and Babloyantz, A. (1991). Predictability of human EEG: a dynamical approach, *Biol. Cybern.*, **64**, 381–391.
Gallinari, P., Thiria, S. and Fogelman Soulie, F. (1988). Multilayer perceptrons and data analysis, Presented at *IEEE Annual International Conference on Neural Networks*, San Diego, California.
Genesereth, M. R. (1984). The use of design description in automated diagnosis, *Artificial Intelligence*, **24**, 431–436.
Giardina, C. R. and Dougherty, E. R. (1988). *Morhological methods in image and signal processing* (Prentice Hall, Englewood Cliffs, NJ).
Goles, E. and Vichniac, G.Y (1986). Invariants in automata networks, *J. Phys. A: Math. Gen.*, **19**, 961–965.
Goles, E. (1985). Comportement dynamique de réseaux d'automates, Thése d'Etat, Grenoble.
Goles-Chacc, E., Fogelman Soulie, F. and Pellegrin, D. (1985). Decreasing energy functions as a tool for studying threshold networks, *Disc. Appl. Math.*, **12**, 261–277.
Gondran, M. and Minoux, M. (1979). *Graphes et algorithmes* (Eyrolles, Paris).
Goodman, I.R., Nguyen, H.T. (1985). *Uncertainty models for knowledge-based systems* (North-Holland, Amsterdam).
Gorman, R.P. and Sejnowski, T.J. (1987). Learned classification of sonar targets using a massively parallel network, *Neural networks for computing*, Snowbird, Utah.
Gorman, W.M. (1964). More scope for qualitative economics, *Review of economic studies*, **31**, 65–68.
Grossberg, S. and Kuperstein, M. (1986). Neural dynamics and adaptive sensory-motor control (North-Holland, Amsterdam).
Grossberg, S. (1986). *The Adaptive Brain I and II*, (Elsevier/North-Holland, Amsterdam).

Grossberg, S. (1987). Competitive learning: from interactive interaction to adaptive resonnance, *Cognitive Sciences*, **11**, 1987.

Guez, A., Protopopsescu, V. and Bahren, J. (1988). On the stability, storage capacity, and design of continuous nonlinear neural networks, *IEEE Transactions on Systems, Man and Cybernetics*, **18**, 80–87.

Guez, A. and Selinsky, A. (1988). A trainable neuromorphic controller., *Journal of robotic systems.*

Guyon, I., Poujaud, I. and Personnaz, L. (1989). Comparing different neural network architectures for classifying hanwritten digits, Presented at *International Joint Conference on Neural Networks*, (Washington).

Hallan, B. and Mallash, K. (1987). *Advances in artificial intelligence* (Wiley, New York).

Harrison, T.D. and Fallside, F. (1987). A connectionist approach to continuous speech recognition, Presented at *IEEE Conf. on Neural Information Processing Systems*, Boulder. Colorado.

Haton, J.-P. (1991). Le raisonnement en Intelligence Artificielle (Inter Editions, Paris).

Hayes, P.J. (1985). The second naive physics, in *Formal Theories of the Commonsense World*, 1–36.

Hebb, D.O. (1949). *The organization of behavior* (Wiley, New York).

Held, R. and Hein, A. (1963). Movement-produced stimulation in the development of visually guided behavior, *J. Compar. Physiol. Psychol.*, **56**, 872–876.

Heskes, T.M., and Kappen, B. (1991). Learning processes in neural networks, *Physical Review A*, **44**, 4.

Hillis, W.D. (1986). *The connection machine* (MIT Press, Cambridge, MA).

Hinton, G.E. and Anderson, J.A. (1981). *Parallel models of associative memory* (Erlbaum Hillsdale , New Jersey).

Hiriart-Urruty, J.-B. and Lemarechal, C. (1993). *Convex Analysis and Minimization Algorithms* (Springer-Verlag, Berlin).

Hofbauer, J. and Sigmund, K. (1988). *The theory of evolution and dynamical systems* (Cambridge University Press, London).

Hogg ,T. and Huberman, B.A. (1984). Understanding biological computation, *Procedings of the National Academy of Sciences* **81**, 6871–6874.

Hogg, T. and Huberman, A. (1992). Acheving global stability through local controls, Presented at the *1991 IEEE International Sympsium on Intelligent Control.*

Hogge, J.C. (1987). The compilation of planning operators from qualitative process theory models, *Report*, UIUCDCS-R-87-1368.

Hogge, J.C. (1987). Time and tplan user's manual, *Report*, UIUCDCS-R-87-1366.

Hogge, J.C. (1987). Tplan: a temporal interval-based planner with novel extensions, *Report*, UIUCDCS-R-87-1367.

Holland, J. (1989). *Induction* (M.I.T. Press, Cambridge, Massachusetts).

Hopfield, J. J and Tank, D.W. (1985). Simple neural optimization networks, *Biol. Cyber.*, **52**, 141–152.

Hopfield, J. J (1982). Neural networks and physical systems with emergent collective computational abilities, *Proc. Nat. Acad. Sci.*, **79**, 2554–2558.

Hubel, D.H. and Wiesel, T.N. (1970). The period of susceptibility to the physiological effects of unilateral eye closures in kittens, *J. Physiology*,

206, 419–436.
Hubel, D.H. (1978). Effects of the deprivation on the visual cortex of cat and monkey, *The Harvey Lectures*, **72**, 1–51.
Huberman, B.A. and Glance, S. (1992). Diversity and collective action, Report.
Huberman, B.A. and Hogg, T. (1988). *The behavior of computational ecologies, The Ecology of Computation* (North-Holland, Amsterdam).
Huberman, B.A. and Struss, P. (to appear). Chaos, qualitative reasoning and the predictability problem, Preprint.
Ironi, L., Stefanelli, M. and Lanzola, G. (1989). Qualitative models of iron metabolism, in *AIME 89: Proc. Europ. Conference Artificial Intelligence in Medecine*, Ed. J. Hunter, J. Cookson, J. Wyatt (Springer-Verlag, Berlin).
Ironi, L., Stefanelli, M. and Lanzola, G. (1990). Qualitative models in medical diagnosis, *Artificial Intelligence in Medecine*, **2**, 85–101.
Ishida, Y., Adashi, N. and Tokumaru, H. (1981). Some results on qualitative theory of matrix, *Trans. of the Society of Instrument and Control Engineers*, **17**.
Jackendoff, R. (1987). *Consciousness and the computational mind* (M.I.T. Press, Cambridge, Massachusetts).
Jacobs, A.M. and Levy-Schoen, A. (1987). Revues critiques, le contrôle des mouvements des yeux dans la lecture: questions actuelles, *L'Annde Psychologique*, **87**, 55–72.
Jacobs, A.M. and O'Regan, J.K. (1987). Spatial and/or temporal adjustments of scanning behavior to visibility changes, *Acta Psychologica*, **65**, 133–146.
Jakubczyk, B. (1989). Stochastic stability of nonsymmetric threshold networks, *Complex Systems*, **3**, 437–470.
Jannerod, M. (1986). Mechanisms of visuomotor coordination: a study in normal and brain-damaged subjects, *Neuropsychologia*, **24** 41–78.
Jeffries, C., Klee, V. and Van Den Driessche, P. (1977). When is a matrix sign stable?, *Can. J. Math.*, **29**, 315–326.
Kandel, E.R. and Schwartz, J.H. (1985). *Principles of neural science* (Elsevier, Amsterdam).
Kaufmann, A., and Gupta, M.M (1985). Introduction to fuzzy arithmetic. Theory and applications (Van Nostrand Reinhold, New York).
Kent, E.W. (1981). *The brains of men and machines* (McGraw-Hill, New York).
Kerszberg, M. and Mukamel, D. (1988). Dynamics of simple computer networks, *Journal of Statistical Physics*, **51**, 777–795.
Kirkpatrick, S., Gelatt, C.D. and Vecchi, M.P. (1983). Optimization by simulated annealing, *Science*, **220**, 671–680.
Kodratoff, Y. (1983). Generalizing and particularizing as the techniques of learning, *Computers and Artificial Intelligence*, **2**, 417–441.
Kodratoff, Y. (1986). *Leçons d'apprentissage symbolique automatique* (Cepadues, Paris).
Kohlas, J. (1987). Modeling uncertainty with belief functions in numerical models, *Rep. # 41*, Instit. for Automation and Operations Research, Univ. of Fribourg.
Kohonen, T., Makisara, K. and Saramaki, T. (1984). Phonotopic maps-insightful representation of phonological features for speech

recognition, *IEEE 7th International Conference on Pattern Recognition*, 182–185.

Kohonen, T. and Oja, E. (1987). Computing with neural networks, *Science*, **235**, 1227.

Kohonen, T. (1974). An adaptive associative memory principle, *IEEE Trans. Computers.*

Kohonen, T. (1984). *Self-organization and associative memory* (Springer-Verlag, Berlin).

Kohonen, T. (1986). Dynamically expanding context, with application to the correction of symbol strings in the recognition of continuous speech, *IEEE*, 1148–1151.

Kohonen, T. (1987). Adaptive, associative, and sel-organizing functions in neural computing, *Applied Optics*, **26**, 4910–4918.

Kohonen, T. (1987). State of the art in neural computing, *ICNN Plenary Speech*, 77–90.

Kohonen, T. (1988). An introduction to neural computing, *Neural Networks*, **1**, 3–16.

Kohonen, T. (1988). The "neural" phonetic typewriter, *Computer*, 11–22.

Kolmogorov, A.N. (1957). On the representation of continuous functions of many variables by superposition of continuous functions of one variable and addition, *Dokl. Akad. Nauk. SSSR*, **114**, 953–956 (in russian); *Amer. Math. Soc. Trans.* (in English), **2**, 55–59.

Krauth, W. and Mezard, M. (1987). Learning algorithms with optimal stability in neural networks, *J. Phys. A*, **20**, 745.

Krone, G., Mallot, H. and Palm, G. (1986). Spatiotemporal receptive fields: a dynamical model derived from cortical architectonics, *Proc. R. Soc. Lond. B* **226**, 421–444.

Kuipers, B.J. and Kassirer, J.P. (1984). Causal reasoning in medicine: analysis of a protocol, *Cognitive Science*, **8**, 363–385.

Kuipers, B.J. (1984). Commonsense reasoning about causality: deriving behavior from structure, *Artificial Intelligence*, **24**, 169–203.

Kuipers, B.J. (1986). Qualitative simulation, *Artificial Intelligence*, **29**, 289–338.

Kuipers, B.J. (1987). Qualitative simulation as causal explanation, *IEEE Trans. on Systems, Man, and Cybernetics,* **17**.

Kuipers, B.J. (1989). Qualitative reasonning with causal models in diagnosis of complex systems, in *Artificial Intelligence, Simulation and Modeling*, ed. L. Widman, K. Loparo, and N. Nielsen (Wiley, New York).

Lady, M.G. (1983). The structure of qualitatively determinate relationships, *Econometrica*, **51**.

Laird, J.E., Newell, A. and Rosenbloom, P.S. (1987). Soar: an architecture for general intelligence, *Artificial Intelligence*, **33**, 1–64.

Lancaster, K.J. (1964). Partitionable systems and qualitative economics, *Review of Economic Studies*, **31**, 69–72.

Lancaster, K.J. (1965). The theory of qualitative linear systems, *Econometrica*, **33**, 395–408.

Lancaster, K.J. (1966). The solution of qualitative comparative static problems, *Quaterly Journal of Economics*, **80**, 278–295.

Lanzola, G., Stefanelli, M. and Barosi, G. (1989). A knowledge system architecture for diagnostic reasoning, ed. J. Hunter, J. Cookson and J. Wyatt, *AIME 89: European Conf. Artificial Intelligence in Medicine*

(Springer-Verlag, Berlin).

Le Cun, Y. and Fogelman Soulie, F. (1987). Modèles connexionnistes de l'apprentissage, in *Apprentissage et Machine*, (Intellectica, Paris). Ed, 114–143

Le Cun, Y. (1985). A learning scheme for asymmetric threshold network, *Proc of Cognitiva, 85*, 599–604.

Le Ny, J.-F. (1989). *Science cognitive et compréhension du langage* (Presses Universitaries de France, Paris).

Lee, W.W. and Kuipers, B.J. (1988). Non-intersection of trajectories in qualitative pjase space: a global constraint for qualitative simulation, in *Proc. Nat. Conf. Artificial Intelligence AAAI 88*, 286–290.

Lepetit, M. and Vernet, D. (1987). Qualitative physics applied to a depropanizer in process control, Presented at *Workshop on Qualitative Physics*, Urbana-Champaign, Illinois.

Lepetit, M. (1986). Physique qualitative: le système physical, LRI Univerité de Paris Sud.

Little, W.A. (1974). Existence of persistent states in the brain, *Math. Biosc.*, **19**, 101–120.

Lord, E. A. and Wilson, C. B. (1984). *The Mathematical Theory of Shape and Form* (Halsted Press, New York).

MacGregor, R. J. (1988). *Neural and Brain Modeling* (Academic Press, Orlando, Florida).

Mackey, M.C. and An Der Heiden, U. (1984). The dynamics of recurrent inhibition, *J. Math. Biology,* **19**, 211–225.

Maderner, N. (1992). Regulation of control systems under inequality viability constraints, *J. Math. Anal. Appl.*, **170**, 591–599.

Marr, D. (1988). *Vision,* (Freeman, San Francisco).

Mattioli, J. and Schmitt, M. (1991). Shape recognition combining mathematical morphology and neural networks, *Proceedings of Applications of Artificial Neural networks*, April 1991, Orlando, Florida.

Mattioli, J. and Schmitt, M. (1993). Discrimination de formes en morphologie mathématique, *C.R. Acad. Sci.*, **317**, 807–810.

Mattioli, J. and Schmitt, M. (1993). An efficient algorithm for computing the antigranulometry, Presented at *SPIE: Application of Artificial Neural Network*, San Diego, CA July 1993.

Mattioli, J. and Schmitt, M. (to appear). On the information contained in the erosion curve, Presented at *Shape in Picture 92*, Driebergen, The Netherlands, September 1992.

Mattioli, J., Schmitt, M., Pernot, E., and Vallet, F. (1991). Shape discrimination based on mathematical morphology and neural networks, Presented at *International Converence on Artificial Neural Networks*, Helsinki, June 1991.

McClelland, J.L. and Rumelhart, D.E. (1985), Distributed memory and the representation of general and specific information, *J. Exp. Psychology,* **114**, 159–188.

McClelland J. L., and Rumelhart, D. E. eds. (1986) . *Parallel distributed processing (Vol.I)* (M.I.T. Press, Cambridge, Massachusetts).

Mc Culloch, W.S. and Pitts, W. (1943). A logical calculus of the ideas immanent in nervous activity, *Bull. Math. Biophysics*, **5**, 115–133.

Meyer, J-A. and Guillot, A. (1990). From animals to animats: everything you wanted to know about the simulation of adaptive behavior,

Technical report BioInf-90-1, Ewle Normale Supérieure.
Mezard, M. and Nadal, J-P. (1989). Learning in feedforward layered networks: the tiling algorithm, *J. Phys. A: Math. Gen.* **22**, 2191–2203.
Michalski, R.S., Carbonell, J.G. and Mitchell, T.M. (1986). *Machine learning, an artificial intelligence approach* (Morgan Kaufmann, Los Altos).
Michel, A.N., Miller, R.K. and Nam, B.H. (1982). Stability analysis of interconnected systems using computer generated Lyapunov functions, *IEEE Trans. Circuits and Systems*, **29**, 431–440.
Michel, A.N., Sarabudla, N.R., and Miller, R.K. (1982). Stability analysis of complex dynamical systems, *Circuits, Systems, Signal Processing*, **1**, 171–202.
Minsky, M. and Papert, S. (1969). *Perceptrons: An Introduction to Computational Geometry* (MIT Press, Cambridge).
Miyake, S. and Fukushima, K (1984). A neural network model for the mechanism of feature extraction, *Biol. Cybern.*, **50**, 377–384.
Moore, R.E. (1979). Methods and applications of interval analysis, *SIAM Studies in Applied Mathematics.*
Moore, R.E. (1984). Risk analysis without monte-carlo methods, *Freiburger Intervall-Berichte*, **1**, 1–48.
Mougeot, M., Azencott, R., and Angeniol, B. (1991). Image compression with back propagation: improvement of the visual restoration using different cost functions, *Neural networks*, **4**, 467–476.
Nadal, J-P., Toulouse, G. and Changeux, J.P. (1986). Networks of formal neurons and memory palimpsests, *Europhysics Lett.*, **535**, 1–10.
Nadal, J-P. and Toulouse, G. (1989). Information storage in sparsely coded memory nets, *Network*, **1**, 61–74.
Neisser, U. (1976). *Cognition and reality* (Freeman and Co., San Francisco).
Nerrand, O. , Roussel-Ragot, P. and Personnaz,, L. (to appear) Neural networks and non-linear adaptative filtering : unifying concepts and new algorithms, *Neural Computation.*
Nesterov, Y. (1984). Minimization methods for nonsmooth convex and quasiconvex functions, *Matekon*, **20**, 519–531.
Nicolis, J. S. (1986). *Dynamics of hierarchical systems: an evolutianory approach* (Springer-Verlag, Berlin).
Nilsson, N.J. (1985). *Learning machines* (McGraw Hill, New York).
Norman, D.A and Rumelhart, D.E. (1975). *Explorations in cognition* (Freeman and Co., San Francisco).
Pan, X. (1985). Expérimentation d'automates à seuil pour la connaissance de caractères, Thèse d'Université Grenoble.
Parnas, H., Flashner, M. and Spira, M.E. (1989). Sequential model to describe the nicotinic synaptic current, *Biophy. J.*, **55**, 875–884.
Parnas, H. and Segel, L.A. (1989). On the contribution of mathematical models to the understanding of neurotransmitter release, (Report).
Paugam-Moisy, H. (1992). Optimisation des réseaux de neurones artificiels, Thèse Ecole Normale Supèrieure de Lyon.
Pearlmutter, B. (1989). Learning state space trajectories in recurrent neural networks, *Neural Computation*, 263–269.
Penrose, R. (1989). *The Emperor's New Mind Concerning Computers, Minds and the Laws of Physics* (Oxford University Press, London).

Peretto, P. (1984). Collective properties of neural networks: a statistical physics approach, *Biol. Cybern.*, **50**, 51–62.

Personnaz, L., Guyon, I. and Ronnet, J.C. (1985). Character recognition, neural networks and statistical physics, in *Cognitiva 1985*, (CESTA, Paris)

Petrowski, A., Personnaz, L. and Dreyfus, G. (1989). Parallel implementations of neural network simulations, in *Hypercube and distributed computers*, ed. J-P Verjus, F. Andre (North-Holland, Amsterdam).

Poggio, T., Torre, V. and Koch, C. (1985). Computational vision and regularization theory, *Nature*, **317**, 314–319.

Poliak, B. (1967). On a general method for solving nonlinear optimization problems, *Dokl. A.N. SSSR*, **1**, 33–36.

Popper, K.R. and Eccles, J-C (1977). *The self and its brain* (Springer International, Berlin).

Prager, R.W., Harrison, T.D. and Fallside, F. (1986). Boltzmann machines for speech recognition, *Computer Speech and Language*, **1**, 3–28.

Pribram, K.H. (1971). *Languages of the brain* (Prentice-Hall, Englewood Cliffs, New Jersey).

Quirk, J. and Ruppert, R. (1965). Qualitative economics and the stability of equilibrium, *Review of Economic Studies*, **32**, 311–326.

Raiman, O., Dague, P. and Deves, P. (1986). Raisonnement qualitatif dans le diagnostic de pannes, Presented at *Congrés sur les systèmes experts et leurs applications*, Avignon.

Ritschard, G. (1983). Computable qualitative comparative static techniques, *Econometrica*, **51**.

Ritter, H. and Schulten, K. (1988). Convergence properties of Kohonen's topology conserving maps: fluctuation, stability, and dimension selection, *Biological Cybernetics*, **60**, 57–71.

Rosenblatt, F. (1961). *Principles of neurodynamics: perceptrons and theory of brain mechanisms* (Spartan Books, Washington).

Rumelhart, D.E., Hinton, G.E. and Williams, R.J. (1986). Learning representations by back propagating errors, *Nature*, **323**, 533–536.

Rumelhart, D.E. and Zipser, D. (1985). Competitive learning, *Cognitive Science*, **9**, 75–112.

Sacks, E. (1990). Automatic analysis of one-parameter planar ordinary differential equations by intelligent numeric simulation, *Report* CS-TR-244-90, Princeton University.

Sacks, E. (1990). Automatic phase space analysis of dynamical systems, *Computing Systems in Engineering*, **1**, 607–614.

Schmitt, M. and Mattioli, J. (1993). *Eléments de Morphologie Mathématique* (Masson, Paris).

Schmitt, M. and Vincent, L. (1993). *Morpholigical Image Analysis. Practical and Algorithmic Handbook* (Cambridge University Press, London).

Seal, J., Commenges, D. and Salamon, R. (1983). A statistical method for the estimation of neuronal response latency and its functional interpretation, *Brain Research*, **278**, 382–386.

Seal, J., Gross, C. and Bioulac, B. (1982). Activity of neurons in area 5 during a simple arm movement in monkeys before and after deafferntation of the trained limb, *Brain Research*, **250**, 229–243.

Segel, L. A. (1984). *Modeling dynamic phenomena in molecular and cellular biology* (Cambridge University Press, London).

Seikkala, S. (1987). On the fuzzy initial value problem, *Fuzzy Sets and Systems Journal*, **24**, 319–330.

Sejnowski, T.J., Kienker, P.K. and Hinton, G.E. (1986). Learning symmetry groups with hidden unit: beyond the perceptron, *Physica D*, **22**, 175–191.

Sejnowski, T.J. and Hinton, G.E. (1985). Separating figure from ground with a Boltzmann machine, *Vision, Brain and Cooperative Computation* (MIT Press, Cambridge, Massachusetts).

Seube, N. (1990). Construction of learning rules in neural networks that can find viable regulation laws ro control problems by self-organization, *Proceedings of INNC 90*, Paris.

Seube, N. (1990). Learning rules for viable control laws, *Proceedings of AINN 90*, Zürich.

Seube, N. (1990). Règles d'apprentissage de lois de contrôle viables, *Revue Technique, Thompson CSF*, **112**.

Seube, N. (1991). Apprentisage de lois de contrôle régulant des contraintes sur l'état par réseaux de neurones, *C.R. Acad. Sci.*, **312**, 446–450.

Seube, N. (1992). Régulation de systèmes contôlés avec contraintes sur l'état par réseaux de neurones, Thèse Université de Paris-Dauphine.

Seube, N. (1991). A Neural Network Approach for AUV Control Based On Viability Theory, in *7th International Symposium on Unmanned Untethered Submersible Technology* (Durham, NC), 191–202.

Seube, N. (to appear). Neural Network Learning Rule For Control: Application To AUV Tracking Control, in *Neural Network for Ocean Engineering Workshop*, (Washington, DC), 185–196.

Seube, N. (1991). Non Linear Feedback Control Law Learning by Neural Networks: A Viability Approach. Presented at *2nd ICIAM Conference.*

Seube, N. (1992). Neural Networks Learning Rules for Non Linear System: Application to Underwater Vehicles Controls in *Non Linear Control System Design Symposium* (NOLCOS 92).

Seube, N. (1993). A State Feedback Control Law Learning Theorem, in *Neural Network in Robotics* (Klüwer Academic Publisher).

Seube, N. and Macias, J.-C. (1991). Design of Neural Network Learning Rules for Viable Feedback Laws, European Control Conference, Grenoble.

Shafer, G. (1976). *A mathematical theory of evidence* (Princeton University Press).

Shallice, (1988). *From neuropsychology to mental structure* (Cambridge University Press).

Sharkey, ed. (1986). *Advances in Cognitive Sciences* (Wiley, New York).

Shillings et al. ed. (1987). *Cognitive science: an introduction* (M.I.T. Press, Cambridge, Massachusetts).

Simon, H.A and Iwasaki, Y. (1986). Causality in device behavior, *Artificial Intelligence*, **29**.

Simon, H.A and Iwasaki, Y. (1986). Theories of causal ordering: reply to De Kleer and Brown, *Artificial Intelligence*, **29**.

Smolensky, P. (1987). A method for connectionist variable binding, *Technical Report* CU-CS-356-87, Institute of Cognitive Science, Univ. of Colorado.

Smolensky, P. (1987). *On the proper treatment of connectionism, Technical Report* CU-CS-359-87, Institute of Cognitive Science, Univ. of Colorado.

Sontag, D. and Sussmann, H.J. (1989). Backpropagation can give rise to spurious local minima even for networks without hidden layers, *Complex Sytems Publications.*

Sontag, D. and Sussmann, H.J. (1990). Back propagation separates where perceptrons do, *Neural networks,* **4**.

Sontag, D. (1990). Feedback stabilization using two-hidden-layer nets, *Rutgers Center for Systems and Control.*

Sontag, D. (1990). Feedforward nets for interpolation and classification, *Rutgers Center for Systems and Control.*

Sowa (1984). *Conceptual structures* (Addison-Wesley, Reading, Massachusetts).

Strat, T. (1984). Continuous belief functions for evidential reasoning, *Proc. of the Nat. Conf. on Artificial Intelligence,* Austin, 213–219.

Struss, P. (1988). Mathematical aspects of qualitative reasoning, *Int. J. Artificial Intelligence in Engineering,* **2**, 156–169.

Sussmann, H. (1993). Uniqueness of the weights for minimal feedforward nets with a given input-ouptput map, *Rutgers Center for Systems Control.*

Todd, P.H (1986). *Intrinsic Geometry of Biological Surface Growth* (Springer-Verlag, Berlin).

Trave, L. and Dormoy, J-L (1988). Qualitative calculus and applications, *12th IMACS World Congress on Scientific Computation,* Paris.

Trave, L. (1988). Qualitative pole assignability in decentralized control, *IFAC/IMACS International Symposium on "Distributed Intelligence Systems - Methods and Applications,* Varna, Bulgaria.

Tuckwell, F.C. (1988). *Introduction to Theoretical Neurology, Vol. 1, Linear Cable Theories and Dendritic Structures* (Cambridge University Press, London).

Tuckwell, F.C. (1988). *Introduction to Theoretical Neurology, Vol. 2, Nonlinear and Stochastic Theories* (Cambridge University Press, London).

Turing, A. (1952). The chemical basis of morphogenesis, *Phil. Trans. Royal Soc.,* **327**, 37.

Turing, A. (1963). *Computing machinery and intelligence, Computers and Thought* (McGraw-Hill, New York).

Varela, F.J., Sanchez-Leighton, V. and Coutinho, A. (1988). Adaptive strategies gleaned from immune networks: viability theory and comparison with classifier systems, in *Evolutionary and Epigenetic Order form Complex Systems: A Waddington Memorial Volume,* ed. B. Goodwin and P. Saunders (Edimburgh University Press).

Varela, F.J. (1981). Living ways of sense-making: a middle path for neuroscience, *Disorder and Order,* 208–221.

Vergis, A., Steiglitz, K. and Dickinson, B. (1986). The complexity of analog computation, *Mathematics and Computers in Simulation,* **28**, 91–113.

Vichniac, G.Y., Lepp, M. and Steenstrup, M. *A neural network for the optimization of communications network design,* (BBN Communications Corporation, Cambridge).

Vichniac, G.Y. (1986). Cellular automata models of disorder and organization, in *Disordered systems and biological organization* (Springer-Verlag, Berlin).

Vicsek, T. (1989). *Fractal growth phenomena* (World Scientific, New York).

Virasoro, M.A. (1988). The effect of synapses destruction on categorization by neural networks, *Europhysics Letters*, **7**, 293–298.

Von Der Malsburg, C. and Bienenstock, E. (1986). Statistical coding and short-term synaptic plasticity: a scheme for knowledge representation in the brain, in *Disordered Systems and Biological Organization*, NATO ASI Series, F20, 247–272.

Von Der Malsburg, C. and Bienenstock, E. (1987). A neural network for the retrieval of superimposed connection patterns, *Europhysics Letters*, **3**, 1243–1249.

Von Neumann, J. (1966). *Theory of self-reproducing automata* (Univ. of Illinois Press, Urbana-Champaign).

Waltz, D.L. and Pollack, J.B. (1985). Massively parallel parsing: a strongly interactive model of natural language interpretation, *Cognitive Science*, **9**, 51–74.

Weigend, A.S., Huberman, B.A. and Rumelhart, D.E. (1990). Predicting the future: a connectionnist approach, *International Journal of Neural Systems*.

Weisbuch, G. and Fogelman Soulie, F. (1985). Scaling laws for the attractors of Hopfield networks, *J. Phys. Lett.*, **46**, 623–630.

Weisbuch, G. (1989). *Dynamique des systèmes complexes. Une introduction aux réseaux d'automates*, (Inter-Editions/Editions du CNRS).

Widrow, B., Stearns, S.D. (1985). *Adaptive signal processing* (Prentice-Hall, Englewood Cliffs, New Jersey).

Widrow, B. and Nguyen, D. (1989). *The truck backer-upper: an example of self-learning in neural networks.*, Presented at *Joint International Conference on Neural Networks.*

Widrow, B. and Hoff, M.E (1960). Adaptive switching circuits, *IRE WESCON Conv. Record*, Part 4, 96–104.

Williams, B.C. (1984). Qualitative analysis of MOS circuits, *Artificial Intelligence*, **24**, 281–346.

Williams, B.C. (1987). *Beyond 'qualitative' reasoning*, Presented at *Workshop on Qualitative Reasoning*, Urbana-Champaign, Illinois.

Yager, R.R (1986). The entailment principle for Dempster-Shafer granules, *Int. J. Intelligent Systems*, **1**, 247–262.

Yip, E.L. and Sincovec, R.F. (to appear). Controllability and observability of continuous descriptor systems, *IEEE Transactions on Automatic Control.*

Zadeh, L.A. (1965). Fuzzy sets, *Information and Control*, **8**, 338–353.

Zadeh, L.A. (1975). Calculus of fuzzy restrictions *Fuzzy Sets and their Applications to Cognitive and Decision Processes*, **1**.

Zadeh, L.A. (1978). Fuzzy sets as a basis for a theory of possibility, *Fuzzy Sets and Systems*, **1**, 3.

Zhao, Jian (1990). Application de la théorie de la viabilité à la commande de robots, Thèse de doctorat Université Pierre et Marie Curie.

Index

Printed in the United States
By Bookmasters